Biochemistry of the Algae and Cyanobacteria

PROCEEDINGS OF THE PHYTOCHEMICAL SOCIETY OF EUROPE

23. **The Genetic Manipulation of Plants and its Application to Agriculture**
    Edited by P.J. Lea and G.R. Stewart

24. **Membranes and Compartmentations in the Regulation of Plant Functions**
    Edited by A.M. Boudet and others

25. **The Biochemistry of Plant Phenolics**
    Edited by C.F. Van Sumere and P.J. Lea

26. **Plant Products and the New Technology**
    Edited by K.W. Fuller and J.R. Gallon

27. **Biologically Active Natural Products**
    Edited by K. Hostettmann and P.J. Lea

28. **Biochemistry of the Algae and Cyanobacteria**
    Edited by L.J. Rogers and J.R. Gallon

PROCEEDINGS OF THE
PHYTOCHEMICAL SOCIETY OF EUROPE

Biochemistry of the Algae and Cyanobacteria

Edited by

L. J. ROGERS

*Department of Biochemistry,
University College of Wales,
Aberystwyth*

and

J. R. GALLON

*Department of Biochemistry,
University College of Swansea,
Swansea*

CLARENDON PRESS · OXFORD
1988

Oxford University Press, Walton Street, Oxford OX2 6DP
Oxford New York Toronto
Delhi Bombay Calcutta Madras Karachi
Petaling Jaya Singapore Hong Kong Tokyo
Nairobi Dar es Salaam Cape Town
Melbourne Auckland
and associated companies in
Berlin Ibadan

Oxford is a trade mark of Oxford University Press

Published in the United States
by Oxford University Press, New York

British Library Cataloguing in Publication Data
Biochemistry of the algae and cyanobacteria.
—— Proceedings of the Phytochemical Society of Europe, 0309-9393. V.28
1. Algae, Biochemistry
I. Rogers, L.J. II. Gallon, J.R. (John R.)
589.3'192
ISBN 0-19-854239-9

Library of Congress Cataloging in Publication Data
Biochemistry of the algae and cyanobacteria / edited by L.J. Rogers and J.R. Gallon.
(Proceedings of the Phytochemical Society of Europe; v. 28)
Includes index.
1. Algae—Physiology—Congresses. 2. Cyanobacteria—Physiology—Congresses. 3. Algae—Biotechnology—Congresses. 4. Cyanobacteria—Biotechnology—Congresses. 5. Botanical chemistry—Congresses. 6. Microbiological chemistry—Congresses. I. Rogers, L.J. II. Gallon, J.R. (John R.) III. Series.
QK565.B56 1988 589.3'1—dc19 88-12554
ISBN 0-19-854239-9

Set by Colset Private Limited, Singapore
Printed and bound in Great Britain by
Biddles Limited, Guildford and King's Lynn

Contents

Contributors

D. V. Amla: National Botanical Research Institute, Lucknow 226001, India

P. Böger: Lehrstuhl für Physiologie und Biochemie der Pflanzen, Universität Konstanz, D-7750 Konstanz, Germany

D.M. Butler: Department of Plant Sciences, University of Leeds, Leeds LS2 9JT, UK

N.G. Carr: Department of Biological Sciences, University of Warwick, Coventry CV4 7AL, UK

A.E. Chaplin: Centre for Continuing Education and Training, University College of Swansea, Singleton Park, Swansea SA2 8PP, UK

G.A. Codd: Department of Biological Sciences, University of Dundee, Dundee DD1 4HN, UK

A.J. Darling: MRC Virology Unit, Institute of Virology, Church Street, Glasgow G11 5JR, UK

Anthony G. Davies: Marine Biological Association, Citadel Hill, Plymouth PL1 2PB, UK

A.E. Douglas: Department of Zoology, University of Oxford, South Parks Road, Oxford, OX1 3PS, UK

L.V. Evans: Department of Plant Sciences, University of Leeds, Leeds LS2 9JT, UK

G.E. Fogg: School of Ocean Sciences, University College of North Wales, Menai Bridge, Anglesey, Gwynedd LL59 5EH, UK

J.R. Gallon: Department of Biochemistry, School of Biological Sciences, University College of Swansea, Singleton Park, Swansea, SA2 8PP, UK

J.L. Harwood: Department of Biochemistry, University College, Cardiff CF1 1XL, UK

A.K. Jones: Department of Botany and Microbiology, University College of Wales, Aberystwyth, Dyfed SY23 3DA, UK

A.L. Jones: Department of Biochemistry, University College, Cardiff CF1 1XL, UK

N.W. Kerby: AFRC Research Group on Cyanobacteria and Department of Biological Sciences, University of Dundee, Dundee DD1 4HN, UK

R.A. Lewin: Scripps Institute of Oceanography, A-002, University of California, La Jolla, California 92093, USA

M.J. Lubberding: Laboratorium voor Microbiologie, Faculteit Wiskunde en Naturwentschappen, Katholieke Universiteit Nijmegen, Toernooiveld, 6525 ED Nijmegen, The Netherlands

H.C.P. Matthijs: Laboratorium voor Microbiologie, Universteit van Amsterdam, Nieuwe Achtergracht 127, 1018 WS Amsterdam, The Netherlands

T.P. Pettitt: Department of Biochemistry, University College, Cardiff CF1 1XL, UK

G.K. Poon: Department of Biological Sciences, University of Dundee, Dundee DD1 4HN, UK

R.H. Reed: Department of Biological Sciences, University of Dundee, Dundee DD1 4HN, UK

H. Riege: Lehrstuhl für Geomikrobiologie, Universität Oldenburg, D-2900 Oldenberg, Germany

L.J. Rogers: Department of Biochemistry, University College of Wales, Aberystwyth, Dyfed SY23 3DD, UK

P. Rowell: AFRC Research Group on Cyanobacteria and Department of Biological Sciences, University of Dundee, Dundee DD1 4HN, UK

S. Scherer: Lehrstuhl für Physiologie und Biochemie der Pflanzen, Universität Konstanz, D-7750 Konstanz, Germany

A.J. Smith: Department of Biochemistry, University College of Wales, Aberystwyth, Dyfed SY23 3DD, UK

R.J. Smith: Department of Biological Sciences, University of Lancaster, Bailrigg, Lancaster LA1 4YQ, UK

A.C. Stewart: Trends in Genetics, Elsevier Publications Cambridge, 68 Hills Road, Cambridge CB2 1LA, UK

W. D. P. Stewart: Agriculture and Food Research Council, 160, Great Portland Street, London, W1N 6DT, UK

P. J. Syrett: Plant and Microbial Metabolism Research Group, School of Biological Sciences, University College of Swansea, Swansea SA2 8PP, UK

Preface

Subscribers to this series of books, the Proceedings of the Phytochemical Society of Europe, may be forgiven if they form the impression that the Society is mainly, if not exclusively, interested in higher plants. Nevertheless, such an impression is erroneous. As proof, in this volume, which is a record of the Phytochemical Society meeting held in Aberystwyth in April 1987, attention is concentrated on the algae.

From the earliest days of plant biochemistry, the algae have played an important role. For example, in 1777, Priestley observed that green matter, deposited on the walls of his water vessels, formed bubbles of 'dephlogisticated air'. More recently, Calvin, Benson, and others used *Chlorella* in the elucidation of the pathway of carbon dioxide assimilation in photosynthesis, an excellent example of how the use of algae has furthered our understanding of plant biochemistry.

The algae comprise a wide range of organisms, from the tiny picoplankton, about a micrometer in size, to Shelley's 'sapless foliage of the ocean', which includes the giant kelp, *Macrocystis*. Specimens of this alga, one of the largest plants known, have been reported to reach almost 50 m in length. In this volume, attention is given to the whole range of algae, both microscopic and macroscopic, including green, red, brown, and blue-green algae. These last organisms are now almost universally known as cyanobacteria but occupy an unusual taxonomic position between bacteria and algae, though there is no doubt that they are prokaryotes. That they should receive special mention in the title of this volume stems from several rather lively discussions during the early stages in the planning of the meeting from which this book originates, when the view was expressed that cyanobacteria could not be included in a meeting entitled 'The Biochemistry of the Algae'. As editors and conference organizers, we felt it important that the cyanobacteria should be included and we hope that their specific mention in the title of this book will serve to deflect the wrath of the taxonomic purists.

Like all plants, the algae are primary converters of solar energy, not only into biomass but also into a large range of higher value, potentially useful products. Furthermore, the microalgae, in particular, offer a potential for cell culture under much less demanding conditions than those needed for cell culture of higher plants. However, efficient exploitation presupposes a detailed understanding of the biochemistry of the algae as well as a knowledge of how they interact with other organisms. In this collection, therefore, an up-to-date account of research into the algae is given under the general part headings of metabolism, biosynthesis, bioenergetics, regulation, interactions, and biotechnology. Each part consists of between two and four review chapters followed by a number of abstracts of posters, relevant to that part, which were presented at the meeting. Together, these contributions illustrate the current state of research into algal biochemistry in Europe.

Our thanks are due to the authors, who were also the speakers at the conference, and to the poster contributors. In addition, we also thank those members of the Department of Biochemistry at the University College of Wales (UCW), Aberystwyth, who worked so tirelessly and efficiently on our behalf. No conference can possibly succeed without the help of such volunteers and the value of their contribution is immeasurable. We are also grateful to Amersham International, Glaxo Group, Midland Bank, Sigma Chemical Company, Tate and Lyle Group, Wellcome Biotech, UCW Aberystwyth, Biotechnology Centre Wales, Mid-Wales Development, and the Wales Tourist Board for financial and other support.

Finally, what of the editors themselves? We have tried to be sparing in our use of the editorial pen yet we are acutely aware that to all contributors editors are a nuisance, altering as they do, apparently at random, lines of original text that were the product of many hours' careful thought. Those alterations we did make were all with a view to improving clarity and continuity within this volume, but it is well known that the road to Hell is paved with good intentions and, if there is a circle in Hell for editors (perhaps among the sowers of discord and the falsifiers), there are probably those who would cheerfully consign us there.

1987
Swansea

J.R.G.
L.J.R.

Part 1: Metabolism

1 The flexibility and variety of algal metabolism

G.E. FOGG

School of Ocean Sciences, University College of North Wales, Menai Bridge, Anglesey, Gwynedd LL59 5EY, UK

Introduction

The papers in this volume give ample evidence that the metabolism of algae, whether we consider individual strains or the group in general, shows great plasticity in pattern and variety in end-products. To put these papers in perspective it may be useful to review the development of our knowledge of this field, which is not only fascinating in itself but has obvious applications in biotechnology. In doing this, attention will be mainly concentrated on three organisms with which I am particularly familiar: *Chlorella, Anabaena*, and *Botryococcus*.

Early studies

For over 200 years there has been some realization that the algae are more active and diverse chemically than other kinds of plants. In 1779 both Priestley and Ingenhousz, the co-discoverers of photosynthesis, described experiments on the gas exchanges brought about by the 'green matter', presumably consisting of *Chlorella*-like forms, which grew in their apparatus as it always does in such unsterilized systems. Priestley observed that it was particularly active in producing 'dephlogisticated air' and that its growth was promoted by organic matter. Ingenhousz was similarly struck by its apparently inexhaustible capacity for producing dephlogisticated air, and remarked:

> *I should rather incline to believe, that that wonderful power of Nature, of changing one substance into another, and of promoting perpetually that transmutation of substances, which we may observe every where, is carried on in this green vegetable matter in a more ample and conspicuous way.*

It can only have been an inspired guess that led Ingenhousz to express our theme so happily, and the next stage in the recognition of the special characters of algal metabolism was likewise more intuitive than scientific. In 1836 Harvey (see Smith 1951) divided algae into three subclasses on the basis of colour—green, brown, and red—a division which remains essentially the same today. It was not until the 1950s that the use of chemical characteristics became respectable amongst higher plant taxonomists, and Harvey had no chemical evidence to justify using colour as a taxonomic criterion. This evidence was rather slow in coming; a water-soluble blue pigment was obtained from blue-green algae by von Eisenbeck in 1836 and a similar red pigment from red seaweeds by Kützing in 1843, and Kützing gave the two pigments the names *phycocyanin* and *phycoerythrin*, respectively (see Rabinowitch 1945). The physicist Stokes, who in 1864 was the first to show that the colour of green leaves comes from a mixture of pigments, demonstrated that brown algae have a chlorophyll different from those in higher plants, and Willstätter and Stoll in 1913 took the first steps in providing a firm chemical background to the idea that the major algal groups are characterized by different photosynthetic pigments (see Rabinowitch 1945). This idea was expanded to a generalization, implicit in the classificatory systems of phycologists such as Bohlin and Oltmanns at the turn of the century, that the algal groups differ chemically as much as morphologically. This was supported by miscellaneous observations and discoveries, such as that of alginic acid, the cell wall component unique to brown seaweeds, by Stanford in 1883.

Phenotypic plasticity—*Chlorella*

Observations such as those just mentioned depended on finding reasonably pure growths of microalgae in Nature or on using macroalgae that could be sorted by hand. The way was opened for a more systematic approach when microalgae were obtained in pure culture by techniques akin to those which had been developed by bacteriologists. This was first achieved, with the common unicellular green alga *Chlorella vulgaris*, by Beijerinck in 1890. This alga favours habitats rich in organic matter, and Beijerinck showed that it is capable of heterotrophic growth in the dark. He also experimented with other sorts of algae and interpreted the accumulation of fats in diatoms as a diversion of intermediates from protoplasmic synthesis that takes place in cells unable to divide but still capable of photosynthesis (Beijerinck 1904). This was an important idea to which we shall return later; for the moment we must follow the career of *Chlorella* as a laboratory organism. This began when the biochemist Warburg, seeking a plant which he could obtain in continuous supply under standard physiological conditions in the laboratory, asked the advice of his uncle, a botanist. A *Chlorella* species was selected and provided the material for a series of fundamental researches on photosynthesis, the first of which was published in 1919. The advantages of *Chlorella* for such work are that it is robust and easily cultured in uniform suspension, that it has the simplest possible life cycle, and that its small size results in rapid exchange of materials between cell and medium. For perhaps another 40 years *Chlorella* was almost the only alga known to the average biochemist. Besides being used in photosynthesis studies, it provided material for work on respiration, mineral nutrition, nitrogen assimilation, and the relation of metabolism to growth; it was the first of any micro-organism to be grown in synchronous culture (Tamiya *et al.* 1953).

From such studies it became clear that, in terms of the proportions of different end-products of metabolism, *Chlorella* is an extremely plastic organism. When rapidly growing, as much as 60 per cent of the cell material is protein, with carbohydrate and fat making up 35 per cent and 5 per cent, respectively. Cells which have been incubated for some time in the light in the absence of a nitrogen source, on the other hand, contain as little as 9 per cent of protein with 6 per cent of carbohydrate and as much as 85 per cent of fat (Spoehr and Milner 1949). After the Second World War a global fat shortage was feared and a patent on this finding was applied for and obtained, but not actually put to any use. Myers (1949) developed the idea that photosynthesis and nitrogen metabolism share common intermediates and intermesh at some stage prior to the formation of carbohydrate, the main path along which carbon flows depending on the previous history of the cell and the conditions to which it is exposed. Myers was influenced both by Beijerinck (see Myers 1980) and by his colleague in the University of Texas, the microbiologist J. W. Foster, who looked on fungal metabolism as a flexible system reacting to excessive supply of a particular substrate by shunts and overflows which result in accumulation of particular internal reserves or extracellular products (Foster 1949). Applying this to a photosynthetic micro-organism one could visualize in those innocent days in the 1950s, when the complexity that was to be reached in present-day biochemical pathway charts was scarcely foreseen, a system of interlocking reversible reactions which might be made to yield a desired product by judicious regulation of conditions or by employment of inhibitors. As we have seen, this concept seems to apply when protein synthesis is prevented by deficiency of nitrogen and intermediates are diverted into fat synthesis, but even here it soon became obvious that conditions that induce fat accumulation are only operative at a particular stage in the cell division cycle (Fogg 1959). Another instance for which it seems to provide an explanation, at least superficially, is that of release of extracellular glycollate by algae.

Tolbert and Zill (1956) were the first to observe that *Chlorella* supplied with ^{14}C-bicarbonate in the light may release much of the ^{14}C fixed into the medium in this form. We now know that glycollate is produced by the oxidation and cleavage of ribulose bisphosphate by RuBP carboxylase-oxygenase, but it appears that in many algae whether the glycollate is released or metabolized further depends on whether or not photosynthesis is supplying intermediates in excess of the capacity of the cell to take them up in growth. Thus, both in the laboratory and in the natural environment, high irradiation and/or deficiency of mineral nutrients induce substantial glycollate release, sometimes to as much as 95 per cent of the total carbon assimilated, and it seems valid to look on the release as an overflow (Fogg *et al.* 1965; Fogg 1983).

The production of glycerol as an organic osmoticum in saline media by the green flagellate *Dunaliella* may be mentioned here. This is not an overflow in the sense that glycollate release apparently is, being under the control of an enzyme system the activity of which is regulated in response to the osmotic potential of the external medium (Ben-Amotz 1980). By manipulation of its growth in saline medium *Dunaliella* can be made to yield up to 85 per cent of its organic matter in the form of glycerol, so that it has distinct possibilities for industrial use.

We must return briefly to *Chlorella*. As we have seen, this genus played an enormous part in the development of algal physiology and biochemistry. Indeed, at one time it was thought that it might also provide the answer to the world's energy and food problems. Because *Chlorella* is robust and unexacting it can be handled in suspension culture by the ordinary methods of chemical engineering. It is capable of achieving conversion of radiant energy to potential chemical energy with an efficiency approaching 30 per cent, and its cells have a minimum of cell wall material and a maximum of protein. However, the cost of providing conditions to enable *Chlorella* to grow at its maximum efficiency proved high, so that *Chlorella* protein was more expensive than the best beef steak. Nevertheless there has been a market in Japan for *Chlorella* as a health food—where it can command \$100 kg^{-1} as compared with a production cost of \$5.4–10.8 kg^{-1} (Tsukada *et al.* 1977). Even the cyanobacterium *Spirulina*, which is somewhat easier to grow and harvest on the large scale and which has had more public appeal, does not seem at present to be a realistic means of producing cheap protein (but see Fay 1983; Jeeji Bai 1985). *Chlorella* cells can be immobilized in agar or alginate supports, where they maintain photosynthetic and respiratory activity for up to two weeks and then are still able to revert to their usual physiological state (Lukavský *et al.* 1986). The possible biotechnological applications of such immobilized *Chlorella* cells do not appear to have been explored yet but this is a promising approach with other algae (Kerby and Stewart, this volume, Chapter 17).

Chlorella has some severe limitations as an experimental tool. It is difficult to make cell-free extracts retaining enzymic activity or to isolate reasonably pure and undamaged preparations of organelles from it. Thus Millbank (1957) did not succeed in obtaining homogenates of *Chlorella* capable of oxidizing tricarboxylic acid cycle intermediates, although he tried a variety of different methods. Mitochondrial preparations can be obtained from *Chlorella*, but with more difficulty than from many other organisms (Grant 1978), and intact chloroplast preparations do not appear to have been obtained. The cell wall of *Chlorella* is relatively thick and tough and its plasma membrane seems impermeable to many of the usual cytological stains and to the digestive enzymes of animals. Free protoplast preparations of *Chlorella* have not, so far as I am aware, been reported. Mutants of *Chlorella* may be obtained but their use in the analysis of metabolic sequences is limited because of the absence of sexual reproduction in this genus, and *Chlamydomonas*, which does reproduce sexually, has become the preferred subject for mutant analysis of photosynthesis (Somerville 1986).

Phenotypic plasticity—*Anabaena*

One activity that did not seem to fit into the shunt/overflow pattern was that of nitrogen fixation by cyanobacteria. In the 1950s it appeared that assimilation of N_2 was rigidly linked to growth and that although these organisms released considerable quantities of of nitrogenous extracellular products this phenomenon had no direct relation to the fixation process (Fogg 1956). This inflexibility we now know to be more apparent than real; nitrogen fixation can be obtained in cell-free extracts of cyanobacteria (Schneider *et al.* 1960) and immobilized mutant cells can be made to liberate large quantities of products of nitrogen fixation such as amino acids (Niven *et al.* 1986; Kerby, and Stewart, this volume, Chapter 17).

On the other hand, what was thought to be a fine example of the flexibility of photosynthesis in nitrogen-fixing cyanobacteria proved illusory. Fogg and Than-Tun (1958) found that in the absence of supplied carbon dioxide *Anabaena cylindrica* appeared to be capable of using molecular nitrogen as an acceptor for photochemically produced hydrogen donors with the concomitant evolution of oxygen. Subsequent workers found this difficult to repeat but Cox (1966) showed that it happens in nitrogen-starved cells. She suggested that in such cells accumulated carbon reserves give rise to pyruvate which acts as a source of energy and reducing power for nitrogen fixation. The carbon dioxide evolved by phosphoroclastic cleavage of pyruvate would be assimilated in oxygenic photosynthesis, thus giving a stoichiometric relation between nitrogen fixation and oxygen evolution. The realization a year or two later (Fay *et al.* 1968) that nitrogenase is inactivated by free oxygen and that under aerobic conditions nitrogen fixation is confined to heterocysts, in which it is spatially separated from oxygenic photosynthesis, confirmed this view.

The segregation of nitrogenase in heterocysts, however, provides its own opportunities for manipulation of the metabolism of cyanobacteria. The photoproduction of hydrogen by algae as a means of capturing and utilizing solar energy has attracted much attention, but so far has proved impracticable because of the sensitivity of hydrogenase to the concomitantly produced oxygen. Nitrogenase also acts as a hydrogenase and in the heterocyst is protected from oxygen, so that Benemann and Weare (1974) were able to obtain the simultaneous photoproduction of hydrogen and oxygen by actively growing cultures of *A. cylindrica*. However, hydrogen production is strongly inhibited by molecular nitrogen so that growth cannot continue for long under the conditions used by Benemann and Weare. Supply of nitrogen in the combined form, of course, suppresses the formation of nitrogenase. Nevertheless, the potentialities of heterocysts for hydrogen production deserve to be explored further. Ochiai *et al.* (1980) have shown that immobilized filaments of the thermophilic cyanobacterium *Mastigocladus laminosus* can function as an anodic photoelectrode for periods of up to 20 days and thus might be used as an electron donor for the production of hydrogen. This would have to be used in conjunction with a separate hydrogenase component and so would be different from the heterocyst system.

Genotypic variety

The metabolic pattern of a given algal strain is plastic within limits and can be manipulated, although not so easily as so far to have been used extensively in biotechnology. The genetic constraints are considerable and at the moment it is perhaps easier to explore the vast diversity of biochemistry presented by the group than to attempt to tailor individual strains to produce particular products. The idea that the algal groups differ in important biochemical ways was for a long time an article of faith backed by precariously few proper chemical studies and by a large number of observations made with often unspecific staining techniques. Surveying the position in 1974 Lewin concluded that, in spite of the pitfalls and rash generalizations that have abounded, there is good correspondence between the biochemical taxonomy of the algae and that based on fine structure and morphology. With the increasing application of new techniques a bewildering mass of information on algal chemistry is becoming available. This seems to call for the establishment of a data bank and computer techniques for extracting general patterns.

There has also been an extension of the biochemical approach in the treatment of the smaller taxa, particularly those such as the genera *Chlorella* and *Scenedesmus* in which there is little of a morphological nature for the taxonomist to get hold of. Kessler (1982) has reviewed chemotaxonomy in the Chlorococcales. Nutritional and biochemical features useful in characterizing bacteria have proved of little use with these algae, but the presence or absence of hydrogenase, starch hydrolysis, secondary carotenoids, and characteristics such as DNA base ratios, tolerance of high concentrations of sodium chloride, acid conditions, and high temperatures allow the recognition of a number of species which can also be distinguished by refined morphological study. Examination of a large number of strains of some of the species shows remarkable uniformity, giving welcome assurance that biochemical diversity does not proliferate indefinitely. It is also reassuring that mutations do not appear frequently enough to be a danger in biochemical systematics. DNA base ratios and serological studies show what one would suspect, that some '*Chlorella*' species are rather distantly related to the others. *Hydra viridissima* shows great discrimination in entering into symbiosis with *Chlorella* strains, accepting only those which release detectable amounts of maltose (McAuley and Smith 1982). Unfortunately Kessler's (1982) studies do not seem to have included such strains, so that we cannot tell whether he discriminates between these algae in a corresponding way to *Hydra*.

A biochemically extreme alga—*Botryococcus*

In seeking particular biochemical or physiological qualities it may thus be best to look among the vast variety of algae existing in Nature. Algae are not always easy to obtain in laboratory culture but there seem to be few that remain obdurate when modern media and culture methods are intelligently and carefully employed. In particular, selective continuous culture is a technique that should not be overlooked where species accustomed to oligotrophic conditions are sought. As an example of the biochemically bizarre algae that may be found and which can be grown in mass culture we may consider *Botryococcus*. This freshwater planktonic chlorophycean forms relatively large colonies in the matrix of which there can be massive accumulation, up to 75 per cent of the dry weight, of hydrocarbons. *Botryococcus* has attracted attention because it seems to be the major component of certain coals (Blackburn 1936). Accumulation of this alga to form such massive deposits does not depend on high rates of primary production but simply on mechanical concentration of the floating colonies against a lee shore.

At the time when algae were being examined as possible sources of edible fats *Botryococcus* was investigated but deemed unsuitable because it grows slowly and most of the lipid which it accumulates is unsaponifiable (Belcher 1968). More recently interest in *Botryococcus* as a possible renewable source of liquid hydrocarbons has stimulated further research. Two types of hydrocarbons are produced—dienic unbranched hydrocarbons with an odd carbon number ranging from C_{23} to C_{31}, and branched unsaturated hydrocarbons with a carbon number ranging from C_{30} to C_{37} which are specific to *Botryococcus* and are called botryococcenes. Which of these types predominates depends both on the strain of *Botryococcus* (Berkaloff *et al.* 1984) and on the conditions of growth (Wolf and Cox 1981; Wolf *el al.* 1985). *Botryococcus* grown in culture usually only achieves a doubling in 72 hours, but if supplied with air enriched with 0.3 per cent of carbon dioxide this is reduced to 40 hours (Wolf *et al.* 1985). Botryococcene accumulation, unlike that of total lipids in other algae, is not induced by nitrogen or phosphorus deficiency and can occur when the organism is actively growing. This gives hope of a reasonably high conversion efficiency of radiant into potential chemical energy.

Uptake of carbon dioxide and other solutes would be expected to be limiting for an alga with bulky colonies and a presumably impermeable matrix such as is possessed by *Botryococcus*. Growth and hydrocarbon production in batch culture have been found to be strongly influenced by associated bacteria, in either an antagonistic or beneficial manner, but no effect on the chemical nature of the hydrocarbons was observed (Chirac *et al.* 1985). Organic substrates increase the rate of cell division and allow higher total biomass to be reached but do not affect oil accumulation (Weetall 1985). A peculiar feature of the hydrocarbon accumulation, revealed by electron microscope studies, is that the hydrocarbons are retained in the matrix by the outer cell wall or *trilamellar sheath* (Wolf and Cox 1981; Berkaloff *et al.* 1984). The internal structure of the cells themselves appears rather constant. The structure of the matrix is extremely variable but this variation was not found to correlate with hydrocarbon composition. There has been a suggestion that the terminal stages in synthesis of botryococcenes take place extracellularly, but this seems improbable, and Wolf *et al.* (1985) are inclined to think that, although there is a preponderance of C_{30} botryococcene in the cell and of higher botryococcenes outside, methylation takes place intracellularly near the periphery of the protoplast.

What biological sense can be made of this? There is no indication that the extracellular hydrocarbons ever act as food reserves but it could be that their accumulation is another example of a protective overflow mechanism. Large colonies such as those of *Botryococcus*, dependent on molecular diffusion rather than turbulence for uptake of solutes, are usually nutrient-limited, and since this particular alga floats at the water surface it must necessarily absorb more radiant energy than can be utilized in providing for growth. The excretion of hydrocarbon may be a way of disposing of surplus energy and thus avoiding damage by photo-oxidation.

Among lipid-accumulating algae *Botryococcus* seems at the moment to offer the best prospect of an economically viable means of utilizing solar energy in the production of hydrocarbons (Ben-Amotz *et al.* 1985). It is a robust and relatively easily grown form and its slow growth is not necessarily a disadvantage. In a dense culture the rate of production is determined by the flux of radiant energy and not by the growth rate and it may be that hydrocarbon is produced with comparatively high photosynthetic efficiency. In having large floating colonies which can be simply skimmed off the culture, *Botryococcus* seems to have a great advantage over smaller forms, such as those studied by Lewin (personal communication), which may have to be harvested by centrifugation. Be this as it may, amongst the tremendous variety of the algae there are undoubtedly many more forms to be found of potential use in biotechnology.

References

Beijerinck, M.W. (1890). *Bot. Ztg* **48**, 725.
Beijerinck, M.W. (1904). *Rec. Trav. bot. néerl.* **1**, 28.
Belcher, J.H. (1968). *Arch. Mikrobiol.* **61**, 335.
Ben-Amotz, A. (1980). In *Biochemical and photosynthetic aspects of energy production*, (ed. A. San Pietro), p. 191. Academic Press, New York.
Ben-Amotz, A., Tornabene, T.G., and Thomas, W.H. (1985). *J. Phycol.* **21**, 72.
Benemann, J.R. and Weare, N.M. (1974). *Science, New York* **184**, 174.
Berkaloff, C., Rousseau, B., Couté, A., Casadevall, E., Metzger, P., and Chirac, C. (1984). *J. Phycol.* **20**, 377.
Blackburn, K.B. (1936). *Trans. R. Soc. Edinb.* **58**, 841.
Chirac, C., Casadevall, E., Largeau, C., and Metzger, P. (1985). *J. Phycol.* **21**, 380.
Cox, R.M. (1966). *Arch. Mikrobiol.* **53**, 263.
Fay, P. (1983). *The blue-greens.* Edward Arnold, London.
Fay, P., Stewart, W.D.P., Walsby, A.E., and Fogg, G.E. (1968). *Nature, Lond.* **220**, 810.
Fogg, G.E. (1956). *A. Rev. Plant. Physiol.* **7**, 51.
Fogg, G.E. (1959). *Symp. Soc. exp. Biol.* **8**, 106.
Fogg, G.E. (1983). *Botanica Marina* **26**, 3.
Fogg, G.E. and Than-Tun (1958). *Biochim. Biophys. Acta* **30**, 209.
Fogg, G.E., Nalewajko, C., and Watt, W.D. (1965). *Proc. R. Soc. Lond. B* **162**, 517.
Foster, J.W. (1949). *Chemical activities of fungi.* Academic Press, New York.
Grant, N.G. (1978). In *Handbook of phycological methods: physiological and biochemical methods* (ed. J.A. Hellebust and J.S. Craigie) p. 25. Cambridge University Press.
Ingenhousz, J. (1979). *Experiments upon vegetables, discovering their great power of purifying the common air in the sunshine, and of injuring it in the shade and at night.* Elmsly & Payne, London.
Jeeji Bai, N. (1985). *Arch. Hydrobiol.* Suppl. **71**, 219.
Kessler, E. (1982). *Prog. phycol. Res.* **1**, 111.
Lewin, R.A. (1974). In *Algal physiology and biochemistry* (ed. W.D.P. Stewart) p. 1. Blackwell Scientific Publications, Oxford.
Lukavský, J., Komárek, J., Lukavska, A., Ludvík, J., and Pokorný, J. (1986). *Arch. Hydrobiol.* Suppl. **73**, 261.
McAuley, P.J. and Smith, D.C. (1982). *Proc. R. Soc. Lond.* B **216**, 7.
Millbank, J.W. (1957). *J. exp. Bot.* **8**, 96.
Myers, J. (1949). In *Photosynthesis in plants* (ed. J. Franck and W.E. Loomis) p. 349. Iowa State College Press, Ames, Iowa.
Myers, J. (1980). In *Primary producitivity in the sea* (ed. P.G. Falkowski) p. 1. Plenum Press, New York.
Niven, G.W., Kerby, N.W., Rowell, P., and Stewart, W.D.P. (1986). *Br. phycol. J.* **21**, 334.
Ochiai, H., Shibata, H., Sawa, Y., and Katoh, T. (1980). *Proc. natn. Acad. Sci. USA* **77**, 2442.
Priestley, J. (1779). *Experiments and observations relating to various branches of*

natural philosophy; with a continuation of the observations on air. J. Johnson, London.
Rabinowitch, E.I. (1945). *Photosynthesis and related processes*, Vol. 1. Interscience, New York.
Schneider, K.C., Bradbeer, C., Singh, R.W., Wang, L.C., Wilson, P.W., and Burris, R.H. (1960). *Proc. natn. Acad. Sci. USA* **46**, 726.
Smith, G.M. (1951). In *Manual of phycology* (ed. G.M. Smith), p. 13. Chronica Botanica, Waltham, MA.
Somerville, C.R. (1986). *A. Rev. plant Physiol.* **37**, 467.
Spoehr, H.A. and Milner, H.W. (1949). *Plant Physiol.* **24**, 120.
Stanford, E.C.C. (1883). *Chem. News* **47**, 254.
Tamiya, H., Iwamura, T., Shibata, K., Hase, E., and Nihei, T. (1953). *Biochim. Biophys. Acta* **12**, 23.
Tolbert, N.E. and Zill, L.P. (1956). *J. biol. Chem.* **222**, 895.
Tsukada, O., Kawahara, T., and Miyachi, S. (1977). In *Biological solar energy conservation* (ed. A. Mitsui, S. Miyachi, A. San Pietro, and S. Tamura) p. 363. Academic Press, New York.
Warburg, O. (1919). *Biochem. Z.* **100**, 230.
Weetall, H.H. (1985). *Appl. Biochem. Biotechnol.* **11**, 377.
Wolf, F.R. and Cox, E.R. (1981). *J. Phycol.* **17**, 395.
Wolf, F.R., Nonomura, A.M., and Bassham, J.A. (1985). *J. Phycol.* **21**, 388.

2 Nitrogen reserves and dynamic reservoirs in cyanobacteria

N.G. CARR

Department of Biological Sciences, University of Warwick, Coventry, CV4 7AL, UK

Reserve materials are frequently located within a micro-organism as relatively large accretions, readily apparent in the electron microscope and sometimes by light microscopy after cytochemical staining. Those accretions whose attributed function is to store carbon, phosphorus, or nitrogen are among the larger of the variety of inclusion bodies that characterize cyanobacteria (Allen 1984; Jensen 1984). This short paper is directed to questioning the principal roles of some cyanobacterial reserves and, possibly contrarily, to emphasizing the importance of others. What do we mean by importance when discussing microbial physiology? The synthesis of many of the most quantitatively minor proteins or other components of a cell has the same absolute importance in terms of cellular maintenance and reproduction as does the synthesis of enzymes of, for example, central metabolic pathways—mutational studies amply confirm this. Thus the term 'importance' is used here to indicate the relative ease by which metabolic role may be interpreted in terms of ecological, and hence evolutionary, advantage.

It has been previously stated that cyanobacteria are unusual among micro-organisms in having specific, insoluble reserves of the element nitrogen and that the possession of these could provide important advantages in the competition with other microbial groups for available nutrients and, in the case of phototrophs, for energy. Such an advantage would obviously be most apparent in an ecosystem that is predominantly nitrogen-limited; the open ocean is certainly the largest example of such. It was for this reason that the discovery some 10 years ago of a widespread population of cyanobacteria belonging to the *Synechococcus* group (see Waterbury *et al.* 1986) and which contributed significantly to total ocean productivity was received as potentially making a profound impact on our understanding of marine microbial population dynamics (Carr and Wyman 1986). Cyanobacteria are reported as possessing two insoluble reserves of nitrogen: cyanophycin, which is a co-polymer of aspartate and arginine; and phycocyanin, which is also a protein component of the light-harvesting apparatus. As mentioned earlier, no other microorganisms are known that possess insoluble nitrogen reserves. Presumably other bacteria, and eukaryotic microorganisms, will expand and contract the metabolic pools of nitrate or ammonia according to the environmental supply; likewise, metabolic pools of amino acids and nucleotide bases will have a degree of flexibility with regard to their size. However, such oscillations have to be carefully limited, otherwise the rates of enzyme reactions which involve the molecules of the expanding and contracting pools will be altered in ways which could not readily be controlled. This constraint becomes more severe if the molecules in question have additional, regulatory functions. There is also the limitation to metabolic pool expansion that is demanded by the necessity to maintain the osmotic balance of the microbial cell and, indeed, the need to prevent nutrients, when in liberal supply, from being lost by diffusion out of the cell. However, there are examples of degrees of nitrogen storage based upon the expansion of intermediary metabolic pools or the transient accumulation of soluble compounds. For instance, the internal concentration of nitrate ions in representative marine phytoplankton differed several fold under conditions of excess and limitation of that nutrient (Dortch *et al.* 1984). Little variation was observed in the size of the ammonium ion pool when the supply of that nutrient was varied. The fluctuation of nitrate pools within macroalgae and the correlation of this with stage of growth is well documented (see Dring 1982). Recently, the existence of a transient pool of what appeared to be a soluble, low molecular weight nucleic acid base complex has been detected in *Paracoccus denitrificans* (Dunstan and Whatley 1986); all the available ammonium ions appeared to be sequestered into this complex at the beginning of a batch culture, and subsequently the nitrogen was distributed from this reserve into nucleic acids and proteins as the growth of the culture proceeded.

Let us now turn to the insoluble nitrogen reserves of cyanobacteria. The chemical nature of cyanophycin [multi-L-arginyl-(polyaspartic acid)] was discovered by Simon (1971) and there is considerable evidence to show that it accumulates within cyanobacteria whenever protein synthesis is curtailed and nitrogen remains available (see Allen 1984). Cyanophycin is non-ribosomally synthesized and, on accumulation in the presence of chloramphenicol, can make up to 9 per cent of total dry weight. The 'structured granules' recognized in electron micrographs are cyanophycin, and these accumulate in non-growing conditions. Cyanophycin, because of the high N to C ratio of its two constituent amino acids, and the fact that nitrogen may be mobilized at the amino level of reduction, clearly has excellent potential as a molecular store of nitrogen. Accordingly, the cyanophycin content of the oceanic *Synechococcus* WH7803 species DC2 was examined, following growth under a variety of conditions that might be expected to maximize content (Newman *et al.* 1987). The surprising result was that no cyanophycin was found, and a rather elaborate isotope dilution experiment of any ^{14}C-labelled cyanophycin with unlabelled cyanophycin from an *Anabaena* species proved that this polymer was not synthesized by *Synechococcus* WH7803. In light of the suggested role of cyanophycin (see below) it is important to qualify this statement. No evidence of insoluble, precipitable, cyanophycin was found in *Synechococcus* WH7803; the existence of a low molecular weight, soluble aspartate–arginine co-polymer could not be excluded. An examination of several other *Synechococcus* species, both marine and freshwater, indicated that cyanophycin was absent from this genus of cyanobacteria altogether. We have already argued that one of the more interesting aspects of the existence of a significant oceanic population of cyanobacteria is the potential of such organisms to store nitrogen (Carr and Wyman 1986). The absence of cyanophycin from *Synechococcus* WH7803 throws into sharper focus the role that the other cyanobacterial nitrogen reserve material, the phycobiliproteins, may play.

Phycocyanin is one of a family of structurally, and in some cases genetically, related proteins (phycobiliproteins) that are central in light harvesting and transmission of light energy to photosystem II (see Glazer 1982; Bryant 1986). In simplistic terms they carry out the process which in higher plant chloroplasts is the responsibility of chlorophyll *b*. Phycobiliproteins are globular proteins, comprised of subunits of around 15 000–20 000 molecular weight, to which are attached one or more chromophore groups constructed of linear tetrapyrroles. The principal classes of phycobiliproteins are phycoerythrin (red in colour, A_{max} approx. 560 nm), phycocyanin (blue, A_{max} approx. 620 nm), and allo-phycocyanin (purple/blue, A_{max} approx. 650 nm); energy can of course only be passed from low to high wavelength centres. Varying proportions of the phycobiliproteins are stacked together to form rods, several of which comprise the phycobilisome, the light-harvesting unit which is revealed on the surface of the thylakoids by electron microscopy (see Bryant, 1986). The phycobiliproteins of cyanobacteria can account for very large proportions, commonly about 30 per cent, of the total soluble protein of the organism. The amount present is usually inversely related to the intensity of utilizable radiation.

There is now considerable evidence that points to phycocyanin (and by implication, phycoerythrin) as having an additional role as a nitrogen reserve (see Allen 1984; Carr and Wyman 1986). The publication by Allen and Smith (1969) provided clear evidence of the preferential breakdown of phycocyanin in nitrogen-limited *Anacystis nidulans* and formed the experimental basis for subsequent discussion of phycobiliproteins as nitrogen reserves. One of the pleasures of studying a diverse group of micro-organism such as cyanobacteria, that has attracted over the years the attention of most disciplines of biology, is that often the literature may be relied upon to contain observations that challenge the accepted order of scientific discovery. We have recently become aware of such an example. Kingsbury (1956) suggested, on the basis of experiments with a species of *Plectonema*, that nitrogen-containing pigments of cyanobacteria are mobilized during nitrogen limitation.

In discussing his experiments Kingsbury points out that Boresch (1910) was probably the first to observe such pigment changes due to nitrogen depletion! The phenomenon of nitrogen reserves in cyanobacteria clearly has a longer history than many of us had previously thought! The effects of nitrogen starvation and its re-supply on the breakdown and re-synthesis of phycobiliproteins of cyanobacteria has been reviewed by Allen (1984), and we now have a considerable amount of information regarding the kinetics of their breakdown, the enzymes involved, and the ways in which environmental factors influence these. Let us now consider what may be the special role of phycobiliprotein in the oceanic *Synechococcus*, which form the greater part of the cyanobacterial picoplankton (Waterbury *et al.* 1986).

Most of our knowledge of the eco-physiology of cyanobacterial picoplankton indicates considerable similarity between isolates from different areas of the ocean. Virtually all of the laboratory-based information has been obtained from *Synechococcus* sp. DC2, described by Waterbury *et al.* (1979) and now referred to as *Synechococcus* sp. WH7803, in the Woods Hole Oceanographic Institution Collection. This has proved to be an amenable laboratory organism, characterized in culture as in Nature by an intense red colour due to the high proportion of phycoerythrin present. This group of micro-organisms was discovered by fluorescent microscopy, the red/orange fluorescence coming from phycoerythrin, and it is this feature that forms the basis of the observations described below. Using continuous cultures of *Synechococcus* WH7803 it became clear that the phycoerythrin content could be varied by both the irradiance supplied and by the extent of availability of nitrogen which was supplied as nitrate (Wyman *et al.* 1985). When cultures that had been maintained under high available nitrogen conditions were deprived of nitrate, growth was maintained for over 24 hours, representing approximately one division time; during this period the phycoerythrin content declined. This was the first formal demonstration that the latter molecule, as well as phycocyanin, apparently acted as a nitrogen reserve. The most interesting phenomenon associated with this process was that related to the fact that *Synechococcus* WH7803 cultured with excess nitrate and a light intensity of 120 μE $m^2 sec^{-1}$ exhibited a degree of phycoerythrin autofluorescence. This means that a proportion of the light energy absorbed by phycoerythrin, instead of being passed through the other components of the phycobilisome to reaction centre chlorophyll, was emitted as phycoerythrin fluorescence. To this extent phycoerythrin was uncoupled from photosynthetic electron transfer; evidence from delayed fluorescence studies supported this interpretation (Wyman *et al.* 1985). The implication of these results was that the accumulation of phycoerythrin by this organism has been such that total light absorbed was greater than could be photochemically converted. Therefore a proportion of light energy was dissipated by autofluorescence of the primary light-harvesting molecule. Presumably this was achieved by the existence of some of the phycoerythrin in a configuration that did not allow energy transduction through the phycobilisome. This interpretation attributes to some of the phycoerythrin a function separate from that of light harvesting for photosynthesis. Not all the phycoerythrin that was used as a nitrogen reserve in the experiments described necessarily exists in this 'uncoupled' state. As in the case of phycocyanin in *A. nidulans*, the mobilization of large amounts of phycoerythrin when a culture of *Synechococcus* WH7803 becomes nitrogen limited would be expected to decrease the light harvesting capacity and hence the photosynthetic rate. Yeh *et al.* (1986) have questioned whether a significant amount of phycoerythrin ever exists in *Synechococcus* WH7803 in the uncoupled form; however, recent picosecond resolved fluorescence data on energy transfer in the organism is in accord with the concept outlined above (Beddard *et al.*, unpublished observations).

We would suggest that the data derived from *Synechococcus* WH7803 show the importance that nitrogen reserves have in a cyanobacterium that cannot fix nitrogen, and which inhabits an environment that is substantially nitrogen limited, and is, at least intermittently, supplied with higher irradiation than can be used in photosynthesis. The available nitrogen held as phycoerythrin allows significant growth and cell division to proceed in the absence of environmentally available nitrogen. The extent to which phycobiliproteins act as carbon reserves is less clear and probably much less important. Certainly the open ocean environment in which *Synechococcus* WH7803 and related species are found will not experience carbon limitation. Phycobiliproteins, in contrast to sugar-containing reserves such as glycogen, cannot easily be thought of as sources of respiratory energy as terminal oxidation of amino acids would largely be prevented by the absence of a complete tricarboxylic acid cycle—still a characteristic feature of all cyanobacteria that have been examined in this respect.

The situation with regard to carbon reserves in cyanobacteria is different in several respects to that of nitrogen, with good evidence that specific reserve macromolecules may act as energy reserves. Like many other microorganisms, cyanobacteria form a polyglucose macromolecule that may loosely be termed glycogen (see Dawes and Senior 1973). A limited number of cyanobacteria will form polyhydroxyacid polymers of high molecular weight (see Allen 1984; Capon *et al.* 1983). These poly-β-hydroxybutyrate-type reserves are characterized by their high level of reduction. Little is known of their metabolism in cyanobacteria and they will not be discussed further. All cyanobacteria possess low rates of oxygen uptake in the dark and these are often, but not always, unaffected by the availability of exogenous substrate. The demonstrations, first that respiratory oxygen uptake was associated with an NADPH oxidase, other respiratory enzymes, and the phosphorylation of ADP, and second that these processes could be associated with the dissimilation of glucose via the oxidative pentose phosphate pathway, established respiratory energy synthesis in these organisms (see Carr 1973; Peshek 1987).

In ^{14}C-bicarbonate-incorporation experiments conducted with *Synechococcus* species (*Anacystis nidulans*) there are indications that the role of glycogen in cyanobacteria may be more than that of a store of energy and carbon for use at some future time. When the proportion of isotope from newly fixed ^{14}C-bicarbonate recovered in isolated glycogen was followed in *Synechococcus* over a 30-min period, and expressed in terms of percentage of total radioactivity incorporated, a peak of some 15–30 per cent was observed at 2–4 min (Carr *et al.* 1982; unpublished observations). The results of this essentially 'Calvin' type of experiment were consistent with a model in which a significant proportion of newly fixed carbon transitorily entered the glycogen pool before flowing into general metabolic synthesis. As would be expected, phosphoglyceric acid was the earliest detected intermediate and a normal Calvin cycle type of fixation was operative. The diversion of the newly fixed carbon into glycogen would occur at the triose level. We suggest that glycogen acts as a dynamic reserve with two functions. The first is its accepted storage role, and the second is one of separation of environmental input of a nutrient from its subsequent metabolic utilization. These two processes have different rates imposed upon them in the integration of the several major nutrients to the ordered macromolecular synthesis necessary for cell growth. It is not only the supply of carbon that may be so regulated, but of all the components (C, N, P, and possibly Fe and S) that must be integrated into the supply, at precise proportions, for the ordered synthesis of cellular macromolecules.

In considering the metabolic role of the aspartate–arginine co-polymer, cyanophycin, its possible dual function is clearly indicated by its distribution. Cyanophycin has been known to exist in heterocysts for many years and this association with nitrogen-fixing cyanobacteria is reinforced by its absence from *Synechococcus* species. None of this genus is capable of nitrogen fixation, although the presence of cyanophycin in examples of non-nitrogen-fixing cyanobacteria, for example *Aphanocapsa* and *Agmenellum*, is well established. The area adjacent to the heterocyst pores—which have connections to the adjoining vegetative cells—contain an electron-dense material that had been earlier shown by Fogg (1951) to stain with the Sakaguchi reagent, this being presumptive evidence of the presence of arginine residues. Much later, cyanophycin was extracted from isolated heterocysts, chemically characterized, and shown to account for about the same percentage of cell dry weight as it did in the vegetative cells (Gupta and Carr 1981*a*). Heterocysts are morphologically and functionally specialized cells that occur in many genera of cyanobacteria and are now established as the location of aerobic nitrogen fixation in those cyanobacteria. Considerable attention has been directed to studies of the development and biochemistry of heterocysts, with particular emphasis on their role in nitrogen fixation and the ways in which cyanobacterial vegetative cell metabolism is modified in heterocysts so as to facilitate nitrogen fixation (for reviews, see Wolk 1982; Haselkorn 1978; Adams and Carr 1981). Heterocysts develop within a cyanobacterial filament only when available nitrogen, usually in the form of nitrate or ammonium ions, becomes limiting, and their formation ceases as soon as this limitation is released. One can fairly say, therefore, that when a cyanobacterial culture is forming heterocysts, nitrogen is the primary limiting nutrient. The question that then arises is: What is the function of a nitrogen reserve, such as cyanophycin, in the cells of an organism that is nitrogen-limited?

In addition to containing cyanophycin, heterocysts also contain all the enzymes for the synthesis of this macromolecule (Gupta and Carr 1981*a*, 1981*b*). The polymerizing enzyme, cyanophycin synthetase, and the initial degrading enzyme, cyanophycinase, are of particular interest. The latter cleaves an aspartate–arginine dipeptide from the cyanophycin molecule. Enzymes that carry out the recycling of nitrogen from arginine and aspartate have been estimated both in vegetative cells and in heterocysts of filamentous cyanobacteria (Gupta and Carr 1981*a*; Carr 1983). The striking feature of these observations is that the cyanophycin polymerase and cyanophycinase enzymes are present in the heterocysts at specific activities 30- and 70-fold greater, respectively, than found in the vegetative cell. These increases in enzymic activity are the largest found of all the enzymes examined so far in heterocysts relative to vegetative cells, other than that of nitrogenase itself. The conclusion that one may draw is that cyanophycin is being rapidly synthesized and degraded within the heterocyst and therefore exists as a dynamic reservoir, rather than as a reserve of nitrogen in the normal sense of that term. The function of cyanophycin in heterocysts would be to separate the processes of nitrogen fixation and utilization. It is known that nitrogen passes from the heterocyst to vegetative cells of *Anabaena* species in the form of glutamine (see Wolk 1982) and this may be considered the major aspect of utilization. We do not know the proportion of total fixed nitrogen that flows through the cyanophycin pool. The *in vitro* rates of cyanophycin synthetase and cyanophycinase are low compared to the actual *in vivo* rate of nitrogen fixation (Gupta and Carr 1981*a*). However, such discrepancies between individual enzymic activities and the overall rate of the process in which the enzymes participate are not unusual, especially when the substrates are macromolecules, whose physical state may be far from optimal for *in vitro* enzymic conversion.

Within the *Synechococcus* group there appears to be more than one aspect of the metabolism of reserve materials not previously recognized. The extent to which these features are general and occur in other cyanobacteria remains to be examined. The role played by nitrogenous reserves as a whole appears to be widespread. Although phycobiliproteins in most cyanobacteria appear to be efficiently coupled to energy transfer, there are some indications that exceptions to this are not limited to the marine *Synechococcus* species discussed here. The maintenance of photosynthesis activity over a period of time during which significant proportions of phycocyanin were mobilized in *Spirulina platensis* under conditions of nitrogen limition have been observed by Boussiba and Richmond (1980). Whether, in addition to any storage function, cyanophycin and glycogen have a widespread role in cyanobacteria as dynamic reservoirs, as has been suggested, also requires investigation. This feature of macromolecular reserves may be a reflection of the autotrophic nature of cyanobacterial nutrition, in which case the examination of the role of glycogen in carbon assimilation of the chemoautotrophs would be worthwhile.

References

Adams, D.G. and Carr, N.G. (1981). *Crit. Rev. Microbiol.* **9**, 45.
Allen, M.M. (1984). *A. Rev. Microbiol.* **38**, 1.
Allen, M.M. and Smith, A.J. (1969). *Arch. Mikrobiol.* **69**, 114.
Boresch, K. (1910). *Lotos (Prague)* **58**, 344.
Boussiba, S. and Richmond, A.E. (1980). *Arch. Microbiol.* **125**, 143.
Bryant, D.A. (1986). *Can. Bull. Fish. Aqua. Sci.* **214**, 423.
Capon, R.J., Dunlop, R.W., Ghisalberti, E.L., and Jefferies, P.R. (1983). *Phytochemistry* **22**, 1181.
Carr, N.G. (1973). In *The biology of blue-green algae* (ed. N.G. Carr and B.A. Whitton) p. 39. Blackwell Scientific Publications, Oxford.
Carr, N.G. (1983). In *Photosynthetic prokaryotes* (ed. G. Papageorgiou and L. Packer) p. 265. Elsevier Biomedical, New York.
Carr, N.G. and Wyman, M. (1986). *Can. Bull. Fish. Aqua, Sci.* **214**, 159.
Carr, N.G., Gupta, M., Rylah, J., and McKie, J. (1982). In *Abstracts IV Int. Symp. Photosyn. Prokaryotes*, p. B12. Bombannes, France.
Dawes, E.A. and Senior, P.J. (1973). *Adv. Microbiol. Physiol.,* **10**, 135.
Dortch, Q., Clayton, J.R., Thoresen, S.S., and Ahmed, S.I. (1984). *Mar. Biol.* **81**, 237.
Dring, M.J. (1982). *The biology of marine plants*. Edward Arnold, London.
Dunstan, R.H. and Whatley, F.R. (1986). *Proc. R. Soc. Lond.* B **227**, 429.
Fogg, G.E. (1951). *Ann. Bot. N.S.* **15**, 23.
Glazer, A.N. (1982). *A. Rev. Microbiol.* **36**, 173.
Gupta, M. and Carr, N.G. (1981*a*). *J. gen. Microbiol.* **125**, 17.
Gupta, M. and Carr, N.G. (1981*b*). *FEMS Microbiol. Lett.* **12**, 179.
Haselkorn, R. (1978). *A. Rev. plant Physiol.* **29**, 319.
Jensen, T.E. (1984). *Cytobios* **39**, 35.
Kingsbury, J.M. (1956). *Biol. Bull.* **110**, 310.
Newman, J., Wyman, M., and Carr, N.G. (1987). *FEMS Microbiol. Lett.* **44**, 221.
Peshek, G.A. (1987). In *The cyanobacteria* (ed. P. Fay and C. Van Baalen) p. 119. Elsevier, Amsterdam.
Simon, R.D. (1971). *Proc. natn. Acad. Sci. USA* **68**, 265.

Waterbury, J.B., Watson, S.W., Guillard, R.R.L., and Brand, L.E. (1979). *Nature, Lond.* **277**, 293.
Waterbury, J.B., Watson, S.W., Valois, F.W., and Franks, D.G. (1986). *Can. Bull. Fish. Aqua. Sci.* **214**, 71.
Wolk, C.P. (1982). In *The biology of cyanobacteria* (ed. N.G. Carr and B.A. Whitton) p. 359. Blackwell Scientific Publications, Oxford.
Wyman, M., Gregory, R.P.F., and Carr, N.G. (1985). *Science* **230**, 818.
Yeh, S.W., Ong, L.J. and Glazer, A.N. (1986). *Science* **234**, 1422.

3 Uptake and utilization of nitrogen compounds

P.J. SYRETT

Plant and Microbial Metabolism Research Group, School of Biological Sciences, University College of Swansea, Swansea SA2 8PP, UK

Introduction

In a microalgal cell growing without nutrient restriction, half of the dry matter is carbon and about one-twelfth is nitrogen (Vaccaro 1965). Oxygen and hydrogen contribute most of the remainder but they are readily available from water. Consequently the determinants for microalgal growth are generally the availabilities of carbon, usually by photosynthesis, of nitrogen, and, to a lesser extent, of phosphorus.

The nitrogen compounds that serve as N sources for the growth of cyanobacteria and algae have been reviewed previously (Syrett 1981). They include N_2 gas (for some cyanobacteria), the inorganic forms NH_4^+, NO_2^-, and NO_3^-, and organic compounds. Many microalgae utilize the amide N of urea and glutamine. Often, amino acids and purines are also good N sources, but pyrimidines are less good.

The uptake of nitrogen compounds

Relatively small uncharged molecules like urea can move passively across plant cell membranes, as can uncharged NH_3 molecules, which are present at higher pH values ($>$pH 8).

Nevertheless, passive entry of nitrogen compounds other than N_2 appears to be insignificant in the utilization of nitrogen by algae. We now have good evidence for the existence of assisted uptake mechanisms for a wide variety of nitrogen compounds, including NH_4^+, NO_2^-, and urea, as well as for NO_3^-, amino acids, and purines.

Studies on uptake by the common laboratory algae *Chlorella* and *Chlamydomonas* and by cyanobacteria have been hampered by the fact that it is virtually impossible in these organisms to demonstrate a significant internal concentration of NO_3^- or NH_4^+; this difficulty appears to be correlated with the absence or sparsity of vacuoles in these organisms. Consequently, while much work has demonstrated the disappearance of NO_3^- and NH_4^+ from the medium around the cells and the appearance of organic N in the cells, it has been difficult to distinguish experimentally between the uptake of a nitrogen compound and its subsequent metabolism, this is particularly true with ionic forms such as NO_3^-. With more highly vacuolate algae, appearance of NH_4^+ and NO_3^- within the cells could be demonstrated more clearly; for example, in the diatom *Ditylum brightwellii* (Eppley and Rogers 1970).

However, with the smaller non- or little-vacuolated algae, several workers have demonstrated that the half-saturation constants for removal of the substrate from the medium (K_s) are often one or two orders of magnitude lower than the Michaelis constant (K_m)of the enzymes involved in the primary metabolism of the substrate (Eppley *et al.* 1969). This difference points to the presence of uptake systems with a high affinity for the substrate. Moreover, with microalgae such as the diatom *Phaeodactylum tricornutum*, despite their small size (volume 150 μm^3), accumulation of metabolites such as NO_3^-, urea, and amino acids within the cells can be demonstrated and it is then possible to measure uptake not only by disappearance from the medium but more unequivocally by following the appearance of the substrate within the cells. In other instances uptake can be followed by use of little- or non-metabolizable analogues of the natural substrate, for example, ^{14}C-methylammonium for NH_4^+ and thiourea for urea. Table 3.1 summarizes the results of some of our studies with *P. tricornutum*. Wheeler (1983) gives an excellent summary of work on the uptake of nitrogenous compounds by marine microalgae as a whole.

One feature of the uptake systems for nitrogenous compounds in *P. tricornutum* is that their activities are relatively low in nitrogen-replete cells. Indeed, cells grown with NH_4^+ as nitrogen source have no ability at all to take up to NO_3^- or NO_2^- and only a low ability to take up urea or amino acids. During a period of about 2 hours of nitrogen deprivation under conditions which allow photosynthesis to continue, the ability to take up NO_3^- and NO_2^- develops and that to take up other nitrogen compounds such as urea and amino acids increases markedly; in particular, the ability to take up methylammonium (an analogue of NH_4^+) increases very greatly—by some 100-fold in 30 hours (Syrett *et al.* 1986).

The increase in the ability to take up nitrogen compounds during a period of N-deprivation is of interest for two quite different reasons. First, it raises the question of the regulatory mechanisms involved. Second, one may ask whether the increases one sees are of ecological significance. Natural phytoplankton populations often occur in waters in which the concentrations of defined nitrogen compounds are very low or undetectable. Have the microalgae in such populations developed an increased ability to take up nitrogen compounds? The work of Glibert and McCarthy (1984), for example, suggests that this is so.

The ability of *P. tricornutum* to take up the basic amino acids, arginine and lysine, differs from the other uptake abilities that we have studied, in that,

Table 3.1 Characteristics of the uptake systems for nitrogenous compounds in *Phaeodactylum*.

Compound	Rate of growth with compound as sole N source	Effect of N deprivation on rate of uptake	Half-saturation constant for uptake (K_s) (μM)
NO_3^-	+ +	0 → + +	7
NO_2^-	not measured	0 → + +	6
NH_4^+	+ +	+ → + + + + + + + + +	<1
$CH_3NH_3^+$	0	+ → + + + + + + + + +	30
Urea	+ +	0 → + +	0.6
Guanine	+ +	0 → + +	0.5
Arginine	+ +	0 → + +	1
Lysine	+	0 → + +	1
		↑ Rate in NH_4^+-grown cells	

All the uptake systems examined are dependent on the presence of Na^+ and show inhibition of rate of uptake by K^+ (Rees *et al.* 1980). Information about other characteristics of these uptake systems is given in references cited in Syrett *et al.* (1986).

not only does it increase during nitrogen deprivation, but it also increases during carbon deprivation. Indeed, the best way to increase the uptake ability for these amino acids is to incubate cells in nitrogen-free medium in darkness (Flynn and Syrett 1985). We do not know how general this response is amongst microalgae; it could be of ecological significance in estuarine habitats where turbulence may not only cut off light from microalgae by resuspending bottom deposits but may also, by disturbing these deposits, release amino acids from the intertidal water where their concentrations are known to be an order of magnitude higher than in the overlying water column (Flynn *et al.* 1987).

With regard to the mechanisms of uptake, entry by some form of assisted transport can be distinguished from entry by passive diffusion by, first, the demonstration of Michaelis–Menten type kinetics with a half-saturation constant which often indicates the presence of a transporter with a high affinity for the substrate (Table 3.1), and, second, by a dependence on a source of metabolic energy, either aerobic respiration or light (Cresswell and Syrett 1981).

Of the details of the transport mechanisms we know little. Falkowski (1975) had evidence for a primary pump for NO_3^-, driven by ATP and located in the plasmalemma of the diatom, *Skeletonema costatum*. Most workers, however, incline to the view that the uptake of nitrogenous compounds is by secondary transport mechanisms driven by gradients of H^+ or Na^+ (Raven 1980). Rees and Syrett (1984) studied the uptake of the urea analogue, thiourea, by *Chlorella* and found evidence for the inward transport of one thiourea molecule with one proton; this is the same stoichiometry as for the entry of glucose, also a neutral compound, into *Chlorella* (Komor and Tanner 1974). Fuggi (1985), in a detailed study of the kinetics of NO_3^- uptake by the acidophilic alga, *Cyanidium caldarum*, found that two protons were taken up with each NO_3^- ion. Moreover, the results indicated that the sequence of binding with the carrier was first the attachment of NO_3^- followed by that of the two protons. Such a mechanism for NO_3^- uptake would imply the entry of $(NO_3, 2H)^+$ consistent with the entry of (thiourea, $H)^+$. However, Schlee *et al.* (1985) argue for the proton-driven entry of NO_3^- into *Chlorella* with a stoichiometry of 1 H^+:1 NO_3^-.

In the marine diatom, *P. tricornutum*, entry of nitrogen compounds is dependent on the presence of Na^+ (Rees *et al.* 1980). The same is true for the entry of glucose and amino acids by another marine diatom, *Cyclotella cryptica* (Hellebust 1978), and of phosphate by the marine fungus, *Thraustochytrium roseum* (Siegenthaler *et al.* 1967). It is tempting to suggest that in marine microalgae, living in water at about pH 8, the uptake pumps are driven by a Na^+ gradient rather than by a H^+ gradient. The Na^+ gradient would then be maintained by a Na^+ pump similar to that so well known in animal cells.

We also know practically nothing about the location of the transporting mechanisms in microalgae. Presumably there must be such a mechanism in the plasmalemma and, where accumulation occurs in vacuoles, in the tonoplast also. With giant algal cells such as those of *Hydrodictyon* evidence for transporters in both boundaries is good. Algae such as *Chlorella* do have vacuoles, albeit small ones. One wonders whether the persistent failure to detect any appreciable quantities of NO_3^- in such cells, in contrast to diatoms such as *Phaeodactylum*, lies with the absence of a NO_3^- transporter in the tonoplast membrane.

The enzymes of nitrogen assimilation

Ammonium (NH_4^+) assimilation

After the demonstration of a ferredoxin-linked glutamate synthase reaction in *Chlorella* and cyanobacteria (Lea and Miflin 1975), the prevailing view has become that NH_4^+ assimilation in microalgae takes place by the incorporation of NH_4^+ into glutamine, catalysed by glutamine synthetase (GS), followed by the reductive reaction of glutamine with α-oxoglutarate to yield two molecules of glutamate, the reaction catalysed by glutamate synthase (for example, see Edge and Ricketts 1978; Cullimore and Sims, 1981).

Two isoforms of GS (E.C. 6.3.1.2) are known, GS_1 and GS_2. In higher plants it is well established that GS_1 is cytosolic in location while GS_2 is chloroplastidic. In microalgae two similar isoforms can be distinguished and, largely from analogy with higher plants, GS_1 is regarded as cytosolic and GS_2 as chloroplastidic (Florencio and Vega 1983; Casselton *et al.* 1986). Whilst several microalgae (e.g. *Chlorella fusca, P. tricornutum*) contain both isoforms, others contain only the chloroplastidic form (e.g. *Chlorella stigmatophora, Euglena gracilis*, and the seaweeds *Fucus vesiculosus* and *Ascophyllum nodosum*); a small group contain only GS_1 (Casselton *et al.* 1986).

Similarly in some eukaryotic algae, perhaps in all, there are two glutamate synthase enzymes, one (E.C. 1.4.1.14.) that uses NADH as reductant and the other (E.C. 1.4.7.1) ferredoxin (Vega *et al.* 1987). The NADH enzyme may be cytosolic and mainly responsible for the assimilation of newly entered NH_4^+; the ferredoxin-dependent enzyme may be chloroplastidic and mainly responsible for the reassimilation of NH_4^+ released during photorespiration. The NADH enzyme from *Chlamydomonas* has an M_r of about 370 000; the ferredoxin-dependent enzyme is smaller and probably contains an iron–sulphur grouping of the (3Fe–XS) type (Vega *et al.* 1987).

The evidence for the main pathway of NH_4^+ assimilation being through GS and glutamate synthase rather than through glutamate dehydrogenase as was thought earlier comes partly from the generally higher affinity of GS (in comparison with glutamate dehydrogenase) for NH_4^+ and partly from the use of methionine sulphoximine (MSX) which is a rather specific inhibitor of GS. Thus, for example, Ahmed *et al.* (1977) examined the NADP-glutamate dehydrogenases from eight marine species of phytoplankton and found the apparent K_m values for NH_4^+ to be in the range 4–10 mM; in contrast the K_m value for glutamine synthase from *Chlamydomonas* is only 0.2 mM (Cullimore and Sims 1981). Cullimore and Sims (1981) also showed that incubation for 2 hours with MSX inhibited completely GS in *Chlamydomonas reinhardii* and also completely stopped NH_4^+ assimilation; indeed, in the presence of MSX *Chlamydomonas* produces NH_4^+, which arises partly from the photorespiratory cycle and partly from protein degradation (Cullimore and Sims 1980; Hipkin *et al.* 1982).

The role of the enzyme glutamate dehydrogenase (GDH) which catalyses the interconversion of α-oxoglutarate plus NH_4^+ and glutamate with electron transfer from either NADH (E.C. 1.4.1.2.) or NADPH (E.C 1.4.1.4.) is now thought of as largely dissimilatory in microalgae; i.e. as catalysing the conversion of glutamate to oxoglutarate and NH_4^+. There is, however, evidence for its role in NH_4^+ assimilation in some species. Ahmad and Hellebust (1985) showed that *Chlorella autotrophica* can assimilate NH_4^+ via NADPH-GDH in the presence of MSX. The best evidence for NH_4^+ assimilation by this pathway comes from studies with the green alga, *Stichococcus bacillaris*. This alga contains a NADPH-GDH with a K_m for NH_4^+ of 1 mM; thus its affinity for NH_4^+ is greater than that of the algal GDH enzymes studied by Ahmed *et al.* (1977). Moreover, the activity of this enzyme is much higher in cells grown on nitrate or low concentrations of NH_4^+ (Everest and Syrett 1983; Ahmad and Hellebust 1986). Whilst 1 mM MSX inhibited almost completely NH_4^+ assimilation by cells grown with ample NH_4^+, it inhibited to a much lesser extent NH_4^+ assimilation by NO_3^--grown cells (Everest and Syrett 1983). Even more striking is the demonstration by Ahmad and Hellebust (1986) that at least some cells in a population of *S. bacillaris* can be trained to grow in the presence of MSX with either NO_3^- or NH_4^+ as a nitrogen source. Such cells have a high NADPH-GDH activity and a low but discernible GS activity; the latter presumably maintains a sufficient supply of glutamine for protein synthesis, although the main pathway of nitrogen assimilation is through the NADPH-GDH pathway (Ahmad and Hellebust 1986).

These observations, together with those on the distribution of the isoforms of GS and glutamate synthase, pose the question of whether assimilation of NH_4^+ by the NADPH-GDH pathway is correlated with an absence from the cytosol of GS and glutamate synthase. It is clear that in *S. bacillaris* and other algae, the levels of the NH_4^+-metabolizing enzymes are dependent on the nature and concentration of the nitrogen source for growth. As yet we lack a comprehensive study in which the activities of the various forms of GDH, GS, and glutamate synthase have all been examined in the same species grown with different nitrogen regimes. Such a study is much needed to clarify the roles of these enzymes.

Nitrate assimilation

In eukaryotic algae, nitrate is reduced to nitrite by nitrate reductase (NR) with NADH or NADPH as electron donor. All the algal enzymes studied can utilize NADH (E.C. 1.6.6.1.) but only some can also utilize NADPH in addition (E.C. 1.6.6.2.); there is no evidence for the occurence of two different enzymes in one organism (Syrett 1981).

The best known NR is that of *Chlorella* (Solomonson and Barber 1987). It is an enzyme of approximate M_r 370 000 made up of four identical subunits each of which contains one FAD molecule, one haem of the cytochrome *b* type, and one molybdenum atom. The molybdenum is present as a pterin–Mo complex which can be removed from the enzyme.

Usually the active NR of eukaryotic algae is soluble and is recovered in the supernatant after cell breakage and centrifugation. There have been suggestions that it is located in the chloroplast or between the chloroplast outer membranes (Grant *et al.* 1970) but direct evidence is lacking and it is generally accepted as being located in the cytoplasm, a view which accords with NADH being the electron donor. Recent immunological evidence has located NR in the chloroplast pyrenoid of several eukaryotic algae (Roldán *et al.* 1987) but it is not certain that this is an active enzyme.

One feature of NR is its complex regulation; it can readily be reversibly inactivated. *In vitro* this is most easily done by reducing it with NADH in the presence of small amounts of HCN (10^{-8} M) which binds to the Mo site; the enzyme is reactivated by mild oxidation with ferricyanide. Irradiation with blue light also reactivates the enzyme, the light being absorbed by the flavin in the enzyme and probably bringing about its oxidation (Maldonado and Aparicio 1987). There is evidence that these changes in activity also occur *in vivo* (Solomonson and Spehar 1977). Certainly nitrate reduction by *Chlamydomonas* is stimulated by blue light (Azuara and Aparicio 1983).

The nitrate reductases of cyanobacteria have received much less detailed study, partly because they are more difficult to solubilize. Those that have been studied differ markedly from the NR of eukaryotic algae. They contain Mo but lack flavin or cytochrome and have an M_r of only 75 000. *In vitro*, reduced ferredoxin, but not pyridine nucleotide, acts as electron donor (Manzano *et al.* 1978).

Nitrite reductase, which catalyses the reduction of NO_2^- to NH_4^+ has also been much less studied than NR. The best known enzyme is from spinach leaf. It is a chloroplast enzyme with a single polypeptide chain with an M_r of 61 000 and contains sirohaem with a tetranuclear iron–sulphur centre. *In vitro*, and presumably *in vivo*, reduced ferredoxin is the electron donor for NO_2^- reduction and no free intermediates between NO_2^- and NH_4^+ are detectable (Kamin and Privalle 1987). Algal nitrite reductases are less well characterized, but from the studies that have been made [e.g. Zumft 1972 (*Chlorella*); Llama *et al.* 1979 (*Skeletonema*); Galván *et al.* 1987 (*Chlamydomonas*)] there is no reason for thinking, at present, that they differ markedly from the spinach leaf enzyme.

Urea assimilation

Many microalgae grow well with urea as sole nitrogen source and most of them contain urease (E.C. 3.5.1.5.), which catalyses:

$$CO(NH_2)_2 + H_2O + 2H^+ \rightarrow CO_2 + 2NH_4^+$$

Urease from jack beans has been purified and been shown to contain nickel (Dixon *et al.* 1975). Again the algal enzymes are less well studied but the dependence of urease-containing algae on the presence of Ni^{2+} in the medium when they are growing on urea (Rees and Bekheet 1982; Oliviera and Antia 1984) indicates strongly that nickel is a constituent of algal urease.

The common green laboratory algae such as *Chlorella* and *Chlamydomonas* grow well on urea but do not contain urease. Instead, these algae, in common with some yeasts, contain two enzymic activities which together are called urea amidohydrolase (E.C. 3.5.1.45.) and which catalyse the same overall reaction as urease but at the expense of one molecule of ATP per urea molecule broken down (Roon and Levenberg 1968). First, a biotin-dependent urea carboxylase catalyses the carboxylation of urea to urea carboxylate (allophanate):

$$CO(NH_2)_2 + HCO_3^- + ATP \rightarrow CO(NH_2)NHCOO^- + ADP + P_i.$$

Allophanate hydrolase then catalyses

$$CO(NH_2)\,NHCOO^- + H_2O + 3H^+ \rightarrow 2NH_4^+ + 2CO_2.$$

Neither of these enzymic activities shows any dependence on Ni^{2+} (Rees and Bakheet 1982). In the green algae, the distribution of urease correlates with that of glycollate oxidase (E.C. 1.1.3.1.) and that of urea amidohydrolase with glycollate dehydrogenase (E.C. 1.1.99.14.). The distribution of the enzyme pairs also correlates with a fundamental division of the green algae made on cytological criteria into a Chlorophyceaen line (including *Chlorella* and *Chlamydomonas*) and a Charaphyceaen line from which the land plants presumably originated (Syrett and Al-Houty 1984). Why some algae should evolve an enzyme system that requires ATP to catalyse a reaction which, with urease, proceeds very well without ATP, and why some yeasts are the only other organisms known to metabolize urea in this way are intriguing questions upon which one may speculate but which lack convincing answers.

Interactions in the assimilation of nitrogen compounds

It is generally not possible to increase significantly the growth rate of an alga growing with NO_3^- or NH_4^+ as sole N source by adding another utilizable nitrogen compound. Interactions must occur, therefore, which, in the long term, equalize the total rate of N uptake and its rate of conversion to new cell material. A good example of such interaction is shown when *P. tricornutum* is incubated with NO_3^-, NO_2^-, and NH_4^+; here NH_4^+ disappears first, then NO_3^-, and finally NO_2^- (Cresswell and Syrett 1982). In this example, NO_2^- assimilation is inhibited by NO_3^-, as has been observed with other diatoms (Eppley and Coatsworth 1968; Serra *et al.* 1978). This inhibition may occur because NO_3^- and NO_2^- are transported across the membrane by a common carrier which has a higher affinity for NO_3^- (Bilbao *et al.* 1981), but this is not proven. Moreover, sometimes the reverse result is found, with NO_2^- addition inhibiting NO_3^- uptake (Syrett and Morris 1963; Thacker and Syrett 1972); such inhibition may then result from NH_4^+ produced from NO_2^- reduction.

Nitrite is probably not a significant natural source of nitrogen for algae and the interaction that has been most studied is that between NO_3^- and NH_4^+ assimilation. In the 1960s we investigated the control mechanisms underlying the preferential assimilation of NH_4^+ by *Chlorella* and found two. First, in the presence of NH_4^+, cells do not make NR; they do make it when NH_4^+ is removed (Morris and Syrett 1963). We can now add that, with NH_4^+, cells also do not have an active uptake mechanism for NO_3^-; this too appears when NH_4^+ is removed (Cresswell and Syrett 1981). Second, the addition of NH_4^+ to cells assimilating NO_3^- rapidly inhibits NO_3^- assimilation (Syrett and Morris 1963). This effect had previously been demonstrated with fungi (Morton 1956); it has now been seen many times with cultures of microalgae.

Later, a third phenomenon was described by the Spanish workers. Addition of NH_4^+ often produces a reversible inactivation of NR in the cells, the inhibition being removed when the NH_4^+ has gone (Losada *et al.* 1970). Lastly, a fourth phenomenon has been described by Deane-Drummond (1985) using the macroalga *Chara*, in which addition of NH_4^+ is followed by an *efflux* of NO_3^- from the cells; the generality of this effect is not known (Larsson and Larsson 1987).

There has been much interest in the inhibition of NO_3^- assimilation which follows NH_4^+ addition, first in attempts to understand the mechanism and second because of its ecological implications. There is evidence from studies of natural populations of marine phytoplankton that they assimilate NH_4^+ rather than NO_3^- when the concentration of NH_4^+ is above about 2 μM (McCarthy *et al.* 1977). It must be added, however, that there are clear examples from natural ecosystems in which both NH_4^+ and NO_3^- are assimilated simultaneously by microalgae. One of the best is in the oyster ponds studied by Maestrini *et al.* (1982).

Earlier interpretations of the rapid inhibition of NO_3^- assimilation following the addition of NH_4^+ tended to implicate inhibition of NR and this explanation was strengthened by the demonstration of a reversible inter-activation of NR in *Chlorella* and *Chlamydomonas* following NH_4^+ addition (Losada *et al.* 1970). We could not demonstrate, however, such inactivation in *Chlamydomonas* grown under our conditions (Syrett and Leftley 1976), even though inhibition of NO_3^- assimilation was clear cut. Moreover, the inactivation of NR is too slow to account for the rapidity with which inhibition by added NH_4^+ occurs (Larsson *et al.* 1985). Attention, therefore, moved to possible effects on NO_3^- transport. This was difficult to investigate with algae such as *Chlorella* and *Chlamydomonas* because no appreciable amount of NO_3^- appears as such in the cells and transport across the cell boundary cannot be distinguished clearly from subsequent metabolism. However, with the diatom *P. tricornutum*, Cresswell and Syrett (1979) were able to measure NO_3^- appearance in the cells and it was clear that NH_4^+ addition not only rapidly inhibited NO_3^- assimilation but it did so by stopping NO_3^- transport.

There appears now to be fairly general agreement that the first effect of NH_4^+ on NO_3^- assimilation is prevention of transport of NO_3^- into the cells; this effect is then followed by effects on NR (Guerrero *et al.* 1981; Ullrich 1983). Opinions differ, however, as to whether inhibition of the NO_3^- transport is caused directly by NH_4^+ or by a product of its metabolism. Syrett and Morris (1963) showed that *Chlorella* cells that had been depleted of available carbon by aeration in darkness, when illuminated in a CO_2-free atmosphere, reduced NO_3^- stoichiometrically to NH_4^+. Because of the absence of available carbon, NH_4^+ was not converted to organic nitrogen, and the addition of NH_4^+ did not inhibit the conversion of NO_3^- to NH_4^+. They argued, therefore, that the inhibitor of NO_3^- assimilation in normal cells was not NH_4^+ but an organic product formed from it. This conclusion is supported by the finding that, in the presence of MSX, which, by inactivating glutamine synthetase stops the formation of organic N from NH_4^+, the addition of NH_4^+ does not inhibit NO_3^- uptake (Flores *et al.* 1981; Di Martino Rigano *et al.* 1982).

It is also noteworthy that the inactivation of algal NR that follows addition of NH_4^+ is prevented by prior incubation with MSX (Rigano *et al.* 1979); again it appears that a product of NH_4^+ assimilation, rather than NH_4^+ itself, is the effector.

Others, however, argue for a direct effect of NH_4^+ on NO_3^- assimilation (Florencio and Vega 1982). One piece of evidence is that methylammonium, an analogue of NH_4^+ which is metabolized either not at all or only as far as γ-N-methylglutamine (Franco *et al.* 1984), also inhibits NO_3^- assimilation (Diez *et al.* 1977). Cresswell and Syrett (1984), however, investigated the inhibition of NO_3^- and NO_2^- uptake by $CH_3NH_3^+$ in *P. tricornutum* and showed that it differed in type from that produced by NH_4^+; inhibition by $CH_3NH_3^+$ took longer to develop and was associated with its accumulation in the cells whereas NH_4^+ inhibition was immediate. NH_4^+ does not always inhibit the uptake of N compounds; for example, it does not inhibit the uptake of arginine by *P. tricornutum* (Flynn and Syrett 1986). Fig. 3.1 shows, however, that, in contrast to 1 mM NH_4^+, the presence of 1mM $CH_3NH_3^+$ does inhibit ^{14}C-arginine uptake by this diatom; we interpret this as additional evidence that NH_4^+ and $CH_3NH_3^+$ do not act similarly.

Ullrich *et al.* (1984) showed that with duckweed (*Lemna*) roots, NH_4^+

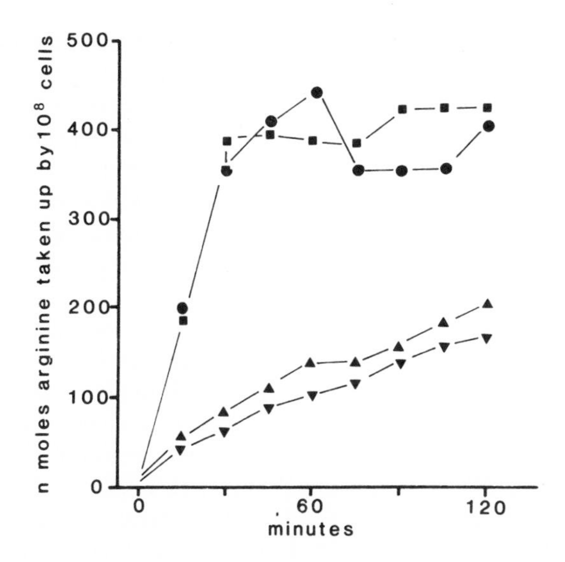

Fig. 3.1. The effect of addition of 1 mM NH_4Cl or 1 mM methylammonium chloride on the uptake of 10 μm ^{14}C-arginine by *P. tricornutum*. ^{14}C-arginine was added at zero time to cultures at pH 8 and 20 °C. Methylammonium was added either 15 min before arginine (-▼-▼-) or simultaneously with it (-▲-▲-). NH_4^+ was added 15 minutes before arginine (-■-■-). Arginine uptake by the control (-●-●-); NH_4^+ addition simultaneously with arginine had no effect on arginine uptake (results not shown). Cell density was 2×10^6 cells ml^{-1} with light energy flux at 32 Wm^{-2}. (Based on unpublished results of K.J. Flynn.)

inhibits both NO_3^- and phosphate uptake. This inhibition is attributed to a depolarization of the membrane potential by NH_4^+ addition (Ullrich 1987). This effect therefore appears to be one of NH_4^+ *per se* and does not act specifically on the transport of NO_3^-. We need to know whether the effects of NH_4^+ studied in microalgae are specific to transport of N compounds or are accompanied by more general effects on transport.

At present, I still favour the view that the inhibition of NO_3^- transport is closely related to the formation of an organic product of NH_4^+ assimilation. This appears to be the simplest way of explaining why NH_4^+ addition does not always inhibit NO_3^- uptake by various microalgae. For example, some algae isolated from the oyster ponds mentioned above show an ability to take up NO_3^- quickly in the presence of concentrations of NH_4^+ as high as 10 or 20 μM, and different species behave differently (Maestrini *et al.* 1986).

If the inhibitor of NO_3^- uptake is related to a product of NH_4^+ incorporation, one might expect the degree of inhibition to depend on the carbon status of the cells and their ability to fix CO_2. Several pieces of evidence point in this direction. For example, 'normal' cells of *Chlamydomonas* do not take up NO_3^- in either light or darkness, but will do so in light if CO_2 is supplied; in contrast, nitrogen-depleted cells, with internal carbon reserves, take up and assimilate NO_3^- in darkness (Thacker and Syrett 1972). The interpretation of this is that normal cells contain a NO_3^- transport inhibitor which is removed when assimilable carbon is readily available and which builds up rapidly when NH_4^+ is added. Larsson *et al.* (1985) showed that the availability of CO_2 during the photosynthetic growth of *Scenedesmus obtusiusculus* altered the NO_3^-/NH_4^+ interaction. With cells grown with air enriched with 3 per cent CO_2, added NH_4^+ was rapidly assimilated and its addition immediately inhibited NO_3^- uptake. With cells grown in air unenriched with CO_2, the effect of NH_4^+ addition was less marked; the rate of NH_4^+ assimilation was low and NO_3^- uptake declined slowly, ceasing completely 40 minutes after NH_4^+ addition. It is likely that the differing responses are related to the carbon status of the cells; organisms grown with 3 per cent CO_2 might well contain carbon reserves that are utilized immediately when NH_4^+ is added.

Work with the cyanobacterium *Anacystis nidulans* (Lara *et al.* 1987) supports these views and, moreover, links the formation of glutamine closely with the inhibition of NO_3^- transport. Glutamine itself, however, appears not to be the inhibitor; rather it is the rate of glutamine formation which is significant. It has been demonstrated recently that the rate of glutamine synthesis is the linking control mechanism that regulates the rate of sucrose catabolism in *Neurospora* (Hernadez and Mora 1986).

The availability of CO_2 as a carbon source is thus important and, using *P. tricornutum*, in which NO_3^- transport into the cell can be demonstrated unequivocally, we have shown that the accumulation of NO_3^- in the cells is dependent upon the presence of CO_2 (Fig. 3.2).

There are, however, as yet some unexplained facets of the interrelationships between NO_3^- transport, presence or absence of CO_2, and pH. Eisele and Ullrich (1977) showed that, in the absence of CO_2, NO_3^- uptake by *Ankistrodesmus braunii* is very slow at pH 6. At pH 8, NO_3^- is taken up more quickly, part of its nitrogen being assimilated but a large portion appearing as NH_4^+. The addition of CO_2 or of glucose allows NO_3^- assimilation (without NH_4^+ appearance) at both pH 6 and pH 8. Di Martino Rigano *et al.* (1985) obtained similar results with a *Chlorella* strain. To explain these observations Eisele and Ullrich (1977) suggested the presence of a NO_3^- carrier which was labile at pH 6. This may be so, but it is worth stressing that in algal cells which are obtaining carbon from carbohydrate, the β-carboxylation reaction that converts C_3 to C_4 compounds is absolutely essential to form the C_4 and C_5 keto acids that are converted to amino acids. Without them NH_4^+ assimilation cannot proceed and feed back mechanisms which prevent entry of NO_3^- must operate to a greater or lesser extent. At pH 6, dissolved inorganic carbon is largely in the form of CO_2. At pH 8 it is largely in the form of

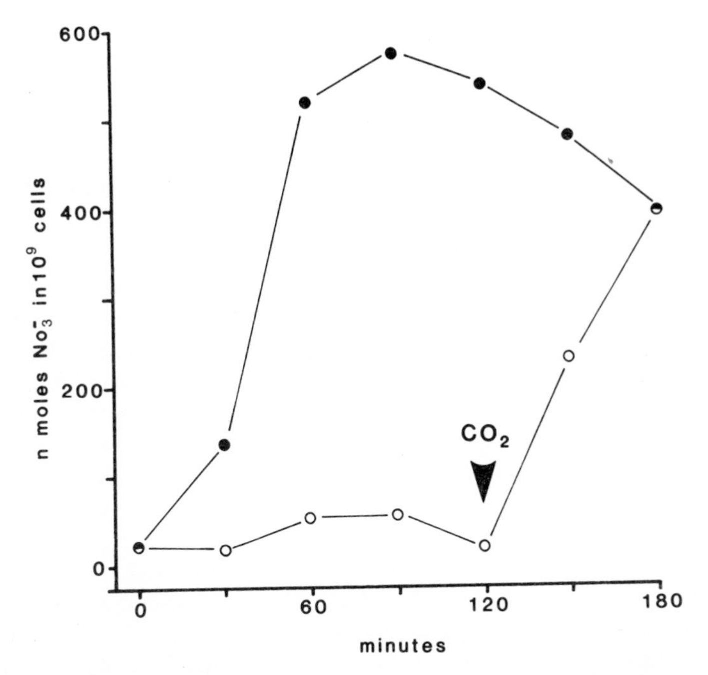

Fig. 3.2. The effect of CO_2 deprivation on NO_3^- accumulation in cells of *P. tricornutum*. Cells were grown with 10 mM NH_4^+ for 3 days and then deprived of nitrogen for 3 hours in order to develop the NO_3^- uptake system (Cresswell and Syrett 1981). The culture was divided into two parts. One part was aerated with CO_2-free air (-○-○-); the other was aerated with air enriched to 0.5 per cent CO_2 (-●-●-). At zero time KNO_3 was added to give a concentration of 300 μM. Samples were taken, cells collected by filtration, and the NO_3^- in them determined (Cresswell and Syrett 1981). After 120 min (arrow), air enriched to 0.5 per cent CO_2 was supplied to the CO_2-deprived culture. Cell density was 1×10^7 cells ml^{-1} with light energy flux at 24 Wm^{-2}. The medium at 20 °C was buffered with 8.2 mM Tris; the initial pH was 7.9 and remained constant when aerated with air minus CO_2, but fell to 7.3 during aeration with air enriched to 0.5 per cent CO_2. (Based on unpublished results of A.M. Peplinska.)

HCO_3^-; CO_2 produced by cellular respiration will then be more difficult to remove and, moreover, being in the form of HCO_3^-, will be a substrate for the carboxylating enzymes responsible for the C_3 to C_4 conversion (Beardall *et al.* 1976; Appleby *et al.* 1980). The differences in response at pH 6 and pH 8 may be in the availability of endogenously produced CO_2 (HCO_3^-) for the β-carboxylation reaction essential to maintain NH_4^+ assimilation. The effect of addition of glucose, which will stimulate respiration and hence provide additional endogenous CO_2 (HCO_3^-), may be explained in this way.

There appears also to be another type of control which links the assimilation of inorganic N and carbohydrate metabolism. It was shown some years ago that *Chlorella* cells, in darkness, cease to assimilate NO_3^- though they still contain appreciable quantities of 'storage' polysaccharide. Addition of glucose allows NO_3^- assimilation to proceed; and, moreever, the cells can use the intracellular 'storage' polysaccharide for NH_4^+ assimilation when NH_4^+ is added (Syrett 1956). These observations have recently been extended with the dinoflagellate *Amphidinium carterae* (Dixon and Syrett 1988; Syrett 1987). *Amphidinium* grows on either NO_3^- or NH_4^+ at about the same rate, but cells growing on NO_3^- always have the ability to utilize added NH_4^+ very quickly. Similarly, cells cease to assimilate NO_3^- in darkness, when they still contain polysaccharide which is utilized immediately once NH_4^+ is added. These findings point to a control on polysaccharide degradation which is released by a product of rapid NH_4^+ assimilation. It is interesting that Harnandez and Mora (1986) have recently shown that, in *Neurospora*, glutamine synthesis provides the regulatory link between carbon and nitrogen metabolism with a high rate of glutamine synthesis stimulating sucrose catabolism. Possibly a similar control operates on polysaccharide catabolism in microalgae. It is noteworthy, too that Hernandez and Mora reached their conclusions from the use of mutants of *Neurospora* with well-defined disabilities in nitrogen metabolism. Until we make use of similar mutants of microalgae our interpretations of regulatory phenomena *in vivo* must remain highly speculative.

Acknowledgement

Our work at Swansea was supported by research grants and studentships from the Natural Environment Research Council.

References

Ahmad, I. and Hellebust, J. A. (1985). *Mar. Biol.* **86**, 85.
Ahmad, I. and Hellebust, J. A. (1986). *New Phytol.* **103**, 57.
Ahmed, S. I., Kenner, R. A., and Packard, T. T. (1977). *Mar. Biol.* **39**, 93.
Appleby, G., Colbeck, J., Holdswoth, E. S., and Wadman, H. (1980). *J. Phycol.* **16**, 290.
Azuara, M. P. and Aparicio, P. J. (1983). *Plant Physiol.* **71**, 286.
Beardall, J., Mukerji, D., Glover, H. E., and Morris, I. (1976) *J. Phycol.* **12**, 409.
Bilbao, M. M., Gabas, J. M., and Serra, J. L. (1981). *Biochem. Soc. Trans.* **9**, 476.
Casselton, P. J., Chandler, G., Shah, N., Stewart, G. R., and Sumar, N. (1986). *New Phytol.* **102**, 261.
Cresswell, R. C. and Syrett, P. J. (1979). *Plant Sci. Lett.* **14**, 321.
Cresswell, R. C. and Syrett, P. J. (1981). *J. exp. Bot.* **31**, 19.
Cresswell, R. C. and Syrett, P. J. (1982). *J. exp. Bot.* **33**, 1111.
Cresswell, R. C. and Syrett, P. J. (1984). *Arch. Microbiol.* **139**, 67.

Cullimore, J.V. and Sims, A.P. (1980). *Planta* **150**, 392.
Cullimore, J.V. and Sims, A.P. (1981). *Phytochemistry* **20**, 933.
Deane-Drummond, C.E. (1985). *Pl. Cell. Envir.* **8**, 105.
Diez, J., Chaparro, A., Vega, J.M., and Relimpio, A. (1977), *Planta* **137**, 231.
Di Martino Rigano, V., Martello, A., Di Martino, C., and Rigano, C. (1985). *Physiol. Plant* **63**, 241.
Di Martino Rigano, V., Vona, V., Fuggi, A., and Rigano, C. (1982). *Physiol. Plant* **54**, 47.
Dixon, G.K. and Syrett, P.J. (1988). *J. Exp Bot*, in press
Dixon, N.E., Gazzola, C., Blakely, R.L., and Zerner, B. (1975), *J. Am. chem. Soc.* **97**, 4131.
Edge, P.A. and Ricketts, T.R. (1978). *Planta* **138**, 123.
Eisele, R. and Ullrich, W.R. (1977). *Plant Physiol.* **59**, 18.
Eppley, R.W. and Coatsworth, J.L. (1968). *J. Phycol.* **4**, 151.
Eppley, R.W. and Rogers, J.N. (1970). *J. Phycol.* **6**, 344.
Eppley, R.W., Rogers, J.N., and McCarthy, J.J. (1969). *Limnol. Oceanogr.* **14**, 912.
Everest, S.A. and Syrett, P.J. (1983). *New Phytol.* **93**, 581.
Falkowski, P.G. (1975). *J. Phycol.* **11**, 323.
Florencio, F.J. and Vega, J.M. (1982). *Phytochem.* **21**, 1195.
Florencio, F.J. and Vega, J.M. (1983). *Z. Naturf.* **38C**, 531.
Flores, E., Guerrero, M.G., and Losada, M. (1981). *Arch. Microbiol.* **128**, 137.
Flynn, K.J. and Syrett, P.J. (1985). *Mar. Biol.* **89**, 317.
Flynn, K.J. and Syrett, P.J. (1986). *Mar. Biol.* **90**, 151.
Flynn, K.J., Syrett, P.J., and Butler, E.I. (1987). *Océanis* **13**, 427.
Franco, A.R., Cardenas, J., and Fernandez, E. (1984). *FEBS Lett.* **176**, 453.
Fuggi, A. (1985). *Biochem. Biophys. Acta* **815**, 392.
Galván, F., Romero, L.C., and Márquez, A.J. (1987). In *Inorganic nitrogen metabolism* (ed. W.R. Ullrich, P.J. Aparicio, P.J. Syrett, and F. Castillo) p. 195. Springer-Verlag, Berlin.
Glibert, P.M. and McCarthy, J.J. (1984). *J. plankton Res.* **6**, 677.
Grant, B.R., Atkins, C.A., and Canvin, D.T. (1970). *Planta* **94**, 60.
Guerrero, M.G., Vega, J.M., and Losada, M. (1981). *A. Rev. Plant Physiol.* **32**, 169.
Hellebust, J.A. (1978). *J. Phycol.* **14**, 79.
Hernandez, G. and Mora, J. (1986). *J. gen. Microbiol.* **132**, 3315.
Hipkin, C.R., Everest, S.A., Rees, T.A.V., and Syrett, P.J. (1982). *Planta* **154**, 587.
Kamin, H. and Privalle, L.S. (1987). In *Inorganic nitrogen metabolism* (ed. W.R. Ullrich, P.J. Aparicio, P.J. Syrett, and F. Castillo) p. 112. Springer-Verlag, Berlin.
Komor, E. and Tanner, W. (1974). *J. gen. Physiol.* **64**, 568.
Lara, C., Romero, J.M., Coronil, T., and Guerrero, M.G. (1987). In *Inorganic nitrogen metabolism* (ed. W.R. Ullrich, P.J. Aparicio, P.J. Syrett, and F. Castillo) p. 45. Springer-Verlag, Berlin.
Larsson, C.M. and Larsson, M. (1987). In *Inorganic nitrogen metabolism* (ed. W.R. Ullrich, P.J. Aparicio, P.J. Syrett, and F. Castillo) p. 203. Springer-Verlag, Berlin.
Larsson, C.M., Larsson, M. and Guerrero, M.G. (1985). *J. exp. Bot.* **36**, 1387.

Lea, P. J. and Miflin, B. J. (1975). *Biochem. biophys. Res. Commun.* **64**, 856.
Llama, M. J., Macarulla, J. M., and Serra, J. L. (1979). *Plant Sci. Lett.* **14**, 169.
Losada, M., Paneque, A., Aparicio, P. J., Vega, J. M., Cardenas, J., and Herrera, J. (1970). *Biochem. biophys. Res. Commun.* **38**, 1009.
McCarthy, J. J., Taylor, W. R., and Taft, J. L. (1977). *Limnol. Oceanogr.* **22**, 996.
Maestrini, S. Y., Robert, J. M., and Truquet, I. (1982). *Mar. Biol. Lett.* **3**, 143.
Maestrini, S. Y., Robert, J. M., Leftley, J. W., and Collos, Y. (1986). *J. exp. mar. Biol. Ecol.* **102**, 75.
Maldonado, J. M. and Aparicio, P. J. (1987). In *Inorganic nitrogen metabolism* (ed. W. R. Ullrich, P. J. Aparicio, P. J. Syrett, and F. Castillo) p. 76. Springer-Verlag, Berlin.
Manzano, C., Candau, P., and Guerrero, M. G. (1978). *Anal. Biochem.* **90**, 408.
Morris, I. and Syrett, P. J. (1963). *Arch. Mikrobiol.* **47**, 32.
Morton, A. G. (1956). *J. exp. Bot.* **7**, 97.
Oliviera, L. and Antia, N. J. (1984). *Br. phycol. J.* **19**, 125.
Raven, J. A. (1980). *Adv. microbial Physiol.* **21**, 47.
Rees, T. A. V. and Bekheet, I. A. (1982). *Planta* **156**, 385.
Rees, T. A. V. and Syrett, P. J. (1984). *FEMS Microbiol. Lett.* **25**, 17.
Rees, T. A. V., Cresswell, R. C., and Syrett, P. J. (1980). *Biochim. Biophys. Acta* **596**, 141.
Rigano, C., Di Martino Rigano, V., Vona, V., and Fuggi, A. (1979). *Arch. Microbiol.* **121**, 117.
Roldán, J. M., Romero, F., López-Ruiz, A., Diez, J., and Verbelen, J. P. (1987). In *Inorganic nitrogen metabolism* (ed. W. R. Ullrich, P. J. Aparicio, P. J. Syrett, and F. Castillo) p. 94. Springer-Verlag, Berlin.
Roon, R. J. and Levenberg, B. (1968). *J. biol. Chem.* **245**, 4593.
Schlee, J., Cho, B-H, and Komor, E. (1985). *Plant Sci.* **39**, 25.
Serra, J. L., Llama, M. J., and Cadenas, E. (1978). *Plant Physiol.* **62**, 991.
Siegenthaler, P. A., Belsky, M. M., and Goldstein, S. (1967). *Science,* **155**, 93.
Solomonson, L. P. and Barber, M. J. (1987). In *Inorganic nitrogen metabolism* (ed. W. R. Ullrich, P. J. Aparicio, P. J. Syrett, and F. Castillo) p. 71. Springer-Verlag, Berlin.
Solomonson, L. P. and Spehar, A. M. (1977). *Nature, Lond.* **265**, 373.
Syrett, P. J. (1956). *Physiol. Plant* **9**, 28.
Syrett, P. J. (1981). In *Physiological bases of phytoplankton ecology* (ed. T. Platt) p. 182. Canadian Bulletin of Fisheries and Aquatic Science, Vol 210, Department of Fisheries and Oceans, Ottawa, Canada.
Syrett, P. J. (1987). In *Inorganic nitrogen metabolism* (ed. W. R. Ullrich, P. J. Aparicio, P. J. Syrett, and F. Castillo) p. 25. Springer-Verlag, Berlin.
Syrett, P. J. and Al-Houty, F. A. A. (1984). *Br. phycol. J.* **19**, 11.
Syrett, P. J. and Leftley, J. W. (1976). In *Perspectives in experimental biology*, Vol. 2 (ed. N. Sutherland) p. 221. Pergamon Press, Oxford.
Syrett, P. J. and Morris, I. (1963). *Biochim. Biophys. Acta* **67**, 566.
Syrett, P. J., Flynn, K. J., Molloy, C. J., Dixon, G. K., Peplinska, A. M., and Cresswell, R. C. (1986). *New Phytol.* **102**, 39.
Thacker, A. and Syrett, P. J. (1972). *New Phytol.* **71**, 423.
Ullrich, W. R. (1983). In *Encyclopedia of plant physiology*, Vol. 15A (ed. A. Lanchli and R. L. Bieleski) p. 376. Springer-Verlag, Berlin.

Ullrich, W.R. (1987). In *Inorganic nitrogen metabolism*, (ed. W.R. Ullrich, P.J. Aparicio, P.J. Syrett, and F. Castillo), p. 32. Springer-Verlag, Berlin.
Ullrich, W.R., Larsson, M., Larsson, C-M., Lesch, S., and Novacky, A. (1984). *Physiol. Plant.* **61**, 369.
Vaccaro, R.F. (1965). In *Chemical oceanography*, Vol. 1 (ed. J.P. Riley and G. Skirrow) p. 365. Academic Press, London.
Vega, J.M., Gotor, C., and Menacho, A. (1987). In *Inorganic nitrogen metabolism* (ed. W.R. Ullrich, P.J. Aparicio, P.J. Syrett, and F. Castillo) p. 132. Springer-Verlag, Berlin.
Wheeler, P.A. (1983). In *Nitrogen in the marine environment* (ed. E.J. Carpenter and D.G. Capone) p. 309. Academic Press, London.
Zumft, W.G. (1972). *Biochim. Biophys. Acta* **276**, 363.

Poster abstracts

Some features of a newly purified *Nodularia* strain

F. Del Compo and S. Sanz-Alferez
Departamento de Biologia, Universidad Autónoma de Madrid, Spain

A *Nodularia* strain, designated M1, has recently been isolated from the bed of a shallow brook, in the Guadarrama Mountains (Spain). Some of its physiological properties include the following:

1. In axenic aerobic cultures actively growing cells are mostly replaced by akinetes. Both the vegetative cells and akinetes, in contrast to the heterocysts, exhibit a high red fluorescence upon irradiation of filaments with green light.

Akinete differentiation seems to be independent of phosphate concentration, nitrogen souce, and light intensity, within the ranges tested, and can be prevented to a significant extent by anaerobiosis. On the other hand, akinetes hardly show up in bacteria-contaminated cultures. These observations suggest that a high oxygen tension, directly or indirectly, is responsible for akinete development. Thus bacteria which are associated with the cyanobacterium in its natural environment could decrease the oxygen levels through respiration.

2. M1 tolerates high water deficits, in such a manner that viability is conserved even in practically dry solid mineral medium. The tolerance was expected, for the sample of soil from which M1 was originally isolated was desiccated.

3. M1 grows with N_2, NO_3^-, or NH_4^+ as N source, at a slower rate than several other *Anabaena* strains assayed. When grown under N_2-fixing conditions, with continuous air bubbling, in a daily light cycle (14 hours light, 10 hours dark), nitrogenase activity reaches its maximum level after some hours of light, and is maintained until the dark phase, when it decreases to a negligible level. The shorter the dark phase the shorter also the light period needed to attain maximum activity. The rise in activity is clearly dependent on non-cyclic photosynthetic electron flow.

The effects of nitrogen depletion/repletion on photosynthesis in *Aphanocapsa* 6308

E. H. Evans
School of Applied Biology, Lancashire Polytechnic, Preston, Lancashire, UK
A. Law and M. M. Allen
Department of Biology, Wellesley College, Wellesley, Mass, USA

When cells of *Aphanocapsa* 6308 are starved for nitrogen, the amount of stored carbohydrate increases, the phycocyanin to chlorophyll *a* ratio decreases, and the rate of oxygen evolution and carbon dioxide fixation decreases. When nitrogen is replenished, the amount of carbohydrate decreases and the rate of oxygen evolution increases immediately and before the increase in phycocyanin or carbon dioxide fixation occurs. The rate of respiration decreases. There is no variable fluorescence in nitrogen-starved cells. Fluorescence was observed only after 24 hours of nitrogen repletion. These data suggest that there is a correlation between carbohydrate level and photosynthetic and respiratory rates, and little correlation between the rate of oxygen evolution and the amount of phycocyanin. Carbohydrate is probably involved both in the respiratory rate and in the formation of cyanophycin granule polypeptide. That carbon dioxide fixation does not increase until 4 to 8 hours after nitrogen is replenished suggests that reducing power may first be needed for some process other than photosynthesis, such as nitrate reduction, within the cell.

The role of intracellular carbonic anhydrase in bicarbonate utilization by marine microalgae

G.K. Dixon and M.J. Merrett
Plant and Microbial Metabolism Research Group, School of Biological Sciences, University College of Swansea, Swansea SA2 8PP, UK

In the marine environment at pH 8 most of the dissolved inorganic carbon exists as bicarbonate (99 per cent), with only 1 per cent present as dissolved CO_2. However, the next fixation of inorganic carbon requires the supply of CO_2 to the active site of Rubisco. Many marine algae, including *Phaeodactylum tricornutum* and *Porphyridium purpureum*, possess appreciable internal carbonic anhydrase but little if any external carbonic anhydrase, so that conversion of bicarbonate to CO_2 external to the plasmalemma does not occur. In these algae higher rates of CO_2-dependent photosynthetic oxygen evolution occur at pH 8 than at pH 5 at constant inorganic carbon concentration, showing that bicarbonate is the preferred form of inorganic carbon from the environment (Patel and Merrett 1986). Photosynthetic oxygen evolution at pH 8 is stimulated by Na^+; in *Porphyridium* Na^+ decreased the $K_{0.5}$ (CO_2) from 220 μM to 57 μM, in agreement with bicarbonate transport occurring by a Na^+/HCO_3^- symport (Katz *et al.* 1986).

A membrane-impermeable inhibitor, acetazolamide (AZ), and a membrane-permeable inhibitor, ethoxzyolamide (EZ) of carbonic anhydrase, were used to investigate the role of intracellular carbonic anhydrase in bicarbonate utilization. The $K_{0.5}(CO_2)$ was increased from 57 μM to 950 μM in *Porphyridium* and from 53 μM to 542 μM in *Phaeodactylum* in the presence of 50 μM EZ at pH 8; in contrast, AZ had little effect. It is concluded that after bicarbonate transport across the plasmalemma intracellular carbonic anhydrase facilitates the conversion of bicarbonate to CO_2 so increasing the steady state flux of CO_2 from inside the plasmalemma to Rubisco.

Patel, B.N. and Merrett, M.J. (1986). *Planta* **169**, 222.
Katz, A., Kaback, H.R., and Avron, M. (1986). *FEBS Lett.* **202**, 141.

Inorganic carbon uptake in a small-celled eukaryotic microalga

J. Munoz and M.J. Merrett
Plant and Microbial Metabolism Research Group, School of Biological Sciences, University College of Swansea, Swansea SA2 8PP, UK

Microalgae, such as *Chlamydomonas reinhardtii*, show a higher affinity for inorganic carbon (HCO_3^- + CO_2) when grown phototrophically on air than when grown with air supplemented with CO_2. This adaption to limiting CO_2 has been correlated with increased carbonic anhydrase activity and the ability of air-grown cells to concentrate inorganic carbon internally to higher levels than could be attained by simple diffusion. The presence of an inorganic carbon concentrating system is thought to be responsible for the $K_{0.5}(CO_2)$ of air-grown cells being appreciably lower than the $K_{0.5}(CO_2)$ of Rubisco from the same cells. We have used a small-celled (2 μM diameter) *Stichococcus* sp. to investigate the effect of the lengths of the diffusion path for CO_2 upon the photosynthetic characteristics of the cell.

Inorganic carbon-dependent photosynthetic oxygen evolution at constant external inorganic carbon concentration is sixfold higher at pH 5 than at pH 8 showing that CO_2 is the preferred inorganic carbon species. Air-grown cells contain appreciable carbonic anhydrase activity but activity is repressed in CO_2-grown cells. The $K_{0.5}(CO_2)$ for air-grown cells at pH 5 was 2.5 μM, but that for CO_2-grown cells was only 1.7 μM. Accumulation of inorganic carbon relative to the external medium occurred in both air-grown and CO_2-grown cells. The short diffusion path facilitates the conversion of CO_2 to HCO_3^- in the cytoplasm, so increasing the HCO_3^- concentration for bicarbonate transport across the chloroplast envelope without the need for carbonic anhydrase to facilitate CO_2/HCO_3^- interconversion.

Ribulose bisphosphate carboxylase-oxygenase of the marine macroalga *Porphyra umbilicalis*

C.M. Hill, A.J. Smith, L.J. Rogers
Department of Biochemistry, University College of Wales, Aberystwyth, Dyfed, SY23 3DD, UK
P. Balding
Sigma Chemical Company, Fancy Road, Poole, Dorset, BH17 7NH, UK

The carboxylase activity of ribulose bisphosphate carboxylase-oxygenase (RUBISCO) in crude extracts of the macroalga *Porphyra umbilicalis* was 18.9 nmol min^{-1} mg $protein^{-1}$; the oxygenase activity was half that of the carboxylase activity.

The enzyme from *P. umbilicalis* was purified in a two-step procedure involving ammonium suphate fractionation and fast protein liquid chromatography. This procedure gave electrophoretically pure RUBISCO. The native enzyme had a molecular weight of 504 000 and a sedimentation coefficient $S°_{20,w}$ of 17.9×10^{-13} sec. SDS-PAGE showed that the molecular weight of the large and small subunits were 51 500 and 17 000, respectively.

Non-denaturing PAGE was used to evaluate the purity of RUBISCO throughout the purification; it was observed that in crude extracts and during purification the native enzyme usually appeared as a doublet rather than as a single band. The possibility of degradation by proteolysis during purification was investigated by purifying the RUBISCO from *P. umbilicalis* in the presence and absence of protease inhibitors. These were added to the extraction medium and enzyme activity was assayed at each stage of the purification procedure. Analysis of the crude extract, the 25–50 per cent $(NH_4)_2SO_4$ fraction and enzyme purified by SDS-PAGE, revealed that in the presence of protease inhibitors the large subunit was always apparent as a single band. However, in the absence of protease inhibitors the large subunit invariably gave two closely related but nevertheless distinct bands, the additional band being of slightly lower molecular weight.

Since this heterogeneity occurred only in the absence of protease inhibitors, the decrease in molecular weight of the large subunit can be ascribed to proteolytic degradation, inferring that in its isolation from *P. umbilicalis* RUBISCO is susceptible to proteolysis. This proteolysis is accompanied by a decrease in the enzymic activity of the homogeneous preparation from 40 nmol min^{-1} mg $protein^{-1}$ in the presence of protease inhibitors to 29 nmol min^{-1} mg $protein^{-1}$ in their absence, though other factors may also contribute. A marked seasonal variation in RUBISCO activity has been observed in extracts and the correspondence of this to protease activity is being investigated.

Hydroxylamine metabolism and its interference with the assimilation of inorganic nitrogen in *Monoraphidium braunii*

T. Balandin and P.J. Aparicio
Centro de Investigaciones Biológicas, Velázquez 144, 28006-Madrid, Spain

Hydroxylamine was reduced to ammonia by *Monoraphidium braunii* cells both in the light and in the dark. In the dark under an argon atmosphere the amount of hydroxylamine that disappeared was stoichiometric with the amount of ammonia released into the medium. The assimilation of ammonia by cells in the light and in the presence of CO_2 was inhibited by hydroxylamine; nitrite and nitrate utilization were also greatly impaired. The inhibition of nitrite reduction was accompanied by loss of the oxygen evolution capacity promoted by hydroxylamine. Nitrate reductase was substantially inactivated *in situ* by hydroxylamine in a similar way to that described for the purified nitrate reductase (Aparicio *et al.* 1985; Balandin *et al.* 1986). The consumption of nitrate was restored when the cell suspensions treated with hydroxylamine were irradiated with blue light, but not with red light.

Photoautotrophic growth of *M. braunii* was completely prevented by hydroxylamine. When the cells had metabolized sufficient hydroxylamine in the medium, vigorous growth resumed, provided a suitable nitrogen source was available.

Hence hydroxylamine appears to behave as an effective phytostatic agent interfering both with inorganic nitrogen metabolism and with the photosynthetic apparatus.

Aparicio, P.J., Balandin, T., Mauriño, S.G., and Maldonado, J.M. (1985). *Photochem. Photobiol.* **42**, 765.
Balandin, T., Fernández, V.M., and Aparicio, P.J. (1986). *Plant Physiol.* **82**, 65.

Tubulin mRNA in the cell cycle of *Chlamydomonas*

D.S.T. Nicholl
Biology Department, Paisley College of Technology, Paisley PA1 2BE, Scotland
J.A. Schloss
University of Kentucky, Lexington, KY 40506-0225, USA
P.C.L. John
The Australian National University, Canberra City, ACT 2601, Australia

Tubulin synthesis during the cell cycle is of interest because of its contribution to the mitotic spindle and cytoskeleton, and the suggestion that tubulin accumulation could play a part in controlling nuclear division (Laffler *et al.* 1981). In *Chlamydomonas reinhardtii* growing under synchronizing conditions, tubulin gene expression is restricted to the period of cell division activity (Ares and Howell 1982). However, under constant environmental conditions tubulin synthesis occurs throughout the cell cycle (Rollins *et al.* 1983).

To investigate tubulin gene expression at the transcriptional level we have used cDNA probes specific for both alpha- and beta-tubulin (Schloss *et al.* 1984) to measure mRNA levels during the cell cycle. Tubulin mRNA was found to accumulate periodically when cells were grown under synchronizing conditions, with a peak just prior to cell division. Under turbidostat conditions, levels remained constant throughout the cell cycle. We conclude that periodic tubulin synthesis in *C. reinhardtii* is not a primary cell cycle event (*i.e.* essential under all conditions) but is an example of a secondary class of event that is necessary for division under certain conditions.

Ares, M. Jr. and Howell, S.H. (1982). *Proc. natn. Acad. Sci. USA* **79**, 5577.

Laffler, T.G., Chang, M.T. and Dove, W.F. (1981). *Proc. natn. Acad. Sci. USA* **78**, 5000.

Rollins, M.J., Harper, J.D.I., and John, P.C.L. (1983). *J. gen. Microbiol.* **129**, 1899.

Schloss, J.A., Silflow, C.D., and Rosenbaum, J.L. (1984). *Mol. cell. Biol.* **4**, 424.

Part 2: Biosynthesis

4 Lipid metabolism

J.L. HARWOOD, T.P. PETTITT, and A.L. JONES

Department of Biochemistry, University College, Cardiff CF1 1XL, UK

Algae are a very varied group of organisms and their lipids reflect this diversity. The subdivision of algae into different classes has been based historically on characteristics such as pigment composition but is a somewhat controversial area and we will not attempt to make definitive statements on this subject. Suffice it to say that if species are taken from the various classes of algae they sometimes show quite different fatty acid compositions. In Table 4.1 the total fatty acid content of some freshwater algae is shown. In general, the same acids which are found in leaves occur, although their relative proportions vary considerably. As a general rule the percentage of α-linolenate is reduced and the amount of hexadecatrienoate is especially variable (cf. *Scenedesmus obliquus*). The decreased amounts of linolenate are probably a reflection of the reduced need for the photosynthetic apparatus in algae growing in media rich in nutrients. It is known that such cultures produce fewer chloroplasts and, hence, because linolenate is concentrated in the thylakoids, this acid is decreased (Hitchcock and Nichols 1971). Like the fatty acids, the complex lipids of freshwater algae are similar to those of higher plants, with such molecules as monogalactosyldiacylglycerol (MGDG) and digalactosyldiacylglycerol (DGDG) and phosphatidylcholine (PC) being predominant.[1] Lipids such as sterol glycosides and cerebrosides are absent or only found in small amounts, and compounds such as the trimethylhomoserine ether lipid, which is characteristic of lower plants (Sato and Furuya 1984), may be of significance.

The marine algae contain a bewildering array of major fatty acids. Particularly noticeable are the large proportions of very long chain polyunsaturated fatty acids; arachidonic and eicosapentaenoic (n-3) acids are usually major components. Some representative compositions for phytoplankton and macroalgae are shown in Table 4.2. Myristate and palmitate are the main saturated acids throughout the phytoplankton, and hexadecenoate (usually

[1]Throughout this paper, the following abbreviations have been adpoted: MGDG, monogalactosyldiacylglycerol; DGDG, digalactosyldiacylglycerol; SQDG, sulphoquinovosyldiacylglycerol; PG, phosphatidylglycerol; PC, phosphatidylcholine; PE, phosphatidylethanolamine; PI, phosphatidylinositol; DPG, disphosphatidylglycerol; PS, phosphatidylserine; MGcDG, monoglucosyldiacylglycerol; PA, phosphatidic acid.

Table 4.1 Fatty acid composition of freshwater algae.

Alga	Fatty acid (per cent total)										
	14:0	16:0	16:1	16:2	16:3	16:4	18:1	18:2	18:3	20:4	Other
Nitella	1	21	2	7	—	—	2	31	17	6	13
Scenedesmus obliquus	1	35	2	tr	tr	15	9	6	30	—	4
Chlorella pyrenoidosa	—	20	3	3	7	—	46	10	12	—	tr
Chlorella vulgaris	2	26	8	7	2	—	2	34	20	—	tr

Data from Hitchcock and Nichols (1971). tr = trace.

n-7) is generally a major constituent. Polyunsaturated C16 acids are often present but in minor proportions. With a few exceptions the C_{18} fatty acids seldom account for more than 10 per cent of the total (Table 4.2), but eicosapentaenoate (usually n-3) is frequently a major component. In the Dinophyceae, docosahexaenoic (n-3) acid is a major constituent. In marine macroalgae the three classes represented in Table 4.2 show certain characteristic compositions. All of the algae contain large amounts of palmitate but, whereas the Phaeophyceae have a high percentage of oleate, the Rhodophyceae have eicosapentaenoate as their major unsaturated fatty acid. Both groups also contain arachidonate in large amounts. In contrast, the Chlorophyceae have only small amounts of C_{20} acids, with linolenate and octadecatetraenoate as major components.

The cyanobacteria (blue–green algae) have a very simple lipid class composition. Their major lipids are MGDG, DGDG, sulphoquinovosyldiacylglycerol (SQDG), and phosphatidylglycerol (PG); this is similar to higher plant chloroplasts. In some species glucose may substitute for galactose (cf. Sato and Murata 1982c). This simple lipid composition is consistent with the postulated evolutionary relationship of cyanobacteria to the chloroplasts of eukaryotes. However, the heterocysts of filamentous cyanobacteria contain other lipid types (see later).

In contrast to other algae, none of the cyanobacteria have been reported to synthesise arachidonate or *trans*-3-hexadecenoate. Moreover, some of the cyanobacteria seem to lack the capacity to make α-linolenate or any appreciable amounts of polyunsaturated fatty acids. In those organisms which lack appreciable linoleate, palmitoleate is prominent. All cyanobacteria contain high levels of palmitate and oleate (Table 4.3).

A discussion of all the lipid classes which have been found in eukaryotic algae is beyond the scope of this short review. Suffice it to say that the three glycosylglycerides (MGDG, DGDG, SQDG) and the phospholipids [PC

Table 4.2 Fatty acid composition of marine phytoplankton[1] and marine macroalgae.

Alga	Fatty acid (per cent total)												
	14:0	16:0	16:1	16:2	16:3	16:4	18:1	18:2	18:3	18:4	20:4	20:5	22:6
Phytoplankton													
Chrysophyceae													
Monochrysis lutheri	10	13	22	5	7	1	3	1	tr	2	1	18	7
Isochrysis galbana	1	16	5	3	2	tr	25	18	11	2	1	4	tr
Xanthophyceae													
Olisthodiscus spp.	8	14	10	2	2	1	4	4	6	18	2	19	2
Monodus subterraneus	2	24	24	—	—	—	9	4	tr	—	5	29	—
Bacillariophyceae													
Lauderia borealis	7	12	21	3	12	1	2	1	tr	—	1	3	—
Thalassiosira fluviatilis	6	11	22	3	9	1	3	1	tr	tr	4	20	—
Dinophyceae													
Amphidinium carterae	3	24	1	1	tr	—	5	1	2	15	—	14	25
Gonyaulax catanella	13	30	3	—	—	—	7	3	6	14	—	1	12
Chlorophyceae													
Dunaliella salina	tr	41	15	tr	—	—	11	8	19	—	—	—	—
Cryptophyceae													
Hemiselmis brunescens	1	13	3	3	tr	tr	2	tr	9	30	tr	14	—
Cryptomonas maculata	5	15	7	1	—	tr	3	—	6	16	1	17	—
Macroalgae													
Phaeophyceae													
Fucus vesiculosus	—	21	2	tr	—	tr	26	10	7	4	15	8	—
Ascophyllum nodosum	—	15	2	1	—	tr	27	11	5	4	18	8	—
Rhodophyceae													
Chondrus crispus	—	34	6	tr	—	—	9	1	1	4	18	22	—
Polysiphonia lanosa	—	32	8	2	—	1	15	4	2	1	6	26	—
Chlorophyceae													
Ulva lactuca	1	18	2	tr	1	18	9	2	17	24	1	2	tr
Enteromorpha intestinalis	—	20	2	4	—	14	12	7	22	13	1	1	—

[1]Data from Pohl and Zurheide (1979). tr = trace.

Table 4.3 Fatty acid composition of some cyanobacteria.

	Fatty acid composition (per cent total)						
	16:0	16:1	16:2	18:0	18:1	18:2	18:3
Anabaena cylindrica	46	6	6	4	6	24	11
Anabaena flos-aquae	40	6	4	1	5	37	11
Anacystis nidulans	47	39	—	1	10	—	—
Chlorogloea fritschii	42	5	tr	6	14	17	16
Mastigocladus laminosus	39	43	—	tr	17	2	—

Data from Hitchcock and Nichols (1971). tr = trace.

and phosphatidylethanolamine (PE)] are usually major lipids. Other phospholipids [PG, diphosphatidylglycerol (DPG), phosphatidylinositol (PI), phosphatidylserine (PS)], triacylglycerols, and fatty acids are usually present in minor amounts. Some Chlorophyceae contain trimethylhomoserine ether lipid (cf. Eichenberger 1982), many freshwater algae the chlorosulpholipids (Mercer and Davies 1979), and Phaeophyceae contain unusual glycolipids (cf. Pham Quang and Laur 1976). The sulphonium analogue of PC has been detected in *Nitzschia alba* (Anderson *et al.* 1978) and also in *Chondrus crispus* and *Polysiphonia lanosa* (T. Pettit and J.L. Harwood, unpublished observations). A review of the lipids of marine algae has been made by Pohl and Zurheide (1979).

In this review we cannot cover all the many diverse aspects of algal lipid metabolism and, therefore, have chosen to survey a few detailed aspects of three quite different classes of algae—namely cyanobacteria, marine macroalgae, and green algae. In this way, we hope to give a representative view of the variety of lipid metabolism occurring in algae in general.

Aspects of lipid metabolism in cyanobacteria

As mentioned above, the cyanobacteria contain a very simple lipid composition with typical 'chloroplast' lipids predominating. The major glycolipids are MGDG and DGDG, but monoglucosyldiacylglycerol (MGcDG) has been reported as a minor component in *Nostoc calcicola* (Feige *et al.* 1980) and *Anabaena variabilis* (Sato and Murata 1982a). A close metabolic relationship between MGDG and MGcDG was indicated by results from pulse-chase experiments and fatty acid labelling patterns. In the latter case it was noted that, whereas saturated fatty acids were labelled at short time intervals, these became progressively more unsaturated during the course of the experiment (Feige *et al.* 1980). Such changes are consistent with the use of the glycosylglycerides themselves as substrates for desaturation in an analogous manner to the use of MGDG for linoleate desaturation in higher plants (cf. Jones and Harwood 1980). The positional distribution of labelled acids between the *sn*-1 and *sn*-2 positions of the glycosylglycerides confirmed the established pattern in cyanobacteria—namely that the distribution was governed by chain length rather than by the degree of unsaturation (Zepke *et al.* 1978). Thus, in both MGDG and MGcDG of *Nostoc* the shorter chain myristate and palmitate were found at the *sn*-2 position, while the C_{18} fatty acids were localized at the *sn*-1 position (Feige *et al.* 1980).

The above distribution of fatty acids was also found in *A. variabilis* (Sato *et al.* 1979; Sato and Murata 1980). When algae were labelled by photosynthetic fixation from $NaH^{14}CO_3$ then it was noted that MGcDG became highly radioactive. During a subsequent pulse-chase, radiolabel was transferred progressively to MGDG and, to some extent, to DGDG (Feige *et al.* 1980; Sato and Murata 1982*a*, 1982*b*). The two other major lipid classes, SQDG and PG, were relatively poorly and relatively highly labelled, respectively (Table 4.4). The total radioactivity in MGDG plus MGcDG remained approximately constant during the pulse-chase in spite of the large change in relative labelling. These findings indicated that most of the MGcDG was converted to MGDG within 1 hour. More detailed analysis of the sugar moieties of the two glycolipids showed a reciprocal relationship between the decline in label of the glucose and the increase in that of the galactose—with the total radioactivity remaining constant. Thus epimerization at C_4 seemed to be taking place.

Table 4.4 Labelling of major lipid classes of *A. variabilis* from $NaH^{14}CO_3$.

Labelling (h)	Radioactivity (per cent total lipids)					
	Chase (h)	MGcDG	MGDG	DGDG	SQDG	PG
0.1	—	58	4	4	6	28
1.0	—	34	30	4	7	25
1.0	10	1	61	7	8	23
Mass		1	54	17	11	17

Data from Sato and Murata (1982*a*).

Sato and Murata (1982*a*) followed the changes in fatty acid patterns which occurred during the pulse-chase experiment and, like Feige *et al.* (1980) using *Nostoc*, obtained evidence for the sequential desaturation of acyl chains while they were still attached to complex lipids. The exact nature of the changes differed for the four lipids studied, but in all cases the starting molecular species was 1-stearoyl, 2-palmitoyl-. The postulated desaturations are shown in Fig. 4.1, and it will be noted that α-linolenate is suggested to be

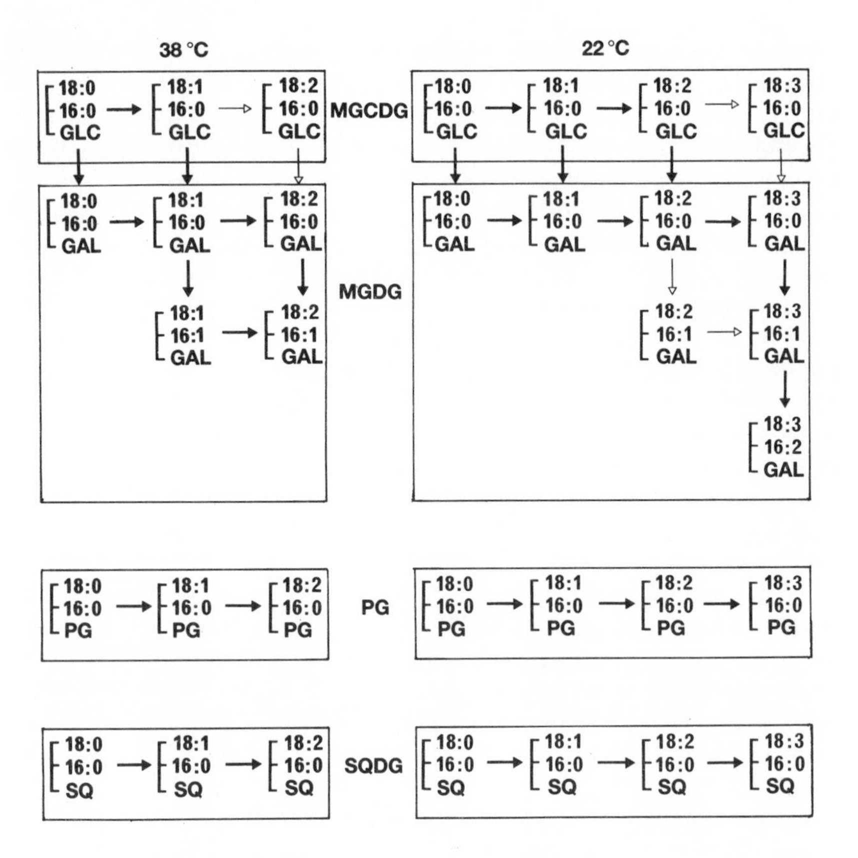

Fig. 4.1. Possible biosynthetic pathways for the molecular species of the acyl lipids of *A. variabilis*. GLC, glucose; GAL, galactose; SQ, sulphoquinovose; DG, diacylglycerol. Data from Sato and Murata (1982*b*) with permission.

formed by sequential desaturation of stearate at the *sn*-1 position of all four lipids. The relative importance of the various pathways for the production of, for example, α-linolenate is not indicated in the figure. However, when the total radioactivity in each lipid and the percentage conversion to linolenate are considered, then MGDG appears to be the major substrate for linoleate desaturation to linolenate (Sato and Murata 1982*b*). This agrees, therefore, with work on higher plants (cf. Harwood 1980; Jones and Harwood 1980).

Labelling studies of complex lipids of *A. variabilis* from $NaH^{14}CO_3$ followed by pulse-chase showed that the primary products of lipid synthesis were, initially, MGcDG, PG, and SQDG (Table 4.4). They also indicated that MGcDG was converted to MGDG by epimerization of glucose to a galactose moiety. Galactosylation could then take place to yield DGDG (Sato and Murata 1982*a*). *In vitro* experiments with *A. variabilis* showed that MGcDG was synthesized by the transfer of glucose from UDP-glucose to diacylglycerol. The glucosyltransferase responsible was membrane-localized. In contrast, UDP-galactose:diacylglycerol galactosyltransferase activity was extremely low in any fraction tested. This confirmed the *in vivo* labelling experiments in that MGDG appeared to be formed via the monoglucosyl-derivative (Sato and Murata 1982*c*). In this respect *A. variabilis* contrasted clearly with the situation in higher plants (Douce and Joyard 1980).

Cyanobacteria are obviously convenient organisms with which to carry out experiments designed to test the effect of environmental parameters. Temperature adaptation experiments have been carried out with *A. variabilis* with regard to changes in both fatty acid and molecular species compositions (Sato *et al.* 1979; Sato and Murata 1980). As expected, the result of shifting growth from 38°C to 22°C was to increase the amount of fatty acyl unsaturation. However, it was noted that only the C_{18} acids and not the C_{16} acids responded to the growth temperature (Table 4.5). Further experiments involving temperature shifts showed that a rapid conversion of palmitate to palmitoleate was seen which was reversed after about two days. In contrast, the increase in C_{18} desaturation was larger and was maintained up to the end of the experiment (200 hours). Although the desaturation of C_{18} acids was found in all lipid classes, the changes in C_{16} acids were confined to MGDG. Sato and Murata (1980), therefore, concluded that the desaturation of palmitate at the *sn*-2 position of MGDG could play a central role in regulating membrane fluidity during rapid thermo-adaptation. Studies of the molecular species of MGDG showed that the most saturated species (1-oleoyl, 2-palmitoyl-) was reduced from 25 to 6 per cent of the total in the first

Table 4.5 Fatty acid composition of lipids of *A. variabilis* grown at 38 °C and 22 °C.

Lipid	Temp.	Fatty acid composition (per cent total)					
	(°C)	16:0	16:1	16:2	18:1	18:2	18:3
MGDG	38	12	14	1	12	12	1
	22	12	11	2	2	6	17
DGDG	38	10	16	1	10	14	1
	22	9	11	4	1	5	20
SQDG	38	27	2	1	14	6	0
	22	27	2	1	5	6	10
PG	38	25	1	0	16	8	0
	22	26	1	0	2	7	15

Data from Sato and Murata (1980).

10 hours following the shift to 22°C. This corresponded to a significant change (14°C to 10°C) in the phase separation temperature for thylakoid lipids and suggested a physiological role for that molecular species.

In spite of the relatively simple composition of cyanobacterial lipids, there have been few *in vitro* studies for these organisms. Two recent reports relate to fatty acid synthesis by *A. variabilis*. Stapleton and Jaworski (1984) found that the fatty acid synthetase showed typical type II properties with requirements for added NADPH and ACP and, like the plant system, was inhibited by CoA. Palmitate was the major product with stearate also formed in significant quantities. At short incubation times two polar derivatives, tentatively identified as ketoacids, were found. These two workers also reported the purification and kinetic properties of the malonyl-CoA:ACP transacylase. Similar results for the properties of the fatty acid synthetase were also obtained by Lem and Stumpf (1984*a*), who used an 80 per cent $(NH_4)_2SO_4$ precipitated fraction of the soluble extract. They also studied the transfer of newly synthesized fatty acids into complex lipids and concluded that CoA-thioesters did not seem to be involved (Lem and Stumpf 1984*b*).

Mention should also be made of the specialized lipids found in the heterocysts of filamentous cyanobacteria. These have been shown to include a specialized class of glycolipids (cf. Bryce *et al.* 1972). The synthesis of such lipids has been studied in *Anabaena cylindrica* during heterocyst differentiation induced by L-7-azatryptophan. During such studies the occurence of a glycoside of hexocosane-1, 25-diol as a normal component of the glycolipid fraction was also reported (Mohy-ud-Dhin *et al.* 1982).

Lipid metabolism in marine macroalgae

We have recently studied the lipid metabolism in three classes of marine macroalgae: *Fucus vesiculosus, Fucus serratus, Ascophyllum nodosum* (Phaeophyceae); *C. crispus, P. lanosa* (Rhodophyceae); and *Enteromorpha intestinalis* (Chlorophyceae). These algae were chosen as common species which were not seasonal and whose lipid composition was typical for its class. All the algae contained the major glycosylglycerides found in higher plants—MGDG, DGDG, and SQDG (Smith and Harwood 1984*a*; Pettitt and Harwood 1986; Jones and Harwood 1987). In the red algae, other sugars were also found in the monoglycosyl and diglycosyldiacylglycerol fractions. In the brown algae the glycosylglycerides constituted 40–70 per cent and in the red and green algae at least 50 per cent of the total acyl lipids. The major phospholipid in the brown algae was PE and this was the major phospholipid labelled by ^{32}P-orthophosphate (Smith *et al.* 1982) or ^{14}C-acetate (Smith and Harwood 1984*a*). An unidentified lipid was found in all three brown algae (Jones and Harwood 1987) and this was highly labelled from ^{14}C-acetate in *F. serratus* (Smith and Harwood 1984*a*). PC was the major phosphoglyceride in red algae and, interestingly, the red algae appeared to contain phosphatidylsulphocholine (Pettitt and Harwood 1987). PG (the characteristic chloroplast phospholipid) was the most highly labelled in the red algae and *E. intestinalis* (Jones and Harwood 1987). It is noteworthy that this lipid is usually well labelled in the light in freshwater algae such as *Chlorella vulgaris* (Nichols *et al.* 1967) and, indeed, its labelling was stimulated by light in *F. serratus* (Smith *et al.* 1982) and *Chondrus crispus* (Pettitt and Harwood 1987) (Table 4.6).

The labelling of glycosylglycerides was generally poor in relation to their abundance, particularly in brown algae (Smith and Harwood 1984*a*; Pettitt and Harwood 1986; Jones and Harwood 1987). In *E. intestinalis* an unknown lipid was provisionally identified as the trimethylhomoserine ether lipid which has been found in various ferns and green algae (Sato and Furuya 1984).

By ^{35}S-sulphate labelling, a number of sulphur-containing lipids were found in red algae. However, apart from SQDG and phosphatidylsulphocholine the other lipids were minor constituents and were poorly labelled (Pettitt and Harwood 1986). It should be noted in this context that several unusual sulphur-containing lipids were also found in the diatom *N. alba* (Anderson *et al.* 1978).

Light was found to increase the labelling of all the typical chloroplast lipids in red algae (Table 4.6). Furthermore, it also changed the pattern of fatty-acid labelling from ^{14}C-acetate. In *F. serratus* a significant increase in the proportion of unsaturated fatty acids labelled was seen in the light (Table

Table 4.6 The effect of light on the labelling of lipid classes in different marine macroalgae.

Lipid	per cent labelling			
	F. serratus		C. crispus	
	Dark	Light	Dark	Light
PA	1.4	1.5	—	—
DPG	9.8	10.8	—	—
PG	10.3	11.6*	5.4	18.5*
PE	54.3	52.1	—	—
PC	4.9	5.1	27.0	17.6*
PI	19.3	18.9	—	—
MGDG	—	—	10.3	14.3*
DGDG	—	—	1.9	6.9*
SQDG	—	—	9.3	13.0

Labelling in *F. serratus* was with ^{32}P-orthophosphate and in *C. crispus* from ^{14}C-acetate.
*Significantly different by Student's *t*-test for paired samples.

Table 4.7 Changes in fatty acid labelling from ^{14}C-acetate in *C. crispus* and *F. serratus* caused by light.

Species	Percentage of total ^{14}C fatty acids						
	14:0	16:0	16:1	18:0	18:1	18:2	Others
C. crispus[a]							
Light	1 ± 1	19 ± 2	3 ± 2	tr	64 ± 6	10 ± 2	3 ± 5
Dark	6 ± 1	31 + 1*	2 ± 1	7 ± 3*	40 ± 6*	3 ± tr*	11 ± 1
F. serratus[b]							
Light	3 ± tr	32 ± 5	—	10 ± 1	43 ± 6	9 ± 2	3 ± 1
Dark	3 ± 1	48 ± 8*	—	13 ± 2	30 ± 6*	1 ± 1*	5 ± 1

Data from Pettitt and Harwood (1987)[a] and Smith and Harwood (1984*b*)[b].
*Significant at 1 per cent level.

4.7). Similar results were also obtained with *C. crispus* (Table 4.7), and they are consistent with those found with higher plants (Harwood 1975).

It is well known that lowering the growth temperature induces an increase in fatty acid unsaturation in most organisms. However, it is not always clear whether these changes are adaptive alterations to maintain membrane fluidity or merely reflect the increased oxygen solubility at lower temperatures (cf. Harwood 1983). Some changes have been noted in the fatty acyl composition of marine algae collected at various times of the year, with sea temperatures varying by as much as 10°C (Jamieson and Reid 1972; Pohl and Zurheide 1979). Whereas *F. serratus* showed little change in fatty acid composition for different seasons (Smith and Harwood 1984*b*), that of *F. vesiculosus* (Jones and Harwood 1987; Pohl and Zurheide 1979) showed significant increases in unsaturation for winter-collected material. When algae were incubated at various temperatures, the pattern of acids labelled from ^{14}C-acetate was different. *F. serratus* (Smith and Harwood 1984*b*), *C. crispus* (Pettitt and Harwood 1987), and *P. lanosa* (Jones and Harwood 1987) showed significant increases in the labelling of unsaturated moieties (Table 4.8). The enhanced desaturation may be the result of raised O_2 concentrations at lower temperatures (cf. Rebeille *et al.* 1980).

Marine algae are subjected to a great deal of pollution from different sources. One such type is heavy metal contamination. Metals such as copper, lead, and zinc may derive from leaching as well as from factory discharges. On the other hand, metals such as nickel and cadmium are usually only elevated by industrial malpractice. Although cadmium, mercury, and lead are generally assumed to be the most dangerous pollutants for man, for marine organisms other metals including copper and zinc assume considerable importance in some areas by virtue of their abundance. These metals are concentrated from the water by brown algae by factors of 10^3–10^4 (cf. Bryan

Table 4.8 Increased labelling of unsaturated fatty acids from ^{14}C-acetate in marine macroalgae caused by low temperatures.

Species	Temp. (°C)	Percentage of total ^{14}C fatty acids							
		14:0	16:0	16:1	18:0	18:1	18:2	20:0	22:0
F. serratus[a]	4	3	24	tr	20	30	5	7	5
	15	5	42*	1	17	13*	n.d.*	6	10
C. crispus[b]	4	1	28	6	12	53			
	15	10	37	5	11	34*			
P. lanosa[c]	5		18	28	2	42	7		
	15		41	11	7	28	n.d.*		

Data from Smith and Harwood (1984*b*)[a]; Pettitt and Harwood (1987)[b], and Jones and Harwood (1987)[c]. tr = trace.
*Significantly different; n.d. = not detected

1983). The site of accumulation of the metal is not known except in a few cases, but includes physodes (Smith *et al.* 1986) and cell walls (unpublished results) depending on the metal concerned. By such means, it is thought that the tissues of some Phaeophyta are able to tolerate relatively high levels of heavy metals and this increases their potential use as indicator organisms for pollution.

In a series of *in vitro* experiments with *F. serratus* it was found that high concentrations of heavy metals, including Cd^{2+}, Cu^{2+}, Ni^{2+}, Pb^{2+}, and Zn^{2+} affected total lipid synthesis (Smith and Harwood 1984*b*). There was also some correlation between changed lipid synthesis and whether the algae had been collected from sites polluted by heavy metals.

In Cornwall, *F. serratus* and *F. vesiculosus* are found at sites in the Fal Estuary which are heavily polluted by run-off from disused copper mines. These algae have become adapted to the pollution and have tissue levels of

Table 4.9 Changes in fatty acid labelling from ^{14}C-acetate induced by copper *in vitro*.

Species	^{14}C fatty acids (per cent total)						
	14:0	16:0	16:1	18:0	18:1	18:2	Others
F. serratus							
Control	7 ± tr	74 ± 6	2 ± 1	2 ± tr	7 ± 3	8 ± 4	tr
+ Cu^{2+}	4 ± tr*	59 ± tr*	3 ± 1*	3 ± tr*	16 ± 3*	11 ± 3	4 ± 2
F. vesiculous							
Control	13 ± 1	65 ± 3	tr	15 ± 2	7 ± 1	tr	tr
+ Cu^{2+}	8.3	57 ± 2*	tr	16 ± 2	17 ± 4*	1 ± tr	2 ± 1

Data from Smith *et al.* (1984).
*Significantly different by paired *t*-test. Cu^{2+} was at 300 $\mu g l^{-1}$. tr = trace.

copper of up to 200-fold the normal values (Bryan 1980). Such concentrations of copper are known to cause inhibition of growth or death in *F. vesiculosus* obtained from relatively unpolluted areas. Experiments *in vitro* showed that elevated Cu^{2+} increased the relative labelling of lipids from ^{14}C-acetate without altering total uptake. An alteration in the pattern of fatty acids synthesized was also seen with an increase in the relative labelling of oleate and a decrease in that of palmitate (Table 4.9). These differences in metabolism were also seen when algae from sites polluted by copper were compared with samples collected from non-polluted areas (Smith *et al.* 1984). Moreover, the endogenous fatty acid composition for algae obtained from differently polluted sites was also changed (Smith *et al.* 1985). In keeping with the *in vitro* experiments oleate was increased by copper pollution (Table 4.10) but, in addition, there were reductions in polyunsaturated acids such as docosapentaenoic, which were not labelled *in vitro*.

The way in which such changes are regulated has not been studied in detail, although it is possible that copper alters the activity of acyl-transferases. Certainly, recent experiments (unpublished results) show that different marine algal species differ quite markedly in the response of their lipid metabolism to copper pollution (A.L. Jones and J.L. Harwood, unpublished data).

Other responses of algae to pollutants have been described (cf. Pohl and Zurheide 1979). These include changes caused by manganese and nitrogen. Correlation of changed lipid metabolism with peaks in phosphate or nitrate levels has been found for marine red, brown, and green algae. In particular, an increased non-polar lipid content was seen during periods of low growth

Table 4.10 Differences in the proportions of endogenous fatty acids in *Fucus* species collected from sites exposed to various amounts of copper pollution.

Species	Fatty acid composition (per cent total)							
	14:0	16:0	18:1 (n − 9)	18:2 (n − 6)	18:3 (n − 3)	18:4 (n − 3)	20:4 (n − 6)	20:5 (n − 3)
F. vesiculosus								
Non-polluted	12	20	14	7	6	5	20	12
Polluted	11	16*	22*	11	4*	3*	19	7*
F. serratus								
Non-polluted	7	16	20	7	9	7	20	9
Polluted	6	14*	29*	9*	6*	6*	18	6*

Data from Smith *et al.* (1985).
*Significantly different.
Only the main fatty acids are shown. Samples were collected in June. Endogenous copper levels in algae were 6 $\mu g\ g^{-1}$ dry weight for the unpolluted and 25 $\mu g\ g^{-1}$ dry weight for the polluted site.

when the contents of nitrogen and phosphate decreased. Conversely, during high-growth periods, when nitrogen and phosphate were plentiful, polar lipid and polyunsaturated fatty acid contents were increased (Pohl and Zurheide 1979).

Lipid metabolism in green algae

Green algae, particularly *C. vulgaris*, have often been used as model systems for higher plants. The ease of experimental manipulation coupled with the rapid metabolic rates are two obvious advantages. Moreover, given suitable techniques green algae can have their growth synchronized, and developmental phenomena such as thylakoid assembly can be followed (Beck and Levine 1977). Such experiments have shown, for example, that chloroplast membrane lipids are synchronized in a sequential or multistep process. Furthermore, lipids appear to be made and inserted into the chloroplast membrane prior to major increases in photosynthetic capacity.

The unicellular alga *Dunaliella salina* has a number of unique properties which make it a useful experimental system. It grows rapidly under axenic conditions to give populations of homogeneous cells. Since it does not have a cell wall, subcellular fractionation is particularly easy and yields fractions with comparatively little contamination. Moreover, because it can tolerate wide ranges of temperature and salinity it can be used to study the effects of these environmental parameters.

Thompson and co-workers have utilized *D. salina* to study lipid metabolism, particularly in relation to low temperature stress. They found that low temperature (12°C) caused an increase in phospholipid, glycolipid, and protein on a per cell basis compared with growth at 30°C. Different membranes were affected to varying degrees. For example, the total chloroplast membrane was found to show an increase of about 20 per cent, whereas the microsomal membranes showed about a 2.8-fold increase in the levels relative to those in cells grown at 30°C (Lynch and Thompson 1982). Increases in fatty acid unsaturation were seen in phospholipids of both thylakoids and microsomes. Generally, microsomal phospholipids responded more quickly and to a greater extent than did chloroplast phospholipids. However, there was little change in the phospholipid class composition for either membrane (Table 4.11). In contrast, low temperature growth caused an increase in the amount of DGDG relative to MGDG (Table 4.11) with a minimal change in the acyl groups of galactolipids (Lynch and Thompson 1982).

Alterations in the microsomal phospholipids were dominated by large changes in the molecular species of PG and PE which occurred before there were noticeable increases in linolenate (Lynch and Thompson 1984*a*). During acclimation in *Dunaliella*, the amounts of saturated and unsaturated species decreased and those of diunsaturated species increased. These changes in acyl

Table 4.11 Changes in the lipid composition of *D. salina* caused by growth at different temperatures.

Lipid (mol %)	30 °C			12 °C		
	Whole cells	Chl	Mic	Whole cells	Chl	Mic
Phospholipids						
PC	25	31	32	20	24	28
PE	15	15	41	25	17	40
PG	37	46	15	34	47	17
PI	11	8	13	12	12	15
Glycolipids						
MGDG	67	66	—	55	58	—
DGDG	21	19	—	27	28	—
SQDG	12	15	—	18	14	—

Data from Lynch and Thompson (1982).
Abbreviations: Chl = chloroplast membranes; Mic = microsome fraction.

distribution also occur rapidly in other species (cf. Dickens and Thompson 1982) and seem to augment the effects of increased acyl-chain desaturation as a means for restoring appropriate membrane properties. In further studies Lynch and Thompson (1984*b*) found that certain molecular species of PG could also be important in modulating the thermal stability of the photosynthetic membranes. Thus changes in chlorophyll fluorescence closely correlated with changes in the major molecular species of this lipid (18:3/16:1 and 18:2/16:1; Fig. 4.2). This result was notable in that there is much interest currently in the role of molecular species of phosphatidylglycerol in chloroplast function, particularly in relation to chilling sensitivity (cf. Murata 1983).

These compositional studies have been followed up in metabolic experiments. Rapid labelling (2 minutes) with ^{14}C-palmitate, ^{14}C-oleate, or ^{14}C-laurate was followed by a chase of up to 16 hours. These precursors were chosen because, since palmitate and oleate were the usual end-products of *Dunaliella* fatty acid synthesis, they would be expected to be used directly for complex lipid synthesis throughout the cell. In contrast, ^{14}C-laurate would give rise to these products only after elongation by the chloroplast fatty acid synthetase. The experiments revealed a slow transfer of lipids into the chloroplasts from other organelles (Norman *et al.* 1985). When the culture temperature was lowered from 30°C to 12°C, glycerolipid synthesis in chloroplasts was decreased more than that in microsomes. It was suggested that the ability of *Dunaliella* chloroplasts to utilize microsomal lipids might be essential for their acclimation to low temperatures.

In view of the interest in the particular function of PG in controlling mem-

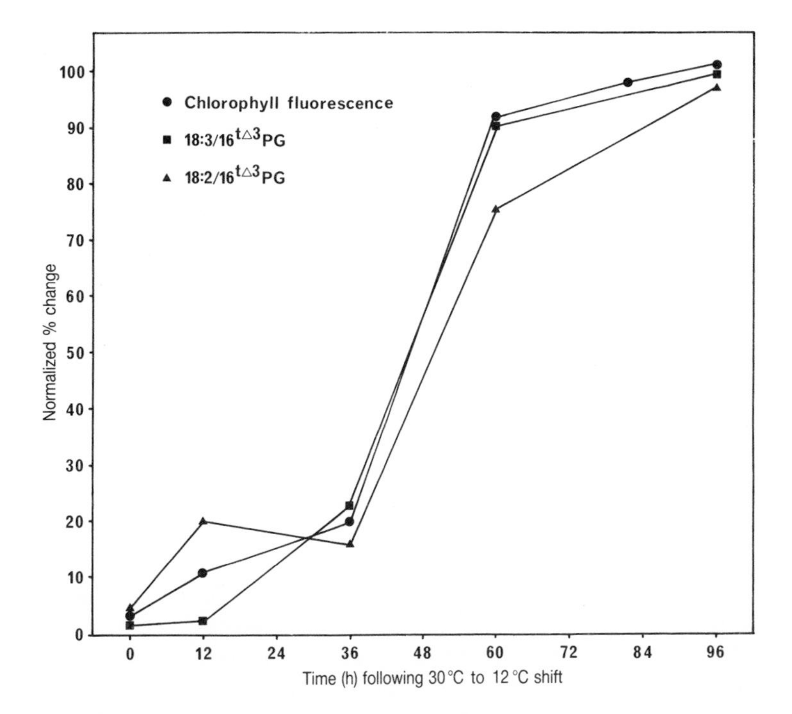

Fig. 4.2. Correlation of changes in chlorophyll fluorescence and the major molecular species of phosphatidylglycerol in chloroplasts from *D. salina*. Data from Lynch and Thompson (1984*b*) with permission.

brane properties (cited above), it was obvious that labelling studies of this particular lipid would be of importance. Moreover, examination of PG from *Dunaliella* had shown that the fatty acid compositions in microsomal and chloroplast fractions were completely different. The microsomal phospholipid had a 'eukaryotic' composition with C_{16} acids at the *sn*-1 position and C_{18} acids at the *sn*-2 position. On the other hand, chloroplast PG was 'prokaryotic' with C_{16} acids enriched at the *sn*-2 position. When these pools of PG were labelled with ^{14}C-palmitate two differences were immediately apparent (Table 4.12). First, the incorporation of ^{14}C-palmitate was 'prokaryotic' in chloroplasts but 'eukaryotic' (i.e. into the *sn*-1 position) in microsomes. Second, the PG of chloroplasts but not that of the microsomes showed a change in positional distribution with time which might have been due to some influx of extra-chloroplastidic PG (Norman and Thompson 1985*a*). These changes were increased by a shift in growth temperature to 12°C. Furthermore, comparison of the labelling of molecular species of chloroplast PG with the endogenous species revealed that high radioactivity was mainly

Table 4.12 Comparison of the distribution of ^{14}C-palmitate in *Dunaliella* chloroplast or microsomal PG.

Time after labelling	Percentage of total PG ^{14}C-palmitate			
	Chloroplasts		Microsomes	
	sn–1	sn–2	sn–1	sn–2
2 min	0–3	97–100	70	30
10 min	35	65	70	30
20 min	50	50	75	25
60 min	50	50	80	20
16 hr	50	50	80	20

Data from Norman and Thompson (1985*a*). Growth of algae was at 30 °C.

associated with the 'eukaryotic' species 1–16:0, 2–18:2 (Table 4.13). In contrast, labelling of individual PG species from ^{14}C-laurate was little affected by temperature. Although the experimental data did not rule out the possibility that retailoring of chloroplast PG occurred exclusively within that organelle, the more likely explanation was that stress induced by growth at 12°C was countered by a low-temperature acclimation followed by the recruitment of key microsomal PG molecular species (Norman and Thompson 1985*a*).

Acyl hydrolases are extremely active in many plant tissues (cf. Burns *et al.* 1979) and could play an obvious role in the retailoring of individual complex lipids. Accordingly, it is interesting that a phospholipid acyl hydrolase (Cho

Table 4.13 Comparison of the radioactive labelling of molecular species of chloroplast PG from ^{14}C-palmitate with the endogenous pattern.

Molecular species	Total PG (wt %)	Labelling (% total radioactivity)	
		0 min chase	20 min chase
14:2/16:0	3	6	3
18:3/*t* 16:1	57	0	0
18:3/16:0, 16:0/18:3	trace	8	7
18:2/*t* 16:1	31	0	0
16:0/18:2, 18:2/16:0	8	83	86
16:0/16:0	trace	3	4

Data from Norman and Thompson (1985*a*).
Labelling for 2 minutes followed by incubation in non-radioactive media for the period shown.

and Thompson 1986) and one which prefentially hydrolysed MGDG (Norman and Thompson 1986) have been described. The former but not the latter enzyme seemed to be activated during acclimation of *Dunaliella* to low growth temperatures. This was in keeping with the general conclusions (above) that chloroplast enzymes are minimally involved in promoting low temperature membrane lipid acclimation.

Before closing this section on *Dunaliella*, mention should be made of its content of the relatively unstudied ether lipid, diacylglyceryltrimethylhomoserine. This compound was first identified in *Ochromonas danica* (Brown and Elovson 1974) and has since been found, sometimes in appreciable amounts, in other algae (e.g. Eichenberger and Boschetti 1978; Janero and Barnett 1983; Eichenberger 1982; and see above). The ether lipid is equivalent to 65 per cent of the total phospholipid of chloroplasts and to 57 per cent of that of the microsomes of *Dunaliella*. *In Chlamydomonas* it has been suggested that it may be involved as a substrate for oleate or linoleate desaturation (Schlapfer and Eichenberger 1983). Oleate desaturation in plants is usually thought to take place with PC as substrate (cf. Harwood 1980) but this lipid is only found in trace amounts in *Chlamydomonas*. Therefore it was interesting that in *Dunaliella*, which contains both lipids, their molar species compositions were very similar, though quite different from other glycerolipids (Norman and Thompson 1985*b*). Furthermore, radiolabelling studies indicated a very close metabolic relationship between the ether lipid and PC, not only during desaturation but also during subsequent transfers of radioactivity between the endoplasmic reticulum and the chloroplasts.

Acknowledgements

We are grateful to the Science and Engineering Research Council and to the Natural Environmental Research Council for financial support of our work on marine algae.

References

Anderson, R., Kates, M., and Volcani, B.E. (1978). *Biochim. Biophys. Acta* **528**, 89.

Beck, J.C. and Levine, R.P. (1977). *Biochim. Biophys. Acta* **489**, 360.

Brown, A.E. and Elovson, J. (1974). *Biochemistry* **13**, 3476.

Bryan, G.W. (1980). *Helgolander Meeresunters* **3**, 6.

Bryan, G.W. (1983). *Science of the Total Environment* **28**, 91.

Bryce, T.A., Welti, D., Walsby, A.E., and Nichols, B.W. (1972). *Phytochemistry* **11**, 295.

Burns, D.D., Galliard, T., and Harwood, J.L. (1979). *Phytochemistry* **18**, 1793.

Cho, S.H. and Thompson, G.A. (1986). *Biochim. Biophys. Acta* **878**, 353.

Dickens, B.F. and Thompson, G.A. (1982). *Biochemistry* **21**, 3604.

Douce, R. and Joyard, J. (1980). In *The biochemistry of plants*, Vol. 4 (ed. P.K. Stump and E.E. Conn) p. 321. Academic Press, New York.
Eichenberger, W. (1982). *Plant Sci. Lett.* **24**, 91.
Eichenberger, W. and Boschetti, A. (1978). *FEBS Lett.* **88**, 201.
Feige, G.B., Heinz, E., Wrage, K., Cochem, N., and Prozelar, E. (1980). In *Biogenesis and function of plant lipids* (ed. P. Mazliak, P. Benveniste, C. Costes, and R. Douce) p. 135. Elsevier, Amsterdam.
Harwood, J.L. (1975). In *Recent advances in the chemistry and biochemistry of plant lipids* (ed. T. Galliard and E.I. Mercer) p. 43. Academic Press, London.
Harwood, J.L. (1980). In *Biogenesis and function of plant lipids* (ed. P. Mazliak, P. Benveniste, C. Costes, and R. Douce) p. 143. Elsevier, Amsterdam.
Harwood, J.L. (1983). *Biochem. Soc. Trans.* **11**, 343.
Hitchcock, C. and Nichols, B.W. (1971). *Plant lipid biochemistry*. Academic Press, London.
Jamieson, G.R. and Reid, E.H. (1972). *Phytochemistry* **11**, 1423.
Janero, D.R. and Barnett, R. (1983). *J. lipid Res.* **23**, 307.
Jones, A.L. and Harwood, J.L. (1987). *Biochem. Soc. Trans.* **15**, 482.
Jones, A.V.M. and Harwood, J.L. (1980). *Biochem. J.* **190**, 251.
Lem, N.W. and Stumpf, P.K. (1984*a*). *Plant Physiol.* **74**, 134.
Lem, N.W. and Stumpf, P.K. (1984*b*). *Plant Physiol.* **75**, 700.
Lynch, D.V. and Thompson, G.A. (1982). *Plant Physiol.* **69**, 1369.
Lynch, D.V. and Thompson, G.A. (1984*a*). *Plant Physiol.* **74**, 193.
Lynch, D.V. and Thompson, G.A. (1984*b*). *Trends biochem. Sci.* **9**, 442.
Mercer, E.I. and Davies, C.L. (1979). *Phytochemistry* **18**, 457.
Mohy-ud-dhin, M.T. Krepski, W.J., and Walton, T.J. (1982). In *Biochemistry and metabolism of plant lipids* (ed. J.F.G.M. Wintermans and P.J.C. Kuiper) p. 209. Elsevier, Amsterdam.
Murata, N. (1983). *Plant Cell Physiol.* **24**, 81.
Nichols, B.W., James, A.T., and Breuer, J. (1967). *Biochem. J.* **104**, 486.
Norman, H. and Thompson, G.A. (1985*a*). *Arch. Biochem. Biophys.* **242**, 168.
Norman, H. and Thompson, G.A. (1985*b*). *Plant Science* **42**, 83.
Norman, H. and Thompson, G.A. (1986). *Biochim. Biophys. Acta* **875**, 262.
Norman, H., Smith, L.A., Lynch, D.V., and Thompson, G.A (1985). *Arch. Biochem. Biophys.* **242**, 157.
Pettitt, T.P. and Harwood, J.L. (1986). *Biochem. Soc. Trans.* **14**, 146.
Pettitt, T.P. and Harwood, J.L. (1987). In *structure and function of plant lipids* (ed. P.K. Stumpf, J.B. Mudd, and W.D. Nes) p. 657. Plenum, New York.
Pham Quang, L. and Laur, M.H. (1976). *Biochimie* **58**, 1367.
Pohl, P. and Zurheide, F. (1979). In *Marine algae in pharmaceutical science* (ed. H.A. Hoppe, T. Levring, and Y. Tanaka) p. 473. Walter de Gruyter, Berlin.
Rebeille, F., Bligny, R. and Douce, R. (1980). *Biochim. Biophys. Acta* **620**, 1.
Sato, N. and Furuya, M. (1984). In *Structure, function and metabolism of plant lipids* (ed. P–A. Siegenthaler and W. Eichenberger) p. 171. Elsevier, Amsterdam.
Sato, N. and Murata, N. (1980). *Biochim. Biophys. Acta* **619**, 353.
Sato, N. and Murata, N. (1982*a*). *Biochim. Biophys. Acta* **710**, 271.
Sato, N. and Murata, N. (1982*b*). *Biochim. Biophys. Acta* **710**, 279.
Sato, N. and Murata, N. (1982*c*). In *Biochemistry and metabolism of plant lipids* (ed. J.F.G.M. Wintermans, and P.J.C. Kuiper) p. 201. Elsevier, Amsterdam.

Sato, N., Murata, N., Miura, Y., and Ueta, N. (1979). *Biochim. Biophys. Acta* **572**, 19.
Schlapfer, P. and Eichenberger, W. (1983). *Plant Sci. Lett.* **32**, 243.
Smith, K.L. and Harwood, J.L. (1984*a*). *Phytochemistry* **23**, 2469.
Smith, K.L. and Harwood, J.L. (1984*b*). *J. exp. Botany* **35**, 1359.
Smith, K.L., Bryan, G.W., and Harwood, J.L. (1984). *Biochim. Biophys. Acta* **796**, 119.
Smith, K.L., Bryan, G.W., and Harwood, J.L. (1985). *J. exp. Botany* **36**, 663.
Smith, K.L., Douce, R., and Harwood, J.L. (1982). *Phytochemistry* **21**, 569.
Smith, K.L., Haan, A.C., and Harwood, J.L. (1986). *Physiol. Plant* **66**, 692.
Stapleton, S.R. and Jaworski, J.G. (1984). *Biochim. Biophys. Acta* **794**, 249.
Zepke, H.D., Heinz, E., Radunz, A., Linscheid, M., and Pesch, R. (1978). *Arch. Microbiol.* **119**, 157.

5 Tetrapyrrole biosynthesis—the C_5 pathway

A. J. SMITH and L. J. ROGERS

Department of Biochemistry, University College of Wales, Aberystwyth, Dyfed SY23 3DD, UK

In cyanobacteria and algae several tetrapyrroles in association with macromolecular polypeptide assemblies play a central role in energy conservation; these include the chlorophylls, phycobilins, and various haems. Tetrapyrroles also have other roles in metabolism including the breakdown of hydrogen peroxide (haem-*b*), the reductive assimilation of nitrite (sirohaem), and C_1 and hydrogen transfers in various biosyntheses (co-enzyme form of vitamin B_{12}). Apart from the phycobilins all of these are cyclic tetrapyrroles (Fig. 5.1).

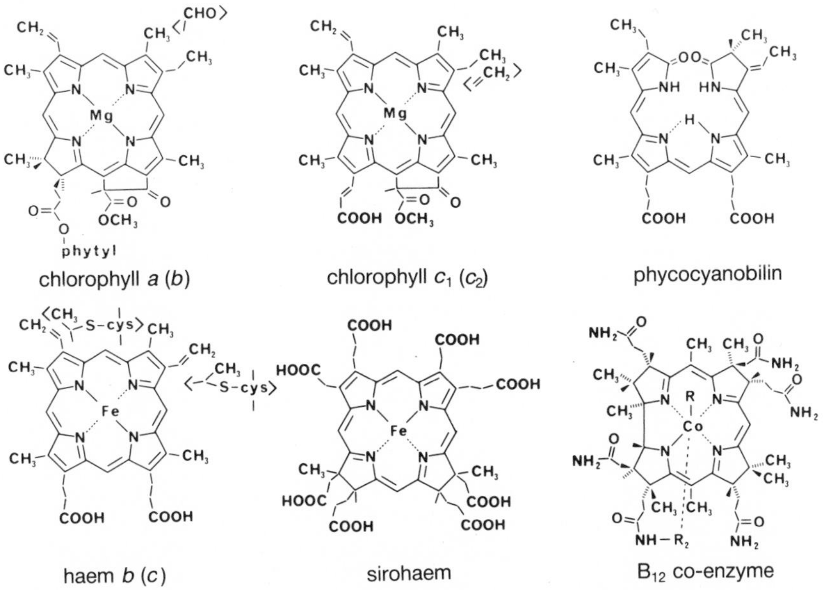

chlorophyll *a* (*b*) — chlorophyll c_1 (c_2) — phycocyanobilin

haem *b* (*c*) — sirohaem — B_{12} co-enzyme

Fig. 5.1. Structures of some representative tetrapyrroles. Although for comparison phycocyanobilin is given in the same conformation it should be noted that this is a linear tetrapyrrole. The variants to the structures are shown in brackets.

The common origin of tetrapyrroles

The idea that the tetrapyrroles share a common biosynthetic pathway grew from the realization of the structural similarities between haems and chlorophylls, and was confirmed by the identification of protoporphyrin and magnesium protoprophyrin as metabolites accumulated by *Chlorella* mutants defective in chlorophyll synthesis (Granick, 1961). The pathway (from precursors) to protoporphyrin was considered to be common to all tetrapyrroles; after this the route to individual tetrapyrroles diverged (Fig. 5.2). This expectation has been confirmed for the major tetrapyrroles. Sirohaem and the tetrapyrrole macrocycle of the B_{12} coenzyme have structural similarities that distinguish them from the other tetrapyrroles (Fig. 5.1), indicating a common origin from a biosynthetic intermediate prior to protoporphyrin.

In all organisms the C_5 compound δ-aminolaevulinic acid (ALA) is a key intermediate in the biosynthesis of tetrapyrroles; one of a number of demonstrations of this relationship was the excretion of porphobilinogen, porphyrins, and phycocyanobilin when the rhodophyte *Cyanidium caldarium* was incubated with ALA (Troxler and Lester 1967). Two molecules of ALA condense to form the monopyrrole porphobilinogen and four molecules of this combine to give a linear type I tetrapyrrole, hydroxymethylbilane (Battersby *et al.* 1982). While other structural isomers may be formed non-enzymically they appear to have no physiological role, and the normal metabolic fate of hydroxymethylbilane is cyclization to produce uroporphyrinogen III, an asymmetric tetrapyrrole. Porphyrins, cobalamines, haems, chlorophylls, and the phycobilin linear tetrapyrroles are all derived from this key intermediate. An outline of this pathway, with an indication of the stoichiometry of the reactions, is summarized in Fig. 5.2. The final steps to chlorophyll have been reviewed recently (Castelfranco and Beale 1983).

In assessing the operation of this pathway, account must be taken of the

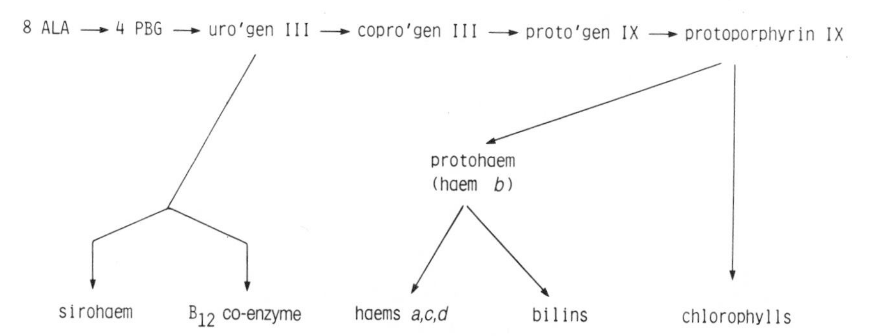

Fig. 5.2. Biosynthetic origin of the tetrapyrroles from δ-aminolaevulinic acid, with an indication of the stoichiometry of the early reactions.

Table 5.1 Cytochrome content of cyanobacterial and algal cells.

Cytochrome	f	b_{559}	b_{563}	aa_3	Reference
Phormidium laminosum	1.1 (1000)	4.3 (260)	2.25 (500)	—	Stewart and Bendall 1980
Synechococcus 6301	—	—	—	0.65 (1700)	Peschek *et al*. 1982
Scenedesmus obliquus	2.3 (480)	3.4 (330)	2.3 (480)	—	Fleischhacker and Senger 1978

Data for these membrane-associated cytochromes are average values (nmol mg^{-1} chlorophyll) of any range quoted; corresponding data for the soluble cyt *c* are not available. Figures in parentheses are corresponding molar ratios of chl/cyt.

types of end-product, their relative proportions, and their location within the cell. In the cyanobacteria, which lack intracellular compartmentation, chlorophyll *a* (Chl *a*) is an integral component of the photosynthetic lamellae, harvesting light at 435 nm and 675 nm *in vivo*. The phycobiliproteins with their phycobilin prosthetic groups function solely as accessory pigments; they absorb light in the mid-region of the visible spectrum at 620–640 nm (phycocyanin) and 560–570 nm (phycoerythrin). The phycobiliproteins are not part of the photochemical reaction centre but are organized in multi macromolecular assemblies, the phycobilisomes, which are peripheral to the photosynthetic lamellae (Glazer 1984). In laboratory cultures of *Synechococcus* 6301 (*Anacystis nidulans*), the cell content of chl *a* is about 10 $\mu g\ mg^{-1}$ dry weight and phycobilin of the order of 5 $\mu g\ mg^{-1}$ dry weight (Hoult *et al.* 1986). In the photosynthetic lamellae, cytochrome *f*(cyt *f*), cyt b_{563}, and cyt b_{559} are integral proteins, the first two being components of the cyt *bf* complex; additionally a soluble cyt *c* may be constitutive or in some species may be induced when cells are grown in copper-deficient medium, here replacing plastocyanin. Cytochromes are also associated with the plasma membrane, cyt *c* being a peripheral protein and cyt aa_3 a membrane-located complex. Amounts of these are of the order shown in Table 5.1. In total, the haem prosthetic groups of the various types of cytochromes represent a minor though qualitatively crucial category of tetrapyrroles.

In the eukaryotic algae, as in higher plants, there is a strict subcellular compartmentation of tetrapyrroles. The chlorophylls, and in red algae the phycobilins, are in the chloroplast, whilst the cytochromes are distributed between the various subcellular organelles. Here also the chlorophylls can be more diverse (Anderson and Barrett 1986); for example, chl *b* (green algae) or chl c_1 and chl c_2 (brown algae) are normally present together with chl *a*. Possession of the additional type of chlorophyll extends the range of incident light which can be utilized for photosynthesis (chl *b*, 480 and 650 nm; chl c_1c_2, 460 nm (major) and 630 nm (minor), *in vivo*).

The origin of ALA in photosynthetic organisms

The diversion of common metabolites to the formation of tetrapyrroles is focused in the formation of ALA, the first compound committed to the synthesis of tetrapyrroles. Cell-free systems from animal tissues (Gibson *et al.* 1958) and from photosynthetic bacteria (Kikuchi *et al.* 1958) contain ALA-synthase, which catalyses the formation of ALA from succinyl-CoA and glycine (Fig. 5.3).

Initial expectations of a common route to ALA in all organisms have been undermined by the inability of cell-free systems from plants and algae to form this key precursor from succinyl-CoA and glycine. In early studies the only instances where the succinyl-CoA pathway appeared to operate were the

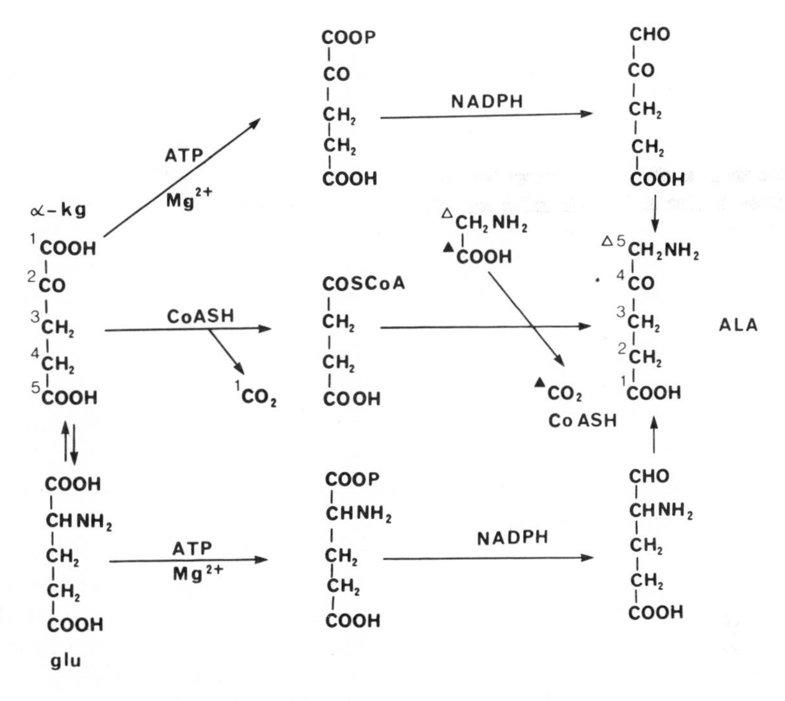

Fig. 5.3. The principal routes proposed to have a role in ALA biosynthesis in photosynthetic organisms. The fate of specific carbons is illustrated. The routes are: dioxovaleric acid pathway (upper); the succinate–glycine pathway (middle); the glutamate 1-semialdehyde pathway (lower).

atypical plant systems such as the greened outer tissues of stored potatoes (Ramaswamy and Nair 1976) and non-greening soybean callus cultures (Wider de Xifra *et al.* 1978). Some indirect evidence, based on the relative incorporation of specifically radiolabelled glycine and succinate into ALA, came principally from studies with the algae *Chlorella fusca* (Porra and Grimme 1974) and *Scenedesmus obliquus* (Klein and Senger 1978) and with etiolated barley (Meller and Gassman 1982). However, there has been no convincing demonstration of the key enzyme, ALA-synthase, in cell extracts from photosynthetically competent cyanobacteria, algae, or plants, with the notable exception of the phytoflagellate *Euglena gracilis* (Dzelzkalns *et al.* 1982; Foley *et al.* 1982). In this organism the enzyme was present in highest activity in achlorophyllous cells, and in greening cells its activity was not related directly to that of chlorophyll formation. This suggested that ALA-synthase was extraplastidic and that the cell requirement of non-plastidic ALA could be met from succinate and glycine when any alternative plastid-located pathway was inoperative (Beale and Foley 1982; Dzelzkalns *et al.*

1982; Foley *et al.* 1982). These and similar studies suggested that an alternative route existed for ALA synthesis in plastids, and that the ALA-synthase pathway, if it occurred at all, only played a role in extraplastidic tetrapyrrole formation. While much of the work directed towards the characterization of this route has centred on plant systems, algal and cyanobacterial studies have also made a significant contribution and these are the prime concern of this review.

An accumulation of ALA has been induced in plants (Harel and Klein 1972), algae (Beale 1970; Richard and Nigon 1973; Oh-Hama and Senger 1975; Jurgenson *et al.* 1976) and cyanobacteria (Meller and Harel, 1978; Kipe-Nolt *et al.* 1978) by laevulinic acid (LA) which is an inhibitor of ALA-dehydratase (Nandi and Shemin 1968). Of particular importance in these studies were the observations that LA also inhibited chlorophyll synthesis and that the ALA which accumulated in the presence of the inhibitor largely accounted for the decrease in chlorophyll compared to untreated controls. This correlation of ALA accumulation on the one hand and inhibition of chlorophyll synthesis on the other did much to restore confidence in ALA as a key intermediate in the synthesis of tetrapyrroles in oxygenic photosynthetic organisms. However, other investigations have indicated that the results of experiments based on the use of LA should be interpreted with caution because, at the high concentrations necessary to inhibit the dehydratase *in vivo*, it may have other effects on metabolism (Klein *et al.* 1975; Konis *et al.* 1978; Oh-Hama and Senger 1975; Owens *et al.* 1978; Schwartzbach *et al.* 1975). In addition, it must be recognized that LA is metabolized extensively (Duggan *et al.* 1981) via the fatty acid oxidation pathway and the tricarboxylic acid cycle (Levasseur and Gassman 1986).

Nevertheless, the accumulation of ALA in the presence of LA has been exploited in isotopic tracer studies of ALA biosynthesis. The degree of incorporation of particular organic compounds into ALA and the location in this molecule of ^{14}C from specifically labelled substrates led to the proposal that either glutamate or α-ketoglutarate was a more direct precursor of ALA than succinate and glycine. In greening plant systems, such as cucumber cotyledons (Beale and Castelfranco 1974) and maize leaves (Meller *et al.* 1975), glycine and succinate were poor precursors compared with glutamate and α-ketoglutarate. Similar observations were made with *Cyanidium caldarium* (Jurgenson *et al.* 1976), *Chlorella* (Meller and Harel 1978), and cyanobacteria (Meller and Harel 1978; Kipe-Nolt and Stevens 1980). In some of these systems, the interpretation of data has been complicated by variation in the efficiency with which organisms can assimilate different organic compounds.

The direct involvement of a C_5 compound in the synthesis of ALA was further investigated by determining the distribution of isotope in ALA obtained from organisms incubated in the presence of LA with specifically labelled glutamate or α-ketoglutarate. This was established by chemical degradation of ALA with alkaline periodate; while succinate and formaldehyde (as the bis-dimedone derivative) can be obtained in high yields (>90 per cent) under optimum conditions (Shemin *et al.* 1955), lower and variable yields have been reported by most groups. Beale *et al.* (1975) reported the demonstration of the incorporation of the intact C_5 skeleton of glutamate into ALA in greening barley. The same conclusion was drawn from similar studies with the eukaryotes *C. caldarium* (Jurgenson *et al.* 1976), *Chlorella vulgaris* (Meller and Harel 1978), and *E. gracilis* (Weinstein and Beale 1983; Gomez-Silva *et al.* 1985), and with the cyanobacteria *Fremyella diplosiphon* (Meller and Harel 1978), *Agmenellum quadruplicatum* (Kipe-Nolt and Stevens 1980), and *Anabaena variabilis* (Avissar 1980). In all of these organisms, C-5 and C-1 of ALA were derived to a significant degree from C-1 and C-5 of glutamate (or α-ketoglutarate), respectively. However, these data were not as clear-cut as they might have been, in all probability because of the scrambling effect of prior metabolism of the labelled substrate on isotope distribution in ALA along with the incomplete recovery of isotope in either succinate or formaldehyde formed from ALA by chemical degradation.

Some of the clearest evidence for the conversion of a C_5 compound such as glutamate to ALA, which has become known as the C_5 pathway, has come from studies exploiting a unique feature of the intermediary metabolism of cyanobacteria. These organisms possess an incomplete TCA cycle blocked at the dehydrogenation of α-ketoglutarate (Smith *et al.* 1967; Hoare *et al.* 1967). A direct consequence of this is the restriction of carbon derived from exogenous substrates such as acetate to a very limited number of metabolites and cell constituents derived from them. Thus isotope from ^{14}C-acetate, which is assimilated more efficiently by cyanobacteria than is isotope from a range of other labelled compounds, is incorporated readily into C_5 compounds such as α-ketoglutarate and glutamate but not directly into C_4 compounds such as aspartate and its metabolites. Consequently, in cyanobacteria, tetrapyrroles derived from ALA formed by the C_5 pathway would be labelled as effectively as glutamate. On the other hand, if cyanobacterial ALA originated from oxaloacetate-derived succinate and from glycine, tetrapyrrole labelling by ^{14}C-acetate would be insignificant.

Experiments on this basis with *Synechococcus* 6301 gave the data presented in Table 5.2 (McKie *et al.* 1981). While aspartate derived from cell protein contained relatively little isotope, glutamate was very heavily labelled by both 1-^{14}C- and 2-^{14}C-acetate; as a consequence of the incomplete TCA cycle this label is located in C-5 and C-4 of glutamate, respectively (Hoare and Moore 1965). Phaeophorbide *a* derived from chl *a* was also heavily labelled by both ^{14}C-acetates. Because certain of the carbons of ALA are lost in the synthesis of chl *a*, the ratio of the specific activities of phaeophorbide *a* to glutamate will differ for organisms grown in the presence of 1-^{14}C- and

76 *A.J. Smith and L.J. Rogers*

Table 5.2 Labelling of aspartate, glutamate, and phaeophorbide-*a* in *Synechococcus* 6301 grown in the presence of 1-^{14}C- or 2-^{14}C-acetate.

Substrate	Specific radioactivity ($10^{-2} \times$ dpm μmol^{-1})			Ratio of specific radioactivities	
	Aspartate	Glutamate	Phaeophorbide	Phaeophorbide *a* / Aspartate	Phaeophorbide *a* / Glutamate
1-^{14}C-acetate	3.74	761	1890	510	2.5
2-^{14}C-acetate	0.76	641	6510	8600	10.2

Chlorophyll *a* was isolated and converted to phaeophorbide *a* for analysis. Aspartate and glutamate were isolated from hydrolysates of total cell protein.

2-^{14}C-acetate; the theoretical values are 2:1 and 8:1, respectively (Fig. 5.4). For the cyanobacterium the experimentally derived ratios were 2.5:1 and 10:1, in excellent agreement given other experimental uncertainties. These data were clearly indicative of the operation of the C_5 route for ALA synthesis with carbons 1 and 2 of ALA derived from carbons 5 and 4 of glutamate (or carbons 1 and 2 of acetate), respectively. They also indicated that the succinate–glycine route does not contribute to ALA synthesis in photoautotrophically grown *Synechococcus* 6301.

The effectiveness of acetate as a precursor of phycocyanobilin in this cyanobacterium was demonstrated by Laycock and Wright (1981). Using the more sophisticated technique of ^{13}C-NMR they identified the location of acetate carbon that had been incorporated into the biliprotein prosthetic group. With 1-^{13}C-acetate only the carboxyl carbons of the propionate side-chains of the pigment were labelled. This could only be accounted for by the synthesis of phycocyanobilin from ALA formed by the C_5 pathway, since C-1 of acetate is restricted to C-5 of glutamate and this, in turn, becomes the C-1 of ALA from which these two carboxyl carbons are exclusively derived (Fig. 5.4).

In further investigations with *Synechococcus* 6301 (Lewis *et al.* 1984) using 1-^{13}C-, 2-^{13}C-, and 1, 2-^{13}C-acetates, the patterns of isotope distribution in chl *a* and phycocyanobilin were again precisely those assumed in the interpretation of the extensive incorporation of acetate carbon into glutamate and chl *a* in cyanobacteria (McKie *et al.* 1981). The alternative that this isotope

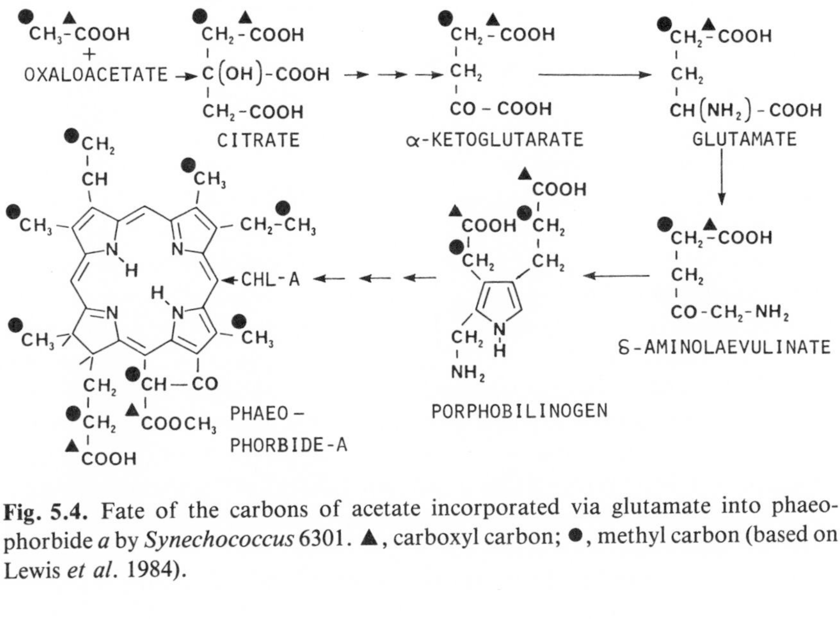

Fig. 5.4. Fate of the carbons of acetate incorporated via glutamate into phaeophorbide *a* by *Synechococcus* 6301. ▲, carboxyl carbon; ●, methyl carbon (based on Lewis *et al.* 1984).

came from labelled acetate via succinate originating from isocitrate by isocitrate lyase could be discounted since label derived from 1-^{13}C-acetate would be randomized between C-1 and C-4 of succinate and would also appear in the pyrrole ring carbons of the bilin. From these observations it can be concluded that chl *a* and phycocyanobilin share a common biosynthetic pathway in which ALA is derived directly from glutamate (Fig. 5.4). These studies with cyanobacteria provide incontrovertible evidence for the synthesis of ALA by the C_5 route in an oxygenic photosynthetic organism. Furthermore, they emphasize the importance of careful design in the application of isotope tracer studies to the elucidation of metabolic pathways.

Exploitation of inhibitors in studies of tetrapyrrole biosynthesis

In recent years several compounds have been shown to block the synthesis of chlorophyll. These inhibitors are of interest in several respects; on the one hand they can be used as tools to further characterize the biosynthetic pathway, while on the other they may serve as the parent compounds for new classes of herbicide.

The first such compound exploited in studying tetrapyrrole biosynthesis was LA, which inhibits ALA-dehydratase, as described earlier. More recently, dioxoheptanoic acid (DA), which inhibits the same enzyme and like LA gives an accumulation of ALA (Meller and Gassman 1981), has also been introduced; it is more potent than LA and has the further advantage of a progressively irreversible effect on ALA-dehydratase (S.A. Pearson, A.J. Smith, and L.J. Rogers, unpublished observations).

Interest has more recently centred on 3-amino 2,3 dihydrobenzoic acid (gabaculin), a much more potent inhibitor of pigment synthesis than either LA or DA. Gabaculin was first characterized as a pyridoxal phosphate antagonist (Rando 1977); in animal cells it inhibited irreversibly several aminotransferases which involve substrates with exchangeable β-protons (Soper and Manning 1982). In higher plants gabaculin blocked chlorophyll synthesis (Flint 1984; Hill *et al.* 1985) and phytochrome formation (Gardner and Gorton 1985; Konomi and Furuya 1986). At the concentration at which it inhibited chlorophyll biosynthesis, gabaculin was considerably more specific in its action than amino-oxyacetate, another pyridoxal phosphate antagonist, as judged by effect on $^{14}CO_2$ assimilation patterns (S.A. Pearson, A.J. Smith, and L.J. Rogers, unpublished observations). Gabaculin also inhibited the phototrophic and chemoheterotophic growth of cyanobacteria, though not of other prokaryotes which are considered to form ALA from succinyl-CoA and glycine (Hoult *et al.* 1986). Exposure of growing cultures of *Synechococcus* 6301 to 50 μM gabaculin resulted in an immediate and complete inhibition of the synthesis of chl *a* and phycocyanin; under these conditions only small amounts of Ehrlich-positive material, presumed to be

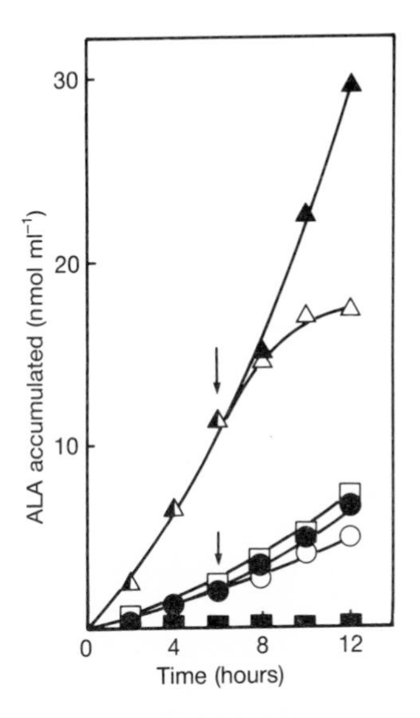

Fig. 5.5. Accumulation of ALA in *Synechococcus* 6301. A cell suspension (0.77 mg dry wt. ml^{-1}) of organisms from the exponential phase of growth was incubated in the light at 30 °C in growth medium with; no addition (■), DA (▲), or gabaculin (●) alone, or both inhibitors (□). After 6 hours a suspension containing DA was supplemented with gabaculin (△), and one containing gabaculin was supplemented with DA (○). Concentrations of DA and gabaculin were 3.2 mM and 50 μM, respectively (after Hoult *et al.* 1986).

ALA, accumulated. With lower concentrations, tetrapyrrole synthesis was suppressed for several hours but then resumed at a lower rate than in untreated organisms. Gabaculin also blocked the accumulation and excretion of ALA in the presence of DA (Fig. 5.5). This suggests that the primary effect of gabaculin is a specific inhibition of the C_5 pathway for ALA formation (Hoult *et al.* 1986). This interpretation is in accord with similar data for this inhibitor obtained with greening barley (Hill *et al.* 1985).

Similar experiments with *Synechococcus* 6301 by Guikema *et al.* (1986) have led to the contrary conclusion that gabaculin blocked chlorophyll biosynthesis after the formation of ALA; this was based on the detection of an accumulation of ALA in the presence of gabaculin, an observation also made by Hoult *et al.* (1986). The differing conclusion drawn by Guikema *et al.* (1986) does not take into account the fact that the amount of ALA excreted in the presence of gabaculin is only a small fraction of that obtained with DA at a concentration that inhibits chlorophyll synthesis to a similar degree. This and the data in Fig. 5.5 make it clear that gabaculin inhibits ALA formation, and not its utilization. It is possible that a contribution to the apparent accumulation of ALA in the presence of gabaculin comes from another

Ehrlich-positive compound, glutamate semialdehyde, which has been reported to accumulate in plants in the presence of this inhibitor (Kannangara and Schouboe 1985) and is a likely intermediate in ALA synthesis in oxygenic photosynthetic organisms; this compound is reported to react with the Ehrlich reagent used in the assay, though the resulting complex has a tenfold lower molar extinction coefficient than that formed with ALA-pyrrole (Kannangara and Schouboe 1985). Although gabaculin was without effect on the catalytic activity of ALA-dehydratase the specific activity of the enzyme was lowered in *Synechococcus* 6301 exposed to the inhibitor (Hoult *et al.* 1986); this may also contribute to the slight accumulation of Ehrlich-positive material seen in gabaculin-treated greening barley (Hill *et al.* 1985; Kannangara and Schouboe 1985) as well as in cultures of cyanobacteria (Hoult *et al.* 1986; Guikema *et al.* 1986). An understanding of this *in vivo* effect on ALA-dehydratase might give an insight into the regulation of the overall pathway. Further information indicating the location of the gabaculin-sensitive site in this section of the pathway has come from studies with cell-free systems (described below).

The exploitation of these inhibitors should enable the co-ordination of the synthesis of apoproteins and their chromophores to be probed. To date, few investigations of this aspect of tetrapyrrole synthesis have been reported, exceptions being the accumulation of a peptide possibly implicated in siro-haem synthesis in LA-treated maize (Schuster and Harel 1985) and the suggestion that apophytochrome and chromophore synthesis may also be out of step in gabaculin-treated pea (*Pisum sativum*) seedlings (Konomi and Furuya 1986).

Studies with cell-free systems

These investigations have for the main part been concerned with higher plants rather than with algae. Plastid preparations from spinach, barley, and maize form labelled ALA from ^{14}C-α-ketoglutarate in a light-stimulated process (Gough and Kannangara 1976; Kannangara and Gough 1977); organelles from immature (greening) leaves were more active than mature chloroplasts. Subsequently L-glutamate was shown to be a more effective precursor than the keto acid (Kannangara and Gough 1977). The maximum rate of ALA formation was equivalent to about 0.3 nmol mg^{-1} protein min^{-1}, a much higher rate than that obtained with cucumber cotyledon plastids (Weinstein and Castelfranco 1978). However, ALA formation at a rate approaching this has been obtained even in the absence of added glutamate (Kannangara and Gough 1977). Further studies with plant systems have established that the enzymes catalysing the conversion of glutamate to ALA are located in the stromal fraction of disrupted chloroplasts (Gough and Kannangara 1977). In such systems the formation of ALA was dependent on addition ATP and NADPH, which accounts for the effect of light on the activity of intact organelles. More recently, intact developing chloroplasts from cucumber cotyledons have been shown to incorporate glutamate through to chlorophyllide (Fuesler *et al.* 1984).

The first algal systems converting glutamate to ALA were obtained from *Chlamydomonas reinhardtii* (Wang *et al.* 1984) and *Chlorella vulgaris* (Weinstein and Beale 1985*a*); like the plant system, ALA formation was dependent on ATP and NADPH. Similar activity has also been found in cell-free preparations from *E. gracilis* and *Cyanidium caldarium* (Weinstein *et al.* 1986). In *Chlorella*, 5 μM gabaculin inhibited the conversion of glutamate to ALA by 50 per cent (Weinstein and Beale 1985*b*).

Taken together, the studies outlined have demonstrated convincingly the direct role of glutamate as an intermediate in tetrapyrrole formation in algae and cyanobacteria, as well as in plants. Amongst possible routes for this C_5 pathway to ALA (Beale *et al.* 1975), two proposals have been the focus of particular interest; these involve DOVA or glutamate 1-semialdehyde, respectively, as the key intermediate (Fig. 5.3). The approach used in the investigation of these alternative routes has exploited radioactive tracer techniques, inhibitors of pigment synthesis, and cell-free systems.

The DOVA pathway

The formation of DOVA from α-ketoglutarate has been reported in maize leaf extracts (Lohr and Friedman 1976). These workers partly purified an enzyme catalysing the conversion of α-ketoglutarate into DOVA in the presence of NADPH. Difficulties in demonstrating this activity were ascribed to the predominance of glutamate dehydrogenase in crude plant extracts. However, the main evidence for the DOVA route has been the demonstration in plants (Salvador 1978*a*) and the algae *C. vulgaris* (Gassman *et al.* 1968) and *E. gracilis* (Foley and Beale 1982) of the transamination of DOVA to ALA.

In an aplastidic mutant of *E. gracilis* both light-grown and dark-grown cells possessed the same levels of DOVA-aminotransferase as wild-type cells and this activity co-purified with glyoxylate-aminotransferase whose relative activity was, however, considerably higher (Foley and Beale 1982). This was also true for a higher plant, *Raphanus sativus* L. (radish), where DOVA-aminotransferase activity did not correlate with chlorophyll synthesis and the apparent subcellular localization of the enzyme was also inappropriate for this role (Shioi *et al.* 1984).

The view that the *in vitro* transamination of DOVA is a consequence of the broad specificity of glyoxylate-aminotransferase is now gaining ground. In *Chlorella regularis* two apparent isoenzymes of DOVA aminotransferase have been partially purified (Shioi *et al.* 1986). However, both proved to be much more active with other substrates, glyoxylate in one case and a number of aldehydes in the other, suggesting that neither of these enzymes play a significant role in ALA formation.

In a *S. obliquus* mutant C-2A′ which forms only traces of chlorophyll in the dark, a light-dependent accumulation of DOVA has been observed in the presence of LA (Dornemann and Senger 1980). Rates of incorporation of isotope from 1-^{14}C-glutamate were low but about the same as the incorporation of label from 2-^{14}C-glycine into DOVA. The conclusion that the labelling of DOVA by glutamate demonstrates that the former is an intermediate in the biosynthesis of ALA does not take into account the possibility that DOVA was a catabolite of ALA (Harel *et al.* 1983). Furthermore, an apparent accumulation of DOVA in cultures of *C. fusca* and *S. obliquus* containing LA (Porra *et al.* 1980) may have been due to an interference by ALA in the assay for DOVA (Porra and Klein 1981). When particular care to exclude this possibility was taken DOVA could not be detected (Meisch *et al.* 1985).

In conclusion, the DOVA pathway as shown by cell extract studies may be a fortuitous association of enzyme activities which, *in vivo*, are located in different cell compartments and serve different functions (Porra and Grimme 1978). The view that ALA is committed solely to tetrapyrrole biosynthesis may be valid in terms of its anabolic role, but under some circumstances its catabolism may be necessary; it is this process that DOVA-aminotransferase activity may possibly initiate. It is known that ALA can be utilized other than in tetrapyrrole synthesis, as shown by the progressive disappearance of ALA that accumulated in experiments using inhibitors of ALA-dehydratase (e.g. see Hill *et al.* 1985); indeed, the green algae *Golenkinia* will grow on ALA as sole carbon source (Ellis and Greenawald 1985).

The glutamate semialdehyde pathway

In recent years a considerable body of evidence for this pathway has accumulated. The observation that ATP and NADPH stimulated ALA formation from glutamate in plastid preparations from greening barley (Kannangara and Gough 1978), as later shown also for maize (Harel and Ne'eman 1983), led to the proposal (Fig. 5.3) that the conversion involved three distinct steps: (1) phosphorylation of glutamate, requiring ATP; (2) reduction of glutamate 1-phosphate to glutamate 1-semialdehyde, requiring NADPH; and (3) an intramolecular transamination of the semialdehyde to ALA. Evidence for involvement of glutamate 1-semialdehyde as an intermediate is its apparent accumulation *in vivo* in the presence of very high concentrations (500 μM) of gabaculin (Kannangara and Schouboe 1985). In these experiments, a rapid decline in glutamate semialdehyde in gabaculin-treated leaves returned to darkness was attributed to its oxidation to glutamate. This was also thought to explain why the amount of semialdehyde that accumulated in the light was not sufficient to account for the difference in chlorophyll compared to the untreated control. In addition, an enzyme catalyzing the conversion of the semialdehyde to ALA has been reported in the soluble protein fraction from barley plastids (Kannangara and Gough 1978). In crude preparations the activity of the enzyme was 12 nmol ALA formed mg^{-1} protein min^{-1}. On purification a fraction was obtained giving five bands on polyacrylamide gel electrophoresis; the specific activity of this preparation was 70 nmol mg^{-1} protein min^{-1}, a 36-fold purification from stromal proteins. A similar activity which has been reported in partially purified extracts of *C. reinhardtii* (Wang *et al.* 1984) has not been characterized in detail.

Although exogenous pyridoxal phosphate was not necessary, and indeed was somewhat inhibitory to the partially purified enzyme (cf. Salvador 1978*b*), partially purified aminotransferase from barley stromal preparations was sensitive to the pyridoxal phosphate antagonist gabaculin (Kannangara and Schouboe 1985); activity decreased to 50 per cent at 2 μM and inhibition was complete at 15 μM gabaculin. This effect was apparently dependent on the simultaneous availability of either the substrate or the product of the enzyme-catalysed reaction. As gel filtration did not reactivate the inhibited enzyme it has been presumed that gabaculin is an irreversible inhibitor; this would be consistent with its effect on γ-aminobutyric acid-aminotransferase and other similar enzymes (Rando 1977). If this is confirmed the rapid recovery of the growth of cyanobacteria and algae following the removal of gabaculin cannot be attributed to a simple reactivation of the enzyme *in vivo* (Hoult *et al.* 1986; Corriveau and Beale 1986). Synthesis of the glutamate semialdehyde-aminotransferase was apparently dependent on chloroplast ribosomes and was induced by a period of exposure to light, possibly through a photosensitive receptor (Gough and Kannangara 1979). Light therefore affects the conversion of glutamate to ALA at three levels: (1) activation of the ALA-synthesizing system in dark-grown plants, possibly by depleting levels of protohaem, which is a feedback inhibitor of ALA synthesis; (2) stimulation of synthesis of aminotransferase on the cytoplasmic ribosomes (this may be the phytochrome-regulated step); and (3) allowing photosynthetic production of ATP and NADPH, the necessary co-factors.

Further studies to establish that glutamate semialdehyde has the properties expected of a biosynthetic intermediate have not been reported. A demonstration of the incorporation of the intermediate into the end-product is crucial and, in addition, the putative intermediate should effectively dilute the incorporation of precurors earlier in the pathway. Because of the nature of these experiments it is likely that a microbial system may be a more appropriate experimental material than plants.

The chemical nature of glutamate semialdehyde and its role as a substrate for the aminotransferase have also been the focus of controversy. The group in Copenhagen have described two procedures for the conversion of N-carbobenzoxy-L-glutamyl 5-benzyl ester to the free semialdehyde or its diethyl acetal with yields up to 11 per cent (Kannangara and Gough 1978; Houen *et al.* 1983). The putative semialdehyde and its derivatives have been characterized by IR, NMR, and mass spectroscopy (Houen *et al.* 1983; Kannangara and Schouboe 1985). The free semialdehyde which has been detected quantitatively using 3-methyl 2-benzothiazolinone hydrazone (Sawicki *et al.* 1961) is stable when in the form of the hydrochloride while the free base is not. Other workers have attempted the synthesis of this compound by several different methods without success. This has been attributed to the spontaneous polymerization of the semialdehyde as it is formed (Meisch and Maus 1983; Kah and Dornemann 1987). Uncertainty will continue to surround this compound until these apparently irreconcilable discrepancies are resolved.

A partial separation of the enzyme activities for the overall conversion of glutamate to ALA first reported for barley (Kannangara and Gough 1979) has also been achieved using extracts prepared from *Chlorella* (Weinstein *et al.* 1987*b*) and *Chlamydomonas* (Wang *et al.* 1984). All of these procedures are based on sequential affinity chromatography using a variety of different ligands. The current view of the proteins required for the conversion of glutamate to ALA is that they include the following (Fig. 5.6): (1) a ligase requiring ATP and Mg^{2+} and a specific RNA molecule to convert glutamate to glutamyl-RNA; (2) a dehydrogenase, dependent on NADPH, to reduce the activated glutamate to the 1-semialdehyde, and (3) an aminotransferase to mediate the intramolecular migration of the amino group in glutamate 1-semialdehyde from C-4 to C-5. Available information about the aminotransferase has been presented, little is known about the dehydrogenase, and the current status of RNA in this process will now be described.

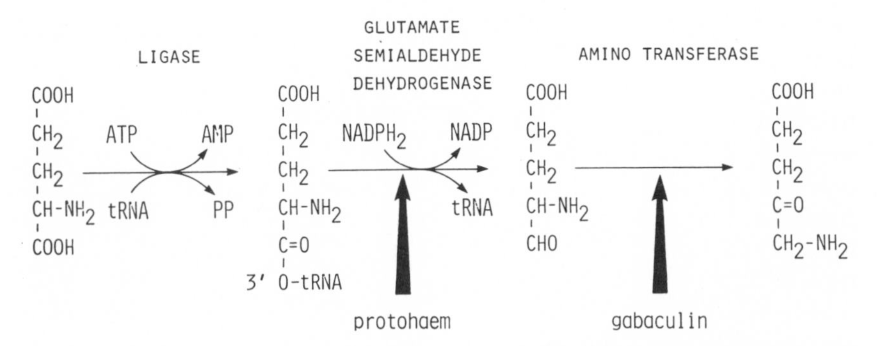

Fig. 5.6. Probable sequence of reactions for the conversion of glutamate to ALA.

The role of RNA in the synthesis of ALA

Initial suggestions of a possible route for the conversion of glutamate to ALA included an activation step yielding glutamyl 1-phosphate (Kannangara and Gough 1978). While a precedent exists for this type of reaction in the biosynthesis of the branched-chain amino acids from aspartate via aspartyl 1-phosphate, no evidence has been obtained for a role for the equivalent glutamyl derivative in ALA synthesis. Instead, data has accumulated over the last few years indicating an involvement of RNA in this process.

The first evidence for a role for RNA in the synthesis of ALA was the demonstration of the sensitivity of this process to treatment with RNase A_1 and snake-venom phosphodiesterase I in stromal preparations from barley chloroplasts (Kannangara *et al.* 1984). With partially fractionated extracts, RNA was detected in material eluted from haem- or chlorophyllin-affinity columns that was essential for the reconstitution of the synthesis of ALA from glutamate. Similar observations have been made with *Chlamydomonas* (Huang *et al.* 1984) and *Chlorella* (Weinstein and Beale 1985*b*). In all three systems the addition of homologous RNA to crude extracts pre-treated to remove endogenous RNA restored the ability to make ALA (Kannangara *et al.* 1984; Weinstein and Beale 1985*b*; Huang and Wang 1986*a*). The level of activity restored was proportional to the amount of RNA added. Moreover, RNA added to a crude extract also stimulated this activity more than six-fold, giving an activity of up to 30 nmol ALA formed mg^{-1} protein $hour^{-1}$ in *Chlorella* preparations. This implies that, in crude extracts, the activity observed is limited by the RNA available. ALA synthesis was also restored in nucleic acid depleted preparations by RNA from cells and tissues competent in oxygenic photosynthesis (Fig. 5.7). Analogous material from non-photosynthetic organisms, including in one study an aplastidic strain of *Euglena*, was essentially inactive (Kannangara *et al.* 1984; Weinstein and Beale 1985*b*; Huang and Wang 1986*a*).

Protein eluted from a Blue-Sepharose affinity column catalysed an ATP- and Mg^{2+} dependent linking of labelled glutamate to homologous and heterologous RNA (Kannangara *et al.* 1984), which is reminiscent of the charging of tRNA molecules in protein synthesis. When RNA isolated from barley stromal preparations was separated by HPLC three distinct fractions were obtained which could be charged with labelled glutamate. However, only one of these RNA fractions was active in a reconstituted system synthesizing ALA (Kannangara *et al.* 1984; Schon *et al.* 1986). This RNA, when isolated and sequenced, was found to contain 76 nucleotides, including the 3′ triplet CCA, and had a sequence compatible with a 'clover-leaf' secondary structure analogous to tRNA molecules. In this structure the anticodon loop contained a modified glutamate anticodon triplet. This species of nucleic acid has been designated $tRNA^{DALA}$ and shown to have a very high sequence

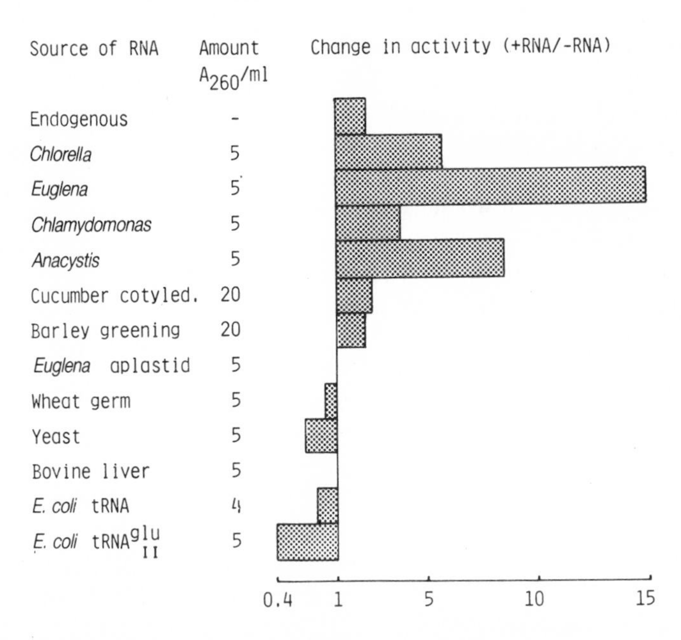

Fig. 5.7. Effect of RNA on the conversion of glutamate to ALA by extracts of *Chlorella* treated with ammonium sulphate to remove endogenous RNA. Values of 1.0 or less indicate that the RNA supplement was without effect or decreased the formation of ALA, and values greater than 1.0 indicate the RNA preparations that stimulated ALA synthesis (based on Weinstein *et al.* 1986).

homology with other glutamate-specific chloroplast tRNAs. In addition, $tRNA^{DALA}$ and fragments obtained from it have been shown to hybridize with chloroplast DNA (Kannangara *et al.* 1984; Schon *et al.* 1986). The covalent linkage between $tRNA^{DALA}$ and glutamate probably involves the 3′OH of the RNA and the carboxyl group at C-1 of glutamate, and evidence to this effect has been presented for a plant system (Kannangara *et al.* 1984).

Despite the detailed information now available about this RNA molecule and its essential role in ALA synthesis in crude preparations, there has been only one preliminary report of the conversion to ALA of glutamate residues linked to tRNA (Kannangara *et al.* 1984); the low efficiency of this conversion, which amounted to only 2 per cent of the glu tRNA, was attributed to the instability of the charged tRNA. While the involvement of tRNA in this type of reaction is most unusual, it is possible that it might facilitate the co-ordinated regulation of pigment and protein synthesis in photosynthetic organisms.

Other metabolic situations where an aminoacylated RNA is involved include a role in bacterial cell wall formation (Chatterjee and Park 1964), synthesis of aminoacyl phosphatidyl glycerol (Nesbitt and Lenartz 1968), ribonuclease P function in *Escherichia coli* (Stark *et al.* 1978), in signal recognition particles in microsomes (Walter and Blobel 1982), and as a component of the ubiquitin proteolytic system in mammalian cells (Ciechanover *et al.* 1985).

The significance of *in vitro* systems for the synthesis of ALA

For any biochemical system the significance of specific transformations demonstrated at the cell-free level must be related to the *in vivo* situation. An assessment of the significance of the C_5 pathway must, therefore, involve a comparison of *in vivo* and *in vitro* rates. Unfortunately, little attempt has been made by investigators to relate the rate of ALA formation from glutamate in cell-free systems to the rate of chlorophyll synthesis *in vivo* either in plants or in algae.

Where ALA synthesis *in vitro* has been assessed radiochemically it is frequently expressed somewhat arbitrarily as cpm incorporated during the period of the incubation. The radiochemical procedure used is often of questionable value for the quantitative assessment of activity because glutamate is commonly used at concentrations well below the apparent K_m of the system. In addition, recent studies have revealed that the endogenous level of acceptor RNA may be insufficient to saturate the system. For the most part, therefore, reported rates of ALA synthesis *in vitro* are at best minimum values. The situation is not much better as far as rates of chlorophyll synthesis *in vivo* are concerned. Where information is provided this is usually in a form that cannot be related to the dry weight or protein content of the experimental material. With micro-organisms chlorophyll content is related to cell density in terms either of unit packed cell volume or of specified cell numbers, and the growth rates of the organism are rarely quoted.

A recent paper (Weinstein *et al.* 1986) has proved relatively more informative in these respects, though even here some useful data for the calculation of *in vivo* rates have been omitted, necessitating assumptions of the dry to wet weight ratio (1:5 assumed) and the protein content of the algal cell material (50 per cent assumed). With intact greening *Chlorella* used in these studies chlorophyll synthesis occurs at a rate of 2 nmol mg^{-1} protein $hour^{-1}$ for a period of 7 hours, which is equivalent to ALA synthesis at 16 nmol mg^{-1} protein $hour^{-1}$. Typical *in vitro* rates of ALA synthesis were in the range 3–6 nmol mg^{-1} protein $hour^{-1}$. Although saturating levels of glutamate were used in these assays, later investigation showed that the observed rates were likely to be significant underestimates of actual capacity because concentrations of acceptor RNA were suboptimal. When this is taken into account, the maximum rate at which glutamate is converted to ALA in the cell-free system from *Chlorella* is likely to be of the order of 34 nmol mg^{-1} protein $hour^{-1}$. Thus in this organism it appears that sufficient activity can be demonstrated in the cell-free system to account for the *in vivo* rates of chlorophyll synthesis. The additional increment needed to take plastid haem synthesis into account will be small and readily accommodated within the observed rates. Although it remains to be seen whether the rates obtained with organisms growing photosynthetically under optimum conditions will be equally compatible, the limited information that is available suggests that the C_5 route via glutamyl-$tRNA^{DALA}$ and glutamate 1-semialdehyde is metabolically credible.

Control of ALA synthesis

Accepting the C_5 route as the sole route in cyanobacteria and the predominant if not the sole route to ALA in the chloroplast, it is necessary to consider the mechanisms that undoubtably exist to co-ordinate the synthesis of this key intermediate with the needs of the organelle. Although the effectiveness of these control mechanisms is apparent from the fact that ALA does not accumulate under physiological conditions, their detail is only now beginning to emerge as a result of the exploitation of biochemical and genetic techniques.

As with a number of other metabolic sequences a possible control via the thioredoxin system has been proposed (Clement-Metral 1979). However, with cell-free preparations converting glutamate to ALA the most relevant observation is a marked sensitivity to protohaem (Wang *et al.* 1984; Weinstein and Beale 1985*a*); at a concentration of 1.5 μM protohaem decreased the synthesis of ALA by 50 per cent. Other compounds tested, including protoporphyrin IX, magnesium protoporphyrin IX, protochlorophyllide, and chlorophyllides *a* or *b*, were substantially less effective with no more than 25 per cent inhibition at 25 μM (Weinstein and Beale 1985*a*). The effect of protohaem has been attributed to an effect on the catalytic activity of the dehydrogenase converting glutamyl-$tRNA^{DALA}$ to glutamate 1-semialdehyde (Huang and Wang 1986*a*). The failure, as yet, to demonstrate effects *in vitro* with intermediates of the chlorophyll branch of the pathway is surprising. It has been suggested from studies with higher plants that protochlorophyllide is only effective as a modulator in combination with its cognate reductase (Stobart and Ameen-Bukhari 1984). Using the mutant approach, Huang and Wang (1986*b*) and Wang *et al.* (1977, 1987) have suggested that protohaem may have the additional role of regulating the synthesis of enzymes involved in the formation of ALA in *Chlamydomonas*. The scheme in Fig. 5.8 incorporates these data and highlights the lack of information on the mechanisms that regulate the flow of intermediates into the several branches of the pathway.

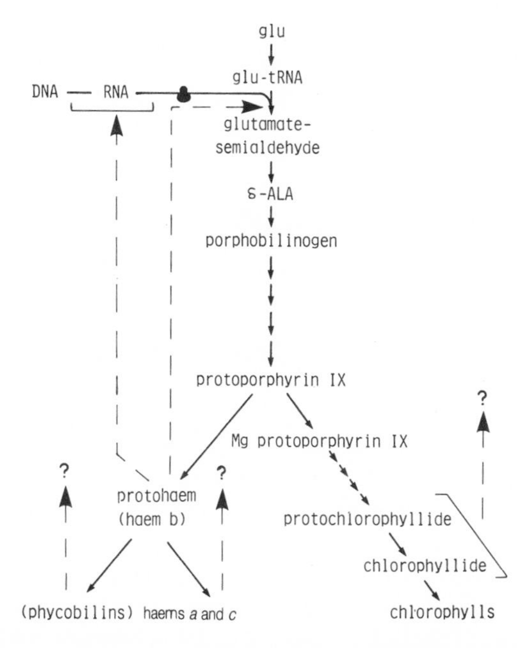

Fig. 5.8. Regulation of the C_5 route for tetrapyrrole biosynthesis in eukaryotic algae (based on references in the text).

One pathway or two?

While there is now a considerable body of evidence that supports the operation of the C_5 pathway as the primary route for tetrapyrrole formation in most photosynthetic organisms, it is still uncertain whether or not ALA-synthase makes a smaller but qualitatively significant contribution to tetrapyrrole formation in these organisms.

In cyanobacteria convincing evidence that the C_5 pathway is the sole provider of ALA has been presented earlier (McKie *et al.* 1981; Laycock and Wright 1981; Lewis *et al.* 1984). Other evidence in support of this conclusion (Einav and Avissar 1984) is less satisfactory since it has been based on the relative degree of incorporation by intact *Spirulina platensis* of isotope from variously labelled substrates into ALA. At best this type of data can only provide qualitative information about the operation of alternative routes since its interpretation is fraught with difficulty.

In an anoxygenic photosynthetic bacterium it appears that only one route is operative, some organisms using the C_5 pathway while others form ALA from succinate and glycine. In the purple non-sulphur bacterium *Rhodopseudomonas sphaeroides*, ^{13}C-NMR studies (Oh-hama *et al.* 1985) suggest that the ALA-synthase pathway operates exclusively. This conclusion is in accord with other data for *Rhodopseudomonas palustris* based on ^{14}C-tracer studies (Andersen *et al.* 1983), and is in accord with the inability of gabaculin to inhibit bacteriochlorophyll formation in *R. sphaeroides* and *Rhodospirillum rubrum* (Hoult *et al.* 1986), though this could also have been due to a lack of permeability of the cell wall and membrane of these bacteria to the inhibitor. In contrast to the ^{13}C-NMR data for the purple non-sulphur photosynthetic bacteria, the comparable study of the purple sulphur bacterium *Chromatium vinosum* (Oh-hama *et al.* 1986) showed that the C_5 pathway operated exclusively in this case. The green phototrophic bacterium *Prosthecochloris aestuarii* (Oh-hama *et al.* 1987) also appeared to utilize the C_5 pathway, confirming earlier conclusions for *Chlorobium limicola* (Andersen *et al.* 1983).

The situation is potentially much more complex in photosynthetic eukaryotes. The mitochondrion has been identified as the location of ALA biosynthesis in animal cells (Patton and Beattie 1973). This is not unexpected because succinyl-CoA, a key intermediate, is produced in this organelle; in addition the mitochondrial inner membrane is a primary, though not the sole, location of tetrapyrroles in the cell. In photosynthetic eukaryotes chlorophyll is, in quantitative terms, the predominant end-product of tetrapyrrole biosynthesis, though is rivalled in some organisms by phycobilins. Here, ALA destined to be converted to chlorophyll is produced via the C_5 route in the chloroplast. However, the possibility must be considered that the alternative route to ALA is associated with the mitochondrion and also contributes to the total cell complement of tetrapyrroles, though a recent report could not detect ALA-dehydratase or porphobilinogen-deaminase in the mitochondrial matrix (Smith 1987). Inevitably, any contribution from the classical succinyl-glycine route would be quantitatively very small, though none the less of qualitative significance in terms of the ultimate role of the end-product in mitochondrial metabolism.

As yet no clear picture has emerged about the relative importance of the two routes to ALA in eukaryotic photosynthetic micro-organisms. In *C. caldarium*, an alga whose taxonomic position has been a matter for considerable debate, 1-^{14}C-glutamate was incorporated into protohaem and haem *a* to a much greater extent than was 2-^{14}C-glycine, even in cells that were unable to form chlorophyll and phycobilins. This suggests that *Cyanidium* makes all its tetrapyrroles, including mitochondrial haems, solely through the C_5 pathway (Weinstein and Beale 1984; Weinstein *et al.* 1987*a*). In the green alga *Sc. obliquus* analyses of incorporation patterns by ^{13}C-NMR (Oh-hama *et al.* 1982) have suggested that only the C_5 pathway contributed to chlorophyll synthesis. However, it has been claimed that ALA can be formed by both the ALA-synthase pathway and the DOVA-based C_5 pathway (Klein and Senger 1978; Dornemann and Senger 1980), though reservations about the interpretation of these data in relation to the significance of the latter route have been outlined earlier.

Most notable in studies with *Euglena* is the report of ALA-synthase in extracts of wild-type and mutant strains and variation in the activity of the enzyme when organisms were transferred to light (Beale *et al.* 1981). These and other studies (Weinstein and Beale 1983; Weinstein *et al.* 1987*a*) have led to the view that in *Euglena* the two biosynthetic routes to ALA are functionally compartmentalized, with chloroplast tetrapyrroles derived via the C_5 pathway and mitochondrial haems formed exclusively via ALA-synthase. However, when chlorophyll formation was inhibited by gabaculin, the level of ALA-synthase, initially some 40 per cent that of untreated cells, was maintained in a succeeding period of illumination, whereas ALA-synthase activity in untreated cells fell dramatically (Corriveau and Beale 1986). This suggested that light activation of the C_5 pathway in *Euglena* was paralleled by a suppression of ALA-synthase, and that the former pathway might be able to contribute to non-plastid tetrapyrrole pools in the light. In investigations with isolated *Euglena* chloroplasts (Gomez-Silva *et al.* 1985) the conversion of glutamate to chlorophyll via the C_5 pathway was concluded to be the sole route to tetrapyrroles in this organelle.

In eukaryotic organisms in particular there is a requirement for close regulation of the synthesis of tetrapyrrole chromophores and their apoproteins so that haem proteins in both mitochondria and etioplasts are produced in the light at approximately the same rate as during dark growth, even though formation of vast amounts of chlorophyll, and phycobiliproteins in some algae, is triggered by illumination. This could be achieved by compartmentation of biosynthetic pathways. However, while it is tempting to envisage the C_5 pathway as being chloroplast-located and the ALA-synthase pathway as being extrachloroplastidic, this has not been verified on experimental grounds, other than in *Euglena*.

An assessment of the situation in higher plants is beyond the scope of this review. Nevertheless, it should be noted that the situation here is no less uncertain. In maize it has been suggested that both ALA-synthase and C_5 routes coexist (Meller and Gassman 1982); a similar conclusion for maize (Klein and Porra 1982) was revised on the basis of later data which suggested sole involvement of the C_5 route in ALA formation (Porra 1986). Studies of the biosynthetic origin of mitochondrial haem *a* in the same tissue suggested that this came mostly, if not entirely, from the C_5 pathway, with little or no contribution via ALA-synthase (Schneegurt and Beale 1986; Weinstein *et al.* 1987*b*). Total cellular protohaem was labelled 30-fold more effectively by 1-^{14}C-glutamate than by 2-^{14}C-glycine, and haem *a* four-times more effectively. However, glycine was incorporated into the farnesyl moiety of haem *a* some 11-times as effectively as glutamate; this portion of the haem *a* originates from mevalonic acid, for which glycine is known to be an effective precursor (Shah and Rogers 1969).

Finally it should be noted that where it is concluded that one pathway operates exclusively this implies that tetrapyrrole formation, if operating in cell compartments other than the site of ALA formation, will require the movement of committed precursors or intermediates late in the pathway across internal membranes to meet these needs. This has obvious implications in the co-ordination of chromophore and apoprotein formation for tetrapyrrole synthesis, particularly where the protein component is encoded in the organelle genome.

The physical and biochemical complexity of the cell in photosynthetic eukaryotes, together with the marked difference in the amount of tetrapyrroles in the two organelles, underlines the need for careful experimental design in order to demonstrate convincingly whether or not both pathways are operating under given conditions, their relative contributions to tetrapyrrole synthesis, and the mechanisms for their regulation. Particular attention must be given to the choice of growth conditions and of experimental organisms. Studies with dark-grown non-pigmented organisms should establish whether or not a distinctive mitochondrial route is operating. It would then be appropriate to extend such investigations to organisms established in photoautotrophic growth. Only at this stage should attention be given to the transitory and largely unnatural situation that is initiated with the transfer of non-pigmented organisms to conditions favouring pigment synthesis. Although in some of these situations the results of labelling studies are likely to be equivocal, an analysis by ^{13}C-NMR spectroscopy of the incorporation of ^{13}C from specifically labelled substrates into organelle-specific tetrapyrroles in order to define isotope distribution may yield more definitive information. Some of the problems mentioned above will undoubtedly be avoided by appropriate studies with organelle preparations.

Further refinement of cell-free systems derived from wild-type and mutant strains, together with the isolation and characterization of the enzymes involved, will provide information necessary to identify the regulatory mechanisms that maintain an appropriate balance between the various end-products and co-ordinate the synthesis of tetrapyrrole prosthetic groups with that of their cognate apoproteins. Studies with algae and cyanobacteria will surely make significant contributions in these areas as they already have in establishing the C_5 pathway as an important route for the synthesis of tetrapyrroles in oxygenic photosynthetic organisms.

References

Andersen, T., Briseid, T., Nesbakken, T., Ormerod, J., Sirevag, R., and Thorud, M. (1983). *FEBS Lett.* **19**, 303.

Anderson, J.M. and Barrett, J. (1986). In *Encyclopedia of plant physiology* Vol. 18, (ed. L.A. Staehelin, and C.J. Arntzen) p. 268. Springer, New York.
Avissar, Y.J. (1980). *Biochim. Biophys. Acta* **613**, 220.
Battersby, A.R., Fookes, C.J.R., Gustafson-Potter, K.E., McDonald, E., and Matcham, G.W.J. (1982). *J. chem. Soc. Perkin Trans.* **1**, 2427.
Beale, S.I. (1970). *Plant Physiol.* **45**, 504.
Beale, S.I. and Castelfranco, P.A. (1974). *Plant Physiol.* **53**, 291.
Beale, S.I. and Foley, T. (1982). *Plant Physiol.* **69**, 1331.
Beale, S.I., Foley, T., and Dzelzkalns, V. (1981). *Proc. Natn. Acad. Sci. USA* **78**, 1666.
Beale, S.I., Gough, S.P., and Granick, S. (1975). *Proc. Natn. Acad. Sci. USA* **72**, 2719.
Castelfranco, P.A. and Beale, S.I. (1983). *A. Rev. plant Physiol.* **34**, 241.
Ciechanover, A., Wolin, S., Steitz, J.A., and Lodish, H.F. (1985). *Proc. Natn. Acad. Sci. USA* **82**, 1341.
Chatterjee, S.N. and Park, J.T. (1964). *Proc. Natn. Acad. Sci. USA* **51**, 9.
Clement-Metral, J.D. (1979). *FEBS Lett.* **101**, 116.
Corriveau, J.L. and Beale, S.I. (1986). *Plant Sci.* **45**, 9.
Dornemann, D. and Senger, H. (1980). *Biochem. Biophys. Acta* **628**, 35.
Duggan, J.X., Meller, E., and Gassman, M.L. (1981). *Plant Physiol.* **68**, 802.
Dzelzkalns, V., Foley, T., and Beale, S.I. (1982). *Arch. Biochem. Biophys.* **216**, 196.
Einav, M. and Avissar, Y.J. (1984). *Plant Sci. Lett.* **35**, 51.
Ellis, R. and Greenawald, M. (1985). *Plant Sci. Lett.* **37**, 213.
Fleischhacker, P. and Senger, H. (1978). *Physiol. Plant.* **43**, 43.
Flint, D.H. (1984). *Plant. Physiol.* **75S**, 170.
Foley, T. and Beale, S.I. (1982). *Plant Physiol.* **70**, 1495.
Foley, T., Dzelzkalns, V., and Beale, S.I. (1982). *Plant Physiol.* **70**, 219.
Fuesler, T.P., Castelfranco, P.A., and Wong, Y-S. (1984). *Plant Physiol.* **74**, 928.
Gardner, G. and Gorton, H.L. (1985). *Plant Physiol.* **77**, 540.
Gassman, M., Pluscec, J., and Bogorad, L. (1968). *Plant Physiol.* **43**, 1411.
Gibson, K.D., Laver, W.G., and Neuberger, A. (1958). *Biochem. J.* **70**, 71.
Glazer, A.N. (1984). *Biochim. Biophys. Acta* **768**, 29.
Gomez-Silva, B., Timko, M.P., and Schiff, J.A. (1985). *Planta* **165**, 12.
Gough, S.P. and Kannangara, C.G. (1976). *Carlsberg. Res. Commun.* **41**, 183.
Gough, S.P. and Kannangara, C.G. (1977). *Carlsberg Res. Commun.* **42**, 459.
Gough, S.P. and Kannangara, C.G. (1979). *Carlsberg Res. Commun.* **44**, 403.
Granick, S. (1961). *J. biol. Chem.* **236**, 1168.
Guikema, J.A., Freeman, L., and Fleming, E.H. (1986). *Plant Physiol.* **82**, 280.
Harel, E. and Klein, S. (1972). *Biochem. biophys. Res. Commun.* **49** 364.
Harel, E. and Ne'eman, E. (1983). *Plant Physiol.* **72**, 1062.
Harel, E., Ne'eman, E., and Miller, E. (1983). *Plant Physiol.* **72**, 1056.
Hill, C.M., Pearson, S.H., Smith, A.J., and Rogers, L.J. (1985). *Bioscience Reports* **5**, 775.
Hoare, D.S. and Moore, R.B. (1965). *Biochim. Biophys. Acta* **109**, 622.
Hoare, D.S., Hoare, S.L., and Moore, R.B. (1967). *J. gen. Microbiol.* **49**, 351.
Houen, G., Gough, S.P., and Kannangara, C.G. (1983). *Carlsberg Res. Commun.* **48**, 567.

Hoult, R.C., Rees, D., Rogers, L.J., and Smith, A.J. (1986). *Arch. Microbiol.* **146**, 57.
Huang, D-D. and Wang, W-Y. (1986*a*). *J. biol. Chem.* **261**, 13451.
Huang, D-D. and Wang, W-Y. (1986*b*). *Mol. gen. Genet.* **205**, 217.
Huang, D-D., Wang, W-Y., Gough, S.P., and Kannangara, C.G. (1984). *Science* **225**, 1482.
Jurgenson, J.E., Beale, S.I., and Troxler, R.F. (1976). *Biochem. biophys. Res. Commun.* **69**, 149.
Kah, A. and Dornemann, D. (1987). *Z. Naturf.* **42c**, 209.
Kannangara, C.G. and Gough, S.P. (1977). *Carlsberg Res. Commun.* **42**, 441.
Kannangara, C.G. and Gough, S.P. (1978). *Carlsberg Res. Commun.* **43**, 185.
Kannangara, C.G. and Gough, S.P. (1979). *Carlsberg Res. Commun.* **44**, 11.
Kannangara, C.G. and Schouboe, A. (1985). *Carlsberg Res. Commun.* **50**, 179.
Kannangara, C.G., Gough, S.P., Oliver, R.P., and Rasmussen, S.K. (1984). *Carlsberg Res. Commun.* **49**, 417.
Kikuchi, G., Kumar, A., Talmage, P., and Shemin, D. (1958). *J. biol. Chem.* **233**, 1214.
Kipe-Nolt, J.A. and Stevens, S.E. (1980). *Plant Physiol.* **65**, 126.
Kipe-Nolt, J.A., Stevens, S.E., and Stevens, C.L.R. (1978). *J. Bact.* **135**, 286.
Klein, O. and Porra, R.J. (1982). *Z. Physiol. Chem.* **363**, 551.
Klein, O. and Senger, H. (1978). *Plant Physiol.* **62**, 10.
Klein, O., Harel, E., Ne'eman, E., Katz, E., and Meller, E. (1975). *Plant Physiol.* **56**, 486.
Konis, Y., Klein, S., and Ohad, I. (1978). *Photochem. Photobiol.* **27**, 177.
Konomi, K. and Furuya, M. (1986). *Plant Cell Physiol.* **27**, 1507.
Laycock, M.V. and Wright, J.L.C. (1981). *Phytochemistry* **20**, 1265.
Levasseur, P.J. and Gassman, M.L. (1986). *Phytochemistry* **25**, 1829.
Lewis, N.G., Walter, J.A., and Wright, J.L.C. (1984). *Phytochemistry* **23**, 1611.
Lohr, J.B. and Friedman, H.C. (1976). *Biochem. biophys. Res. Commun.* **69**, 908.
McKie, J., Lucas, C., and Smith, A.J. (1981). *Phytochemistry* **20**, 1547.
Meisch, H-U. and Maus, R. (1983). *Z. Naturf.* **38C**, 563.
Meisch, H-U., Reinle, W., and Wolf, U. (1985). *Biochim. Biophys. Acta* **841**, 319.
Meller, E. and Gassman, M.L. (1981). *Plant Physiol.* **67**, 728.
Meller, E. and Gassman, M.L. (1982). *Plant Sci. Lett.* **26**, 23.
Meller, E. and Harel, E. (1978). In *Chloroplast development* (ed. G. Akoyunoglou and J.H. Argyroudi-Akoyunoglou) p. 51. Elsevier, Amsterdam.
Meller, E., Belkin, S., and Harel, E. (1975). *Phytochemistry* **14**, 2399.
Nandi, D.L. and Shemin, D. (1968). *J. biol. Chem.* **243**, 1236.
Nesbitt, J.A. and Lennartz, W.J. (1968). *J. biol. Chem.* **243**, 3088.
Oh-hama, T. and Senger, H. (1975). *Plant Cell Physiol.* **16**, 395.
Oh-hama, T., Seto, H., and Miyachi, S. (1985). *Arch. Biochem. Biophys.* **237**, 79.
Oh-hama, T., Seto, H., and Miyachi, S. (1986). *Arch. Biochem. Biophys.* **246**, 192.
Oh-hama, T., Seto, H., and Miyachi, S. (1987). *Prog. photosyn. Res.* **IV**, 445.
Oh-hama, T., Seto, H., Otake, N., and Miyachi, S. (1982). *Biochem. Biophys. Res. Commun.* **105**, 647.
Owens, T.G., Riper, D.M., and Falkowoki, P.G. (1978). *Plant Physiol.* **62**, 516.
Patton, G.M. and Beattie, D.S. (1973). *J. biol. Chem.* **248**, 4467.

Peschek, G.A., Schmetterer, G., Lauritsch, G., Nitschmann, W.H., Kienzl, P.F., and Muchl, R. (1982). *Arch. Microbiol.* **131**, 261.
Porra, R.J. (1986). *Eur. J. Biochem.* **156**, 111.
Porra, R.J. and Grimme, L.H. (1974). *Arch. Biochem. Biophys.* **164**, 312.
Porra, R.J. and Grimme, L.H. (1978). *Ind. J. Biochem.* **9**, 883.
Porra, R.J. and Klein, O. (1981). *Anal. Biochem.* **116**, 511.
Porra, R.J., Klein, O., Dornemann, D., and Senger, H. (1980). *Hoppe-Seyler's Z. Physiol. Chem.* **361**, 187.
Ramaswamy, N.K. and Nair, P.M. (1976). *Ind. J. Biochem. Biophys.* **13**, 394.
Rando, R.R. (1977). *Biochem.* **16**, 4604.
Richard, F. and Nigon, V. (1973). *Biochim. Biophys. Acta* **313**, 130.
Salvador, G.F. (1978*a*). *Plant Sci. Lett.* **13**, 351.
Salvador, G.F. (1978*b*). *C.R. Acad. Sci. Paris* **286**, 49.
Sawicki, E., Hauser, T.R., Stanley, T.W., and Elbert, W. (1961). *Anal. Chem.* **33**, 93.
Schneegurt, M.A. and Beale, S.I. (1986). *Plant Physiol.* **81**, 965.
Schon, A., Krupp, G., Gough, S.P., Berry-Lowe, S., Kannangara, C.G., and Soll, D. (1986). *Nature, Lond.* **322**, 281.
Schuster, A. and Harel, E. (1985). *Plant Physiol.* **77**, 648.
Schwartzbach, S.D., Schill, J.A., and Goldstein, N.H. (1975). *Plant Physiol.* **56**, 313.
Shah, S.P.J. and Rogers, L.J. (1969). *Biochem. J.* **114**, 395.
Shemin, D., Russell, C.S., and Abramsky, T. (1955). *J. biol. Chem.* **215**, 613.
Shioi, Y., Doi, M., and Sasa, T. (1984). *Plant Cell Physiol.* **25**, 1487.
Shioi, Y., Doi, M., and Sasa, T. (1986). *Plant Cell Physiol.* **27**, 67.
Smith, A.G. (1987). *Prog. Photosyn. Res.* **IV**, 453.
Smith, A.J., London, J., and Stanier, R.Y. (1967). *J. Bact.* **94**, 972.
Soper, T.S. and Manning, J.M. (1982). *J. biol. Chem.* **257**, 13930.
Stark, B.C., Kole, R., Bowman, E.J., and Altman, S. (1978). *Proc. Natn. Acad. Sci. USA* **75**, 3717.
Stewart, A.C. and Bendall, D.S. (1980). *Biochem. J.* **188**, 351.
Stobart, A.K. and Ameen-Bukhari, I. (1984). *Biochem. J.* **236**, 741.
Troxler, R.F. and Lester, R. (1967). *Biochemistry* **6**, 3840.
Walter, P. and Blobel, G. (1982). *Nature, Lond.* **229**, 691.
Wang, W-Y., Boynton, J.E., and Gillham, N.W. (1977). *Molec. gen. Genet.* **152**, 7.
Wang, W-Y., Huang, D-D., Stachon, D., Gough, S.P., and Kannangara, C.G., (1984). *Plant. Physiol.* **74**, 569.
Wang, W-Y., Huang, D-D., Chang, T-E., Stachon, D., and Wegmann, B. (1987). *Prog. photosyn. Res.* **IV**, 423.
Weinstein, J.D. and Beale, S.I. (1983). *J. biol. Chem.* **258**, 6799.
Weinstein, J.D. and Beale, S.I. (1984). *Plant Physiol.* **74**, 146.
Weinstein, J.D. and Beale, S.I. (1985*a*). *Arch. Biochem. Biophys.* **237**, 454.
Weinstein, J.D. and Beale, S.I. (1985*b*). *Arch. Biochem. Biophys.* **239**, 87.
Weinstein, J.J. and Castelfranco, P.A. (1978). *Arch. Biochem. Biophys.* **186**, 376.
Weinstein, J.D., Mayer, S.M., and Beale, S.I. (1986). *Plant Physiol.* **82**, 1096.
Weinstein, J.D., Mayer, S.M., and Beale, S.I. (1987*a*). *Prog. photosyn. Res.* **IV**. 435.

Weinstein, J. D., Schneegurt, M. A., and Beale, S. I. (1987*b*). *Prog. photosyn. Res.* **IV**, 431.
Wider de Xifra, E. A., Stella, A. M., and Batlle, A. M. del C. (1978). *Plant Sci. Lett.* **11**, 93.

Poster abstracts

The occurrence and biosynthesis of polyunsaturated fatty acids in the marine dinoflagellate *Crypthecodinium cohnii*

R.J. Henderson
NERC Unit of Aquatic Biochemistry, Department of Biological Science, University of Stirling, Stirling FK9 4LA, UK

J. Leftley
SMBA, Dunstaffnage Marine Research Laboratory, Oban, Argyll, UK

J.R. Sargent
NERC Unit of Aquatic Biochemistry, Department of Biological Science, University of Stirling, Stirling FK9 4LA, UK

The biosynthetic pathways which produce long chain polyunsaturated fatty acids (PUFAs), such as 20:5(n-3) and 22:6(n-3), in marine algae are unknown. In terrestrial plants, phosphatidylcholine (PC) and monogalactosyldiacylglycerol are known to be involved as substrates in the pathway leading to 18:3(n-3) (Gounaris *et al.* 1986). PC was found to be the principal polar lipid in the non-photosynthetic dinoflagellate *Crypthecodinium cohnii*, although triacylglycerols (TAGs) were the most abundant lipid overall. 22:6(n-3) was almost exclusively the only PUFA in the lipids of this organism. A small amount of 20:4(n-6) occurred in phosphatidylinositol. Fatty acids synthesized by the alga from U-^{14}C-acetate were rapidly incorporated into glycerolipids. Their rate of incorporation into TAGs was linear over 6 hours, whereas that into PC was biphasic with a decline occurring after 3 hours. The relative labelling of saturates and monoenes in TAGs was similar to their relative mass, whereas PUFAs in both TAGs and total phospholipids were poorly labelled on this basis. The converse was true of saturates and, particularly, of dienes in total phospholipids.

Fatty acids esterified in phospholipids may act as substrates for the desaturases involved in the formation of 22:6(n-3), and the conversion of dienes to trienes may be the rate-limiting step in this pathway.

Gounaris, K., Barber, J., and Harwood, J.L. (1986). *Biochem. J.* **237**, 313.

Identification of α-tocotrienol in the cyanobacterium *Synechocystis* sp. 6714

C.J. Mullins, T.J. Walton and R.P. Newton
Department of Biochemistry, School of Biological Sciences, University College of Swansea, Swansea SA2 8PP, UK

A.G. Brenton
Mass Spectrometry Research Unit, University College of Swansea, Swansea SA2 8PP, UK

Cyanobacteria, uniquely amongst prokaryotes, synthesize meroterpenoids of the vitamin E series. In addition to the widely distributed α-tocotrienol, the presence of monomethyl and dimethyltocols has also been established (Newton *et al.* 1977; Mullins *et al.* 1985, 1986). To date, tocotrienols, found in eukaryotic plants, have not been reported in cyanobacteria. However, the tocochromanol fraction of *Synechocystis* sp. 6714 isolated from cells cultured autotrophically at high cell density was found to contain, in addition to α-tocopherol, a reducing compound with the chromatographic properties of α-tocotrienol. The 70 eV electron impact mass spectrum of this material showed a prominent molecular ion (m/e 410) and intense fragment ions at m/e 191 and 151, characteristic of the 7,8-dimethyltocochromanol nucleus, and was essentially identical to that of authentic α-tocotrienol. The mass analysed ion kinetic energy spectra obtained from the molecular ions of authentic α-tocotrienol and putative α-tocotrienol (m/e 410) were closely similar containing strong fragment ions (m/e 191 and 151) isobaric with those obtained under similar conditions from the molecular ion of α-tocopherol (m/e 416). The chromatographic and mass spectral data are thus entirely consistent with the identification of α-tocotrienol, representing the first description of a member of the tocotrienol series in cyanobacteria.

Mullins, C.J., Newton, R.P., and Walton, T.J. (1985). *Biochem. Soc. Trans.* **13**, 1242.
Mullins, C.J., Walton, T.J., Newton, R.P., Beynon, J.H., Brenton, A.G., and Griffiths, W.J. (1986). *Biochem. Soc. Trans.* **14**, 696.
Newton, R.P., Walton, T.J., and Moyse, C.D. (1977). *Biochem. Soc. Trans.* **5**, 1486.

Photosynthesis-deficient mutants of *Scenedesmus obliquus*

A. Henry, J. Crofts, R. Powls, and J.F. Pennock
Biochemistry Department, The University of Liverpool, Liverpool L69 3BX, UK

Photosynthesis-deficient mutants of *Scenedesmus obliquus*, obtained by X-ray irradiation or with chemical mutagens, can be grown heterotrophically in the dark and have proven very useful in studies on the biosynthesis of chloroplast pigments. A survey of the amounts of chlorophyll, plastoquinone, α-tocopherol, and phytol allows the classification of mutants into four groups. Group 1 has pigment levels similar to wild type; group 2 has no α-tocopherol and phytol and has chlorophylls with a geranylgeranyl side-chain; group 3 has low chlorophyll; and group 4 has both low chlorophyll and low plastoquinone. Group 2 mutants, which cannot reduce geranylgeranyl pyrophosphate to phytyl pyrophosphate, accumulate geranylgeraniol. These mutants are also deficient in vitamin K_1. Mutant F154 accumulates α-tocotrienol as well as α-tocopherol and may shed some light on the interrelationships of these compounds.

An interesting series of compounds has been found in a mutant blocked in carotenoid biosynthesis such that phytoene accumulates. Tlc of total lipid revealed many UV-absorbing materials, all showing the same absorption spectrum as phytoene. However, their chromatographic properties suggest the presence of polar groups such as epoxides and hydroxyls which are found in the xanthophylls of *S. obliquus* wild-type cells. It appears that the accumulation of phytoene allows epoxidase and hydroxylase enzymes, normally functional in xanthophyll formation, to use phytoene as a substrate. Thus accumulation of intermediates of a biosynthetic pathway in a mutant may clearly give misleading information about that pathway.

Chemical variability in the oil-rich alga *Botryococcus braunii*

P. Metzger and E. Casadevall
Laboratoire de Chimie Bioorganique, ENSCP, 11, rue Curie, 75231 Paris Cedex 05, France

Botryococcus braunii strains were isolated from samplings collected in Bolivia, France, the Ivory Coast, Morocco, Peru, Thailand, and the West Indies. The oils extracted from laboratory cultures, that always contained a hydrocarbon fraction, showed on analyses some important differences. These differences are related to the chemical structure of the hydrocarbons, and in some case to the presence of other neutral lipids predominating in the oil. On the basis of the hydrocarbon chemical structure, we arrived at a subdivision of *B. braunii* into three races:

A race. The hydrocarbons are straight chain alkadienes and trienes, odd-numbered from 23 to 31. This race can be subdivided into four groups depending on the nature of the predominant neutral lipids:

Hydrocarbon strains: hydrocarbons account for 50–87 per cent of the oil;
Alcohol strains: C_{27} secondary fatty alcohols predominate (75–85 per cent of the oil);
Aldehyde strains: even-numbered, C_{44}–C_{68} β-unsaturated aldehydes, derived from an aldol condensation (45–60 per cent of the oil) are present;
Triacylglycerol (TAG) strains: TAG, essentially triolein, and hydrocarbons predominate, each accounting for 36 per cent of the oil.

B race. Here, triterpenoid hydrocarbons C_nH_{2n-10} ($30 \leq n \leq 37$), termed botryococcenes, are always the major components (75 per cent of the oil).

Depending on their origin the strains exhibit a large variation in the number and relative ratio of their hydrocarbons.

T race. Here, only one hydrocarbon has been characterized, a $C_{40}H_{78}$ tetraterpene (lycopodiene) which makes up 2–40 per cent of the oil. Some strains accumulate highly viscous tetraterpenols.

Origin of the variability of non-isoprenoid hydrocarbons in *Botryococcus braunii*—specificity of the elongation-decarboxylation system

J. Templier, C. Largeau, and E. Casadevall
Laboratoire de Chimie Bioorganique et Organique Physique, UA CNRS 456, Ecole Nationale Supérieure de Chimie, 11 Rue Pierre et Marie Curie, 75231 Paris Cedex 05, France

Large amounts of non-isoprenoid hydrocarbons occur in several strains of *Botryococcus braunii*. In some strains the hydrocarbons chiefly correspond to an homologous series of C_{23} to C_{31}, odd, unbranched, dienic products with a terminal unsaturation and a *cis*-9, 10 double bond. Such hydrocarbons derive from oleic acid via an elongation–decarboxylation process (Templier *et al.* 1984). Other strains contain, in addition, a second major series of dienes differing only by the trans-stereochemistry of the internal unsaturation (Metzger *et al.* 1986). Feeding experiments on different strains led to the following conclusions:

(i) Elaidic acid is the precursor of the *trans*-dienes.

(ii) The ability of a given strain to produce either exclusively the *cis* series, or a mixture of *cis*-and *trans*-dienes, is not related to differences in the elongation-decarboxylation system. In fact, this system is not highly specific and can transform both oleic and elaidic acid. The variability of the strains can be attributed to their differences in producing various of these acids.

(iii) While not highly specific, the elongation-decarboxylation systems can probably use a very limited number of acids as precursors. So, linoleic acid (presence of a second unsaturation) and vaccenic acid derivatives (shift of the double bond to the *cis*-11,12 position) are not transformed into the corresponding hydrocarbons.

Templier, J., Largeau, C., and Casadevall, E. (1984). *Phytochemistry* **23**, 1017.
Metzger, P., Templier, J., Largeau, C., and Casadevall, E. (1986). *Phytochemistry* **25**, 1869.

The effect of gabaculin on the photosynthetic activity of *Synechococcus* 6301

R.C. Hoult, L.J. Rogers, and A.J. Smith
Department of Biochemistry, University College of Wales, Aberystwyth, Dyfed SY23 3DD, UK

Recently, we reported that 3-amino 2,3 dihydrobenzoic acid (gabaculin) inhibited the C_5 route of δ-aminolaevulinic acid biosynthesis in cyanobacteria. The present work extends that observation in defining the effect of an inhibition of tetrapyrrole biosynthesis on the photosynthetic activity of *Synechococcus* 6301.

The addition of 50 μM gabaculin to exponentially growing cultures caused *in vivo* rates of oxygen evolution and CO_2 fixation per unit volume of culture to decline progressively over a 60 hour period. When expressed on a chlorophyll basis both activities showed an initial slight increase in rate but declined thereafter, falling to below 50 per cent of the control rates (900 μmol O_2 or 400 μmol CO_2 $mgChl^{-1}$ $hour^{-1}$) after 36 hours of exposure to the inhibitor. After 60 hours, rates of oxygen evolution and CO_2 fixation were 6 and 20 per cent, respectively, of those of the untreated cultures.

Thylakoid preparations obtained from cell-free extracts of organisms exposed to 50 μM gabaculin for 36 hours possessed *in vitro* rates of photosystem I and II activity comparable to control thylakoids. Both preparations evolved oxygen in the presence of 2 mM $K_3Fe(CN)_6$ at a rate of 130 μmol O_2 $mgChl^{-1}$ $hour^{-1}$. Photosystem I activity was assayed by the auto-oxidation of methyl viologen in the presence of DCMU and NaN_3 and rates of 2100 and 2200 μmol O_2 consumed $mgChl^{-1}$ $hour^{-1}$ were obtained for the control and gabaculin-treated thylakoids, respectively.

The progressive decrease in growth rate characteristically observed in cultures exposed to 50 μM gabaculin is likely to be due to the decline in *in vivo* photosynthetic rates. The *in vitro* studies suggest that the chlorophyll associated with the photosystems is still functional after 36 hours of exposure to gabaculin. The decline in whole cell rates of photosynthesis may therefore be the result of impaired function of the cytochrome *b-f* complex, not probed in these *in vitro* assays. Alternatively, a decrease in phycobiliproteins, the major light-harvesting pigments of photosystem II, may seriously impair the channelling of photons into reaction centres, thereby decreasing the activity of the electron transport system as a whole. These two possibilities are currently under investigation.

Ultraviolet inducible proteins in *Chlamydomonas reinhardtii*

P. Nicholson and C. J. Howe
Department of Biochemistry, University of Cambridge, Tennis Court Road, Cambridge CB2 1QW, UK

As part of a study into the effects of DNA-damaging agents on *Chlamydomonas reinhardtii*, cells were exposed to UV light. Protein synthesis after this treatment was examined by labelling *in vivo* with ^{35}S-methionine followed by one-dimensional SDS-polyacrylamide gel electrophoresis. These experiments showed that the synthesis of two high-molecular-weight polypeptides is consistently specifically increased following low level irradiation. The molecular masses of these polypeptides were estimated at approximately 71 kDa and 65 kDa. Inhibitors of protein synthesis on organellar or cytoplasmic ribosomes were used in conjunction with the UV irradiation and ^{35}S-methionine labelling to identify the site(s) of synthesis of these two polypeptides. The inhibitor studies revealed that both polypeptides are synthesized on cytoplasmic rather than on organellar ribosomes.

Standard one-dimensional electrophoresis can detect only relatively major changes in protein synthesis. To study minor alterations in the protein profile a more sensitive two-dimensional electrophoretic analysis was employed. With this technique, synthesis of an additional six polypeptides was observed to be induced, or selectively enhanced, following UV irradiation. The approximate molecular masses of these additional polypeptides were 61 500, 61 000, 50 000, 30 000, 29 000, and 17 500 Da.

This work was supported by the AFRC.

Part 3: Bioenergetics

6 Molecular biology of photosynthetic reaction centres

A.C. STEWART

Trends in Genetics, Elsevier Publications Cambridge, 68 Hills Road, Cambridge CB2 1LA, UK

Introduction

Recent years have seen major advances in our understanding of the composition, structure, and function of photosynthetic reaction centres, the pigment–protein complexes, usually membrane-bound, that carry out the primary conversion of light to chemical energy. Arguably, the most significant breakthroughs have been the crystallization of the reaction centre from purple photosynthetic bacteria, and the isolation and sequencing of a number of the genes coding for reaction centre proteins. Much information has come from a combination of these studies with improved purification procedures for reaction centre complexes from a wide variety of photosynthetic organisms.

In this review, I shall summarize the information such approaches have yielded about the composition and organization of the bacterial reaction centre and the Photosystem I (PSI) and Photosystem II (PSII) reaction centre complexes of higher plants, algae, and cyanobacteria.

The reaction centre of purple photosynthetic bacteria

The purple photosynthetic bacteria have chromatophore membranes containing a single photochemical reaction centre in which the primary electron donor is a specialized bacteriochlorophyll (bchl) dimer. The primary acceptor, a molecule of bacteriophaeophytin (bphaeo), reduces a quinone, Q_A, which is magnetically coupled to a non-haem iron atom. Q_A in turn reduces the secondary quinone, Q_B. Electrons are returned to the oxidized primary donor via a cyclic pathway through a membrane-bound cytochrome *bc* complex and an extrinsic *c*-type cytochrome.

Purified reaction centre complexes from *Rhodopseudomonas (Rps.) viridis* contain three polypeptides designated L, M and H (molecular masses from sequencing data 30.6 kDa, 35.9 kDa, and 28.3 kDa, respectively); one molecule of cytochrome *c*; four bch1, one pair of which forms the primary donor P960; two bphaeo; one non-haem iron; and one menaquinone (Q_A). The secondary acceptor, Q_B (ubiquinone), is lost during purification of the complex (Michel and Deisenhofer 1986). Other purple bacteria, such as *Rhodobacter (Rb.) sphaeroides* and *Rb. capsulata*, contain similar L, M, and H subunits, but the isolated complexes often do not contain a cytochrome molecule.

The reaction centres from both *Rps. viridis* and *Rb. sphaeroides* have recently been crystallized in photochemically active form and their structures determined to 3Å resolution (Deisenhofer *et al.* 1985; Chang *et al.* 1986). The primary donor and electron acceptors lie in a hydrophobic environment associated with the membrane-spanning L and M subunits, and the complex so formed is almost symmetrical about an axis running between these subunits perpendicular to the membrane. The bchl special pair is located near the periplasmic side of the membrane. The porphyrin ring of each member of the special pair is associated with a molecule of accessory bchl, which is in turn in contact with a bphaeo, forming two branches extending across the membrane. However, the presence of Q_A and non-haem iron on only one branch implies that, of the two possible electron transfer pathways, only one is functional.

The genes encoding the L, M, and H subunits of *Rps. viridis, Rb. capsulata* and *Rb. sphaeroides* have been isolated and sequenced (Michel *et al.* 1985, 1986*b*; Youvan *et al.* 1984; Williams *et al.* 1984). The corresponding genes from the three bacteria show significant homology (approximately 50 per cent) and where amino-acid changes occur they are usually conservative ones. Knowledge of the sequences for the *Rps. viridis* proteins has allowed the polypeptide chains and side-groups to be positioned precisely within the atomic model of the reaction centre, and has enabled the identification of individual amino acids and sequences that are closely associated with the various prosthetic groups (Michel *et al.* 1986*a*, 1986*b*).

The L and M subunits each contain five hydrophobic membrane-spanning alpha-helices. In addition to an overall structural similarity, the L and M subunits share significant sequence homologies; for example, in the positions of a number of conserved His residues that are ligands for the special pair, accessory bchl molecules and non-haem iron, and the positions of conserved Trp residues that are hydrogen-bonded to the two bphaeo molecules.

Other residues have been identified that are not shared by L and M, but are conserved among corresponding L and M subunits from different bacteria and may be critical for reaction centre function. For example, interaction with a Glu on the L subunit may distinguish one bphaeo from the other. An overall asymmetry in the distribution of polar and non-polar side-chains in the region of the special pair may contribute to directing electrons along the functional electron transfer branch to Q_A, whose head-group is exclusively associated with residues on the M subunit.

The H subunit, which is not essential for the primary photochemical reaction, is much less hydrophobic than L or M, with only one membrane-spanning segment and the bulk of the protein located on the cytoplasmic side of the membrane (Michel *et al.* 1985). The function of this subunit remains unknown, although roles in reaction centre assembly or interaction with light-harvesting complexes have been suggested (Michel and Deisenhofer 1986).

Also peripheral to the *Rps. viridis* membrane, but on the periplasmic side, is cytochrome *c*. Of its four haem groups, only one appears to be close enough to the special pair to be able to transfer electrons to it.

Organization of reaction centre genes

In *Rb. capsulata, Rb. sphaeroides*, and *Rps. viridis*, the L and M subunits form part of a cluster that also contains the genes for the alpha and beta subunits of the closely associated light-harvesting complex and, in the case of *Rps. viridis*, the cytochrome *c* gene (Williams *et al.* 1984; Youvan *et al.* 1984; Michel *et al.* 1986*b*). It has been suggested that the cluster constitutes a polycistronic operon that may be transcribed from a single oxygen-repressed promoter. It remains to be clarified how the synthesis of the correct ratio of light-harvesting to reaction centre proteins is achieved, and how synthesis of L and M is co-ordinated with that of H, whose gene is located some distance away from the putative operon.

Photosystem II

Photosystem II of oxygenic organisms performs a reaction in many ways analogous to that of the bacterial reaction centre. Light excites a specialized chlorophyll species, P680, which donates an electron, via an intermediate phaeophytin acceptor, to a tightly bound plastoquinone molecule (Q_A) closely associated with a non-haem iron atom. The more loosely bound secondary acceptor, Q_B, also a plastoquinone, acts as a two-electron gate into the plastoquinone pool.

A major difference between PSII and the reaction centre of purple bacteria is that electrons to re-reduce the oxidized primary donor come not from a cyclic electron transfer pathway but from water. A still-enigmatic manganese-containing complex sequentially removes four electrons from two water molecules, yielding a molecule of oxygen and four protons that are released to the lumen of the thylakoid. A component 'Z' with quinone-like properties (see Renger and Govindjee 1985) has been proposed as the intermediate electron donor to P680; the relationship of Z to the reactions of oxygen evolution is still not understood.

The PSII core

Over the last few years, PSII preparations essentially devoid of chlorophyll *b*-containing light-harvesting complexes, PSI, and other components of the electron transfer chain have been obtained from a number of higher plants, algae, and cyanobacteria (for recent reviews, see Arntzen and Pakrasi 1986; Bryant 1986). A consensus has emerged that so-called 'core' preparations which are inactive in oxygen evolution contain six major polypeptides whose approximate molecular masses, based on SDS gels, are 47–51, 43–45, 32–34, 30–32, 10, and 4.5 kDa. Other small polypeptides of 4, 5, and 5.5 kDa may also be present (Ljungberg *et al.* 1986). The preparations contain, per 50 chlorophyll molecules, approximately 10 carotenoid, one P680, one Z, one Q_A, and two phaeophytins.

Until recently, this type of preparation was thought to be the simplest one capable of carrying out the primary photochemistry of PSII. The two largest polypeptides (hereafter referred to as the *psbB* and *psbC* proteins) each bind approximately 20–25 chlorophylls, and there was some evidence that for *psbB* these included P680 (Yamagishi and Katoh 1984; De Vitry *et al.* 1984). The 30–34 kDa polypeptides (referred to as D1 and D2) were proposed to be quinone-binding proteins. D1, a rapidly metabolized light-induced protein, binds the secondary quinone, Q_B, and is the target for high-affinity binding of the herbicides DCMU and atrazine, which compete with Q_B (reviewed by Arntzen and Pakrasi 1986). The physiological function of D2 was unclear, but a possible role on the oxidizing side of PSII had been suggested. The 10 kDa polypeptide was identified as cytochrome b_{559}, a constituent of all active preparations, but one whose function remains unclear. It is now thought that the 4.5 kDa polypeptide is also a subunit of this cytochrome (Hermann *et al.* 1984; Widger *et al.* 1985).

Very recently, this view of the PSII core has been challenged, largely as a result of comparison of PSII with the newly elucidated structure of the *Rps. viridis* reaction centre. This challenge culminated in the isolation of a 'super-core' PSII particle that is capable of photochemical reduction of phaeophytin (Nanba and Satoh 1987). The particle is devoid of both the *psbB* and *psbC* chlorophyll-binding proteins, containing only D1, D2, one to two molecules of cytochrome b_{559}, five chlorophyll *a*, two phaeophytin, and one beta-carotene. It is now proposed that a D1–D2 heterodimer is the binding site for P680.

The genes for PSII proteins

In higher plants and algae, the genes for the 30–32 (D1), 47–51, 43–45, 32–34 (D2), 10, and 4.5 kDa polypeptides are encoded by chloroplast genes denoted *psbA, psbB, psbC, psbD, psbE,* and *psbF*, respectively (Herrmann *et al.* 1985). Recently, two further chloroplast-encoded genes have been assigned to PSII. The *psbG* gene encodes a predominantly hydrophilic 24 kDa protein found in oxygen-evolving PSII preparations but not in PSII 'core' particles (Steinmetz *et al.* 1986). The gene designated *psbH* (Westhoff *et al.* 1986; Hird *et al.* 1986) encodes a 10 kDa phosphoprotein of unknown function.

Westhoff *et al.* (1986) have reported that the spinach *psbH* gene is transcribed as part of an operon comprising the *psbB* gene and the *petB* and *petD* genes of the cytochrome *bf* complex.

The genes for several chloroplast-encoded PSII proteins have been mapped on a number of chloroplast genomes including those of spinach, pea, maize, and wheat (for example, see Courtice *et al.* 1985; Westhoff *et al.* 1986), tobacco (Shinozaki *et al.* 1986), liverwort (Ohyama *et al.* 1986), and the cyanelle genome of *Cyanophora paradoxa* (Lambert *et al.* 1985). The recent complete sequencing of the tobacco and liverwort chloroplast genomes (Shinozaki *et al.* 1986; Ohyama *et al.* 1986) has revealed a number of so far unassigned open reading frames; some of these are co-transcribed with known PSII genes and may prove to encode additional PSII proteins. Fig. 6.1 shows current maps of the wheat and pea chloroplast genomes.

The most frequently studied PSII gene, *psbA*, is highly conserved (85–100 per cent homology) amongst a wide variety of photosynthetic organisms, including higher plants and eukaryotic algae (reviewed by Herrmann *et al.* 1985) and cyanobacteria (reviewed by Bryant 1986). Most higher plant chloroplast genomes contain a single copy of *psbA*, but in some eukaryotic algae the gene is within the inverted repeat region and thus duplicated (Erickson *et al.* 1984), while several cyanobacteria have been found to contain multiple regions of homology to the *psbA* gene (Bryant 1986). Only one of the two *psbA* genes of *Anabaena* PCC 7120 appears to be expressed. Evidence has been obtained for the expression of more than one of the *psbA*-homologous genes of *Anacystis nidulans* R2, though transcripts from one of the three genes predominate, and any one of the genes can produce enough of the D1 protein to support photoautototrophic growth (Golden *et al.* 1986). Such multigene families for *psbA* appear so far to be unique to the cyanobacteria.

The deduced amino acid sequence of the 32 kDa D1 protein predicts a strongly hydrophobic protein that is initially synthesized as a 34 kDa precursor which is post-translationally processed before or during insertion into the thylakoid membrane. Hydropathy analysis (Holschuh *et al.* 1984) and a recent investigation of the spinach protein, making use of antibodies raised against a range of peptides corresponding to different regions of the sequence, suggest that the protein has five transmembrane segments (Sayre *et al.* 1986), with the N- and C-termini on the stromal and lumenal sides of the membrane, respectively. Residues located near the stromal ends of the fourth and fifth helices may partipate in binding non-haem iron, and residues in the surface-located sequence linking these regions might include binding sites for Q_A and Q_B. Within the latter region, point mutations at Val219, Phe225, and Ser264 have all been shown to be associated with herbicide resistance (Arntzen and Pakrasi 1986; Sayre *et al.* 1986).

Overlapping the 3' end of the spinach *psbC* gene by 50 bp, but in a different

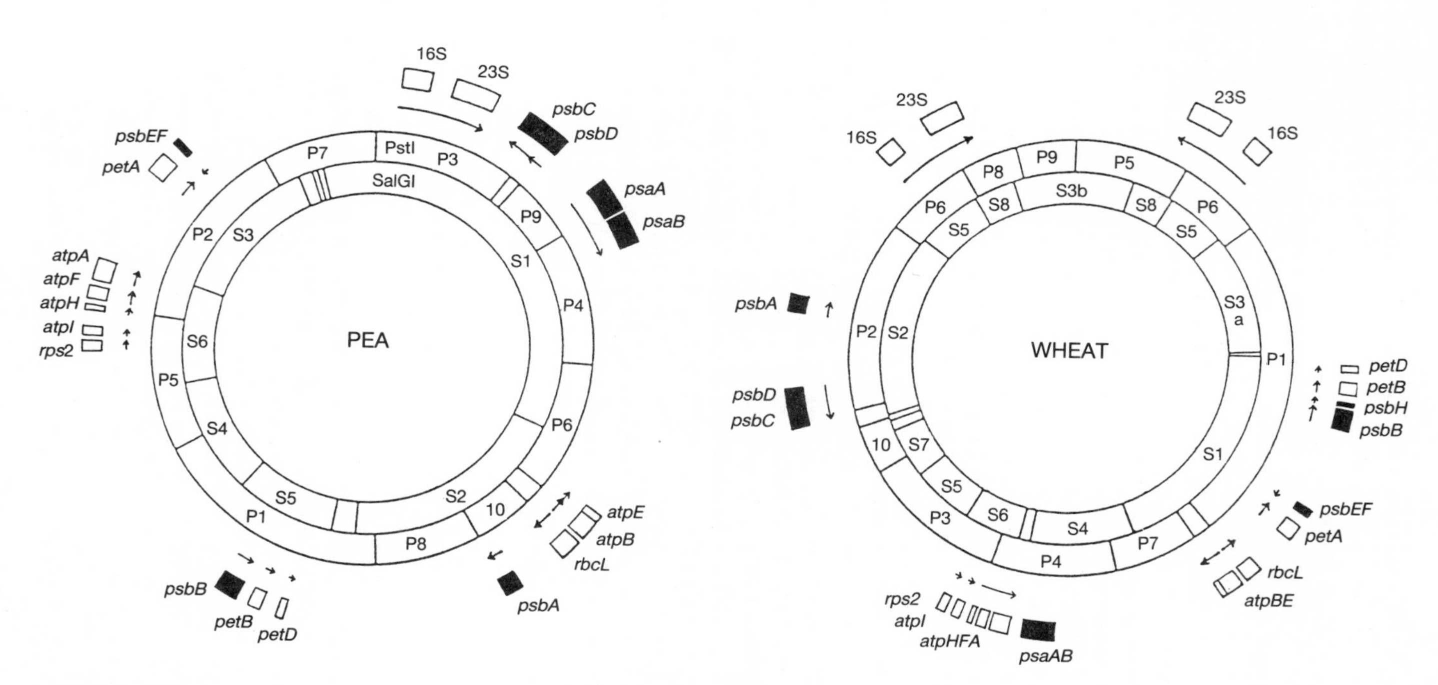

Fig. 6.1. Restriction maps of pea and wheat chloroplast DNA, showing the locations of genes coding for components of the photosynthetic membrane complexes. Genes for PSI and PSII components are shown as filled boxes. Figure kindly provided by Dr. J.C. Gray.

reading frame, is the *psbD* gene (Holschuh *et al.* 1984; Alt *et al.* 1984). The genes are also adjacent and presumably co-transcribed in a number of other higher plants. Interestingly, the cyanobacteria *Synechococcus* 7002 and *Synechocystis* 6803 each contain two copies of *psbD*, one of which is adjacent to *psbC* (Bryant 1986). The *psbD* gene has been sequenced from spinach (Alt *et al.* 1984; Holschuh *et al.* 1984), pea (Rasmussen *et al.* 1984), and *Chlamydomonas* (Rochaix *et al.* 1984), and again is highly conserved. The corresponding protein D2 is highly hydrophobic and its hydropathy profile is strikingly similar to that of D1 (Alt *et al.* 1984). Although the overall sequence homology between D1 and D2 is only about 26 per cent, some highly conserved features are evident, most notably conserved histidines that are also found in the L and M subunits of the bacterial reaction centre (Rochaix *et al.* 1984). These homologies will be discussed further below, in the context of the structural and functional analogies between PSII and the photosynthetic bacterial reaction centre.

The sequences of the *psbB* and *psbC* genes encoding the two large chlorophyll-binding proteins are also highly conserved (Morris and Herrmann 1984; Bookjans *et al.* 1986) and also predict highly hydrophobic proteins. In spinach, the genes encode proteins of 56 and 52 kDa respectively. The *psbB* and *psbC* genes appear to be distantly related to one another, showing not only some sequence homology but also remarkably similar hydropathy profiles. Nether protein shows significant homology with either D1 or D2, or with the L and M proteins of the bacterial reaction centre. On the basis of their hydropathy profiles, the *psbB* and *psbC* proteins have each been predicted to have five (Holschuh *et al.* 1984) or seven (Alt *et al.* 1984) transmembrane helices, with a large hydrophilic region located near the C-terminus. A number of conserved pairs of His residues may participate in binding chlorophyll.

Sequences with a substantial degree of homology (≥ 80 per cent) to *psbB* and *psbC* have also been detected in a number of cyanobacteria. *Synechococcus* 7002 (Bryant 1986) and *Synechocystis* 6803 (Vermaas *et al.* 1986) and the cyanelle genome of *Cyanophora paradoxa* (Bryant 1986) each has a single copy of these genes.

The *psbE* and *psbF* genes are adjacent to one another in higher plants (Herrmann *et al.* 1984; Courtice *et al.* 1985; Carrillo *et al.* 1986), in the cyanobacterium *Synechocystis* 6803, and in *C. paradoxa* cyanelles (Bryant 1986). They encode the apoproteins of the two subunits of cytochrome b_{559}. Hydropathy profiles predict a single transmembrane region in each polypeptide, and it has been suggested that the haem group is co-ordinated between the two conserved His residues in adjacent helices, forming a heterodimer. The function of this intriguing protein remains a mystery.

Structure of the PSII reaction centre

Substantial evidence now favours the model that the PSII reaction centre is fundamentally similar in structure and function to the reaction centre of the purple photosynthetic bacteria (Michel and Deisenhofer 1986; Trebst 1986). Most telling are homologies between D1 and D2 and the bacterial reaction centre subunits, L and M respectively. In particular, the His residues of L and M that are ligands to the bchl special pair are conserved in D1 and D2, as are those that bind the non-haem iron. A Trp residue that forms part of the Q_A binding site is conserved between M and D2. Both L and D1 bind the photo-affinity-labelled herbicide, azidoatrazine, that has been proposed to attack the Q_B binding site. Significantly, a Phe that is conserved between L and D1 forms part of this site, and it has been shown that a mutation of this residue to Tyr in *Chlamydomonas* D1 protein confers atrazine resistance.

A PSII photochemical reaction centre consisting of a D1–D2 heterodimer is consistent with a number of recent reports suggesting functions for D1 on both the reducing and oxidizing side of PSII (Takahashi *et al.* 1986; Metz *et al.* 1986; Ikeuchi and Inoue 1987). As mentioned previously, the recent isolation of a photochemically active PSII preparation containing only D1, D2, and cytochrome b_{559} now seems to provide conclusive support to this argument.

Nevertheless, it should be noted that some experimental findings are still difficult to accommodate within the new model. For example, there are reports that PSII particles depleted of 30–35 kDa proteins show normal photoreduction of Q_A (Yamagishi and Katoh 1984; Satoh 1986). However, no direct evidence is presented that the proteins lost from these particles do indeed correspond to D1 and D2, which have notoriously variable staining properties.

Vermaas *et al.* (1986) have used the novel approach of directed mutagenesis of the transformable cyanobacterium, *Synechocystis* 6803, to study the functions of PSII proteins. A mutant in which the only genomic change was disruption of the *psbB* gene contained non-functional PSII complexes and apparently lacked only the *psbB* protein. However, the possibility cannot be ruled out that lack of the *psbB* protein indirectly affects PSII photochemistry by preventing proper assembly of D1 and D2 to form the reaction centre.

Oxygen evolution

The ability of PSII to generate a potential high enough to oxidize water (approximately +1.0 V) is one property that clearly distinguishes it from the bacterial reaction centre. Mn^{2+}, Ca^{2+}, and Cl^- ions are all required for oxygen evolution (reviewed by Renger and Govindjee 1985), but many questions remain about the nature of the protein components of the water-oxidizing complex.

Substantial progress has followed the isolation of PSII preparations active in oxygen evolution. The first such preparation was obtained from a cyanobacterium (Stewart and Bendall 1979). Since 1981, preparations from other cyanobacteria have been reported, and purified 'PSII membranes' have been obtained from higher plants (reviewed by Bryant 1986). The most highly resolved preparations now available are those from the thermophilic cyanobacterium, *Synechococcus* sp. (Ohno *et al.* 1986) and recent spinach preparations that are free of the light-harvesting chlorophyll *a*/*b* protein complex (Ikeuchi *et al.* 1985; Tang and Satoh 1985). In addition to the components of the PSII 'core', the *Synechococcus* preparation contains a Tris-removable polypeptide of 35 kDa, an 8 kDa polypeptide of unknown function, four Mn, and one Ca, while the spinach preparation contains the core components plus polypeptides of 33 kDa and 22 kDa.

The 33–35 kDa component of these preparations corresponds to the largest of three extrinsic proteins (33, 23, and 16 kDa) that are located at the inner surface of higher plant thylakoids and have been shown to participate in oxygen evolution (reviewed by Andersson 1986). The 33 kDa protein appears to be highly conserved amongst oxygenic organisms; antibodies raised against the spinach protein cross-react with a corresponding protein not only from other higher plants but also from cyanobacteria (Stewart *et al.* 1985*b*). The role of the 33 kDa protein is still unclear. Considerable evidence favours an association with at least two of the four functional Mn in the oxygen-evolving complex (Andersson 1986; Hunziker *et al.* 1987). However, treatment with 1M $MgCl_2$ or $CaCl_2$ removes the 33 kDa protein from PSII without removing Mn, and such preparations can be at least partially reactivated by 150–200 mM Cl^- ions. The spinach protein has now been purified and sequenced (Oh-oka *et al.* 1986) and, significantly, found to contain a region of homology to bacterial Mn-superoxide dismutases.

The 23 and 16 kDa proteins appear to function *in vivo* to mediate high-affinity binding of the Ca^{2+} and Cl^- ions required for oxygen evolution (reviewed by Andersson 1986); *in vitro* they can be replaced by higher, non-physiological concentrations of these ions. The 23 and 16 kDa proteins from spinach have been purified and partially sequenced (Vater *et al.* 1986). These proteins have been much less highly conserved than the 33 kDa protein during evolution. Similar proteins are present in the green alga, *Chlamydomonas* (Bennoun *et al.* 1981), but antibodies to the 23 kDa and 16 kDa proteins from spinach detect no corresponding proteins in cyanobacteria (Stewart *et al.* 1985*b*), even though cyanobacteria require Ca^{2+} for oxygen evolution.

Also apparently missing from cyanobacterial preparations are intrinsic polypeptides of 22 and 24 kDa that, from immunological evidence, appear to be closely associated with the oxygen-evolving complex of spinach (Ljungberg *et al.* 1984*b*). It is not known whether either of these polypeptides is related to the 22 kDa polypeptide in the highly purified oxygen-evolving

```
1                   5                   10
X-Gln-Gln-Phe-Arg-Asn-Ala-Met-Asp-Asp-
                    15                  20
Lys-Leu-Ala-Thr-Asp-Phe-Gly-Lys-Lys-Ile-
                    25                  30
Asp-Leu-Asn-Asn-Thr-Asn-Val-Arg-Ala-Phe
                    35                  40
Met-Gln-Tyr-Pro-Gly-Met-Tyr-Pro-Thr-Leu-
                    45                  50
Ala-Arg-Met-Ile-Leu-Lys-Asn-Ala-Pro-Phe
```

Fig. 6.2. N-terminal sequence of the 9 kDa protein from *P. laminosum* oxygen-evolving PSII particles.

particles described by Tang and Satoh (1985) or to the product of the *psbG* gene.

There may be other, smaller proteins that participate in oxygen evolution. Cytochrome b_{559} is invariably a constituent of active preparations, and Ljungberg *et al.* (1984*a*, 1986) have detected intrinsic polypeptides of 6.5 and 7 kDa, and an extrinsic polypeptide of 10 kDa, that are missing from preparations inactive in oxygen evolution.

Stewart *et al.* (1985*a*, 1985*b*) have reported that removal of an extrinsic 9 kDa polypeptide from oxygen-evolving particles from the cyanobacterium *Phormidium laminosum* is associated with loss of activity, and a polypeptide of similar size has been detected in active PSII particles from *Synechococcus* sp. (Ohno *et al.* 1986). The 9 kDa protein from *P. laminosum* has been purified and partially sequenced (Fig. 6.2) and shows no homology with any of the higher plant proteins that have been sequenced so far. It does not appear to be associated with the Ca^{2+} or Cl^- requirements for oxygen evolution, and its role in PSII remains unknown.

Figure 6.3 presents models for oxygen-evolving PSII in higher plants and in cyanobacteria, based on the evidence summarized above.

Genes coding for the proteins of the oxygen-evolving complex

The 33, 23, and 16 kDa proteins of higher plants and green algae, unlike the components of the PSII 'core', are encoded by the nuclear genome (Westhoff *et al.* 1985) and synthesized as larger precursors that are processed to their final forms during or after translocation to the thylakoid lumen. cDNA clones for all three genes have been obtained from spinach (Tittgen *et al.* 1986) and from pea and wheat (J.C. Gray, personal communication). There is some evidence that small families of genes may encode the 33 and 16 kDa proteins from spinach. Mayfield *et al.* (1987) have identified and sequenced a

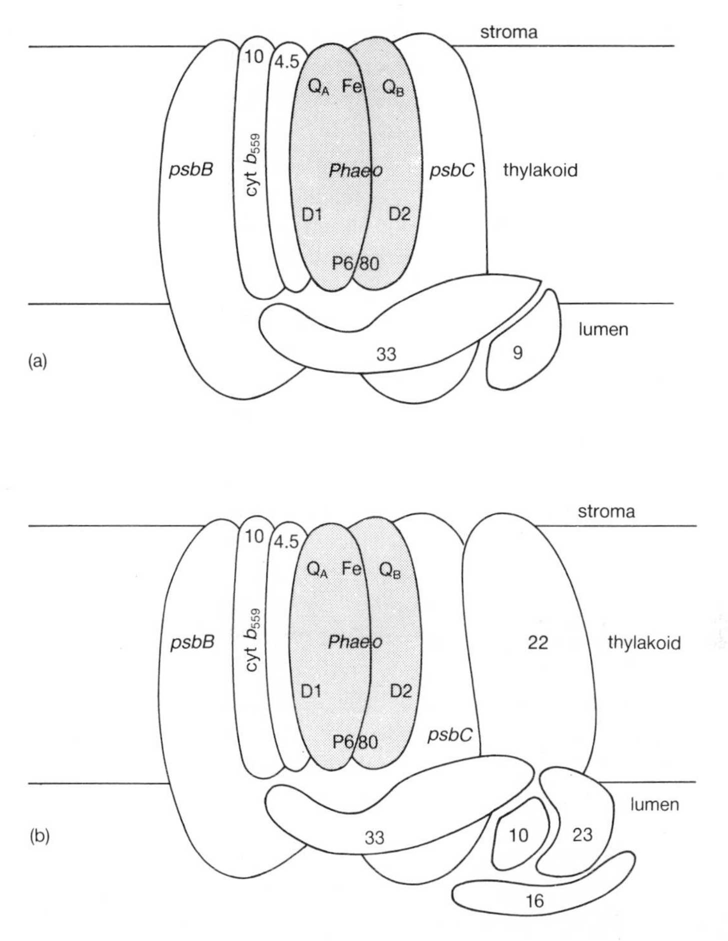

Fig. 6.3. Schematic models for the PSII core and the oxygen-evolving complex from (a) a higher plant or green alga, (b) a cyanobacterium. Components of the PSII reaction centre are shaded.

clone from a *Chlamydomonas* cDNA library coding for Oxygen Evolution Enhancer Protein 2, a 20 kDa protein that is the functional equivalent of the 23 kDa protein from spinach. The protein and the corresponding mRNA were absent from a mutant strain that was almost devoid of oxygen evolution capacity.

Photosystem I

In PSI, light energy generates a reductant strong enough to reduce ferredoxin (Fd), which in turn reduces Fd-NADP reductase and NADP. The primary photochemical donor of PSI is again a specialized chlorophyll species, P700. A number of membrane-bound electron acceptors have been identified between P700 and Fd, mainly by EPR spectroscopy (reviewed by Rutherford and Heathcote 1985). The exact pathway of electrons through these acceptors is not clear, but there is recent evidence, both for higher plants and for cyanobacteria, that the earliest acceptors, A_0 and A_1, are a monomeric form of chlorophyll and vitamin K (phylloquinone), respectively, while subsequent acceptors, A, B, and X, are iron–sulphur centres (Malkin 1986).

In terms of its protein composition, PSI, which at one time was better characterized than PSII, is now in some respects the less well understood of the two. 'Core' PSI preparations (that is, not containing accessory light-harvesting complexes) have been isolated from a number of higher plants, green algae, and cyanobacteria (Wollman 1986; Bryant 1986). These preparations have been reported to comprise between 7 and 16 polypeptides. All contain one or two large polypeptides of an apparent molecular mass (on SDS gels) of 60–70 kDa, plus a variable number of smaller polypeptides ranging from 4–20 kDa. The 60–70 kDa polypeptide is the apoprotein of a 100 kDa chlorophyll protein containing P700 and approximately 40 antenna chlorophyll molecules. The amounts of other components have not been firmly established, but a complex from barley has been reported to contain 10 atoms of iron and 14 of acid-labile sulphur (Høj and Møller 1986), and preparations from cyanobacteria contain two molecules of vitamin K, only one of which appears to be required for photochemical activity (Malkin 1986). Carotenoids are present, but no phaeophytin.

Evidence has recently been presented that one iron–sulphur centre (probably centre X) is associated with the 110 kDa P700 chlorophyll protein (Høj and Møller 1986), but the locations of other iron–sulphur centres are unknown. An 8 kDa polypeptide from spinach and barley PSI preparations is highly enriched in Cys, but no iron has been found associated with this protein.

Immunological evidence suggests that the major polypeptide of the PSI complex has been highly conserved across a wide range of higher plants, the green alga *Chlamydomonas*, and the cyanobacterium *Mastigocladus laminosus* (Bryant 1986).

The genes for PSI proteins

The controversy over the number and nature of high molecular weight polypeptides in PSI has been at least partly resolved by the isolation of two maize and spinach chloroplast genes, *psaA* and *psaB* (alternatively *psaA1* and *psaA2*), respectively, encoding two polypeptides of deduced molecular mass 80–85 kDa; the larger was immuno-precipitated by antibodies raised against the P700 chlorophyll protein (Fish *et al.* 1985; Kirsch *et al.* 1986). Highly homologous genes have been detected in the chloroplast genomes of other higher plants (Courtice *et al.* 1985; Lehmbeck *et al.* 1986), *Chlamydomonas* (Kück, 1985), in the cyanelle genome of *C. paradoxa* (Lambert *et al.* 1985), and in the genome of *Synechococcus* 7002 (Bryant 1986).

The *psaA* and *psaB* genes of maize, spinach, and pea have been sequenced. The corresponding genes are highly conserved (approximately 90 per cent homology, and *psaA* and *psaB* are also related to each other (40 per cent homology). The discovery of yet another pair of homologous genes suggests that gene duplication has played a major part in the evolution of photosynthetic membrane complexes.

In the higher plants and cyanobacteria that have been examined to date the genes are close together. Co-transcription has been shown for the spinach operon, which also contains the gene for a ribosomal protein (Kirsch *et al.* 1986), and it has therefore been suggested that the PSI core is a heterodimer of the *psaA* and *psaB* proteins. However, the genes are much further apart (45 kbp) in the *Chlamydomonas* genome (Kück 1985).

The deduced amino-acid sequences of *psaA* and *psaB* predict highly hydrophobic proteins, possibly with up to 11 transmembrane domains. Each protein contains approximately 40 conserved His residues which may form pairs in adjacent helices to participate in non-covalent binding of chlorophyll. There is also a sequence Asp-Pro-Thr-Thr-Arg that is conserved not only between *psaA* and *psaB*, but is also found in the *psbB* protein of the PSII core. The suggested participation of this sequence in binding the reaction centre chlorophyll species (Fish *et al.* 1985) now seems less likely in view of the accumulating evidence that P680 is not located on the *psbB* protein. The *psaA* protein contains four conserved Cys residues, two of which are also present in *psaB*; it has been suggested that these cysteines may participate in the formation of the iron–sulphur centre(s) associated with the P700 chlorophyll protein.

The complete sequencing of the tobacco and liverwort chloroplast genomes (Ohyama *et al.* 1986; Shinozaki *et al.* 1986) has revealed two conserved open reading frames, which, on the basis of the occurrence and spacing of their Cys residues, seem likely to encode 4Fe–4S proteins. There is a strong possibility that these are components of the PSI complex.

At least three nuclear-encoded proteins also contribute to the PSI complex isolated by Westhoff *et al.* (1983). cDNAs have now been obtained for subunits 2, 5, and 6, encoding proteins of 22, 16, and 12 kDa, respectively. Genomic blots hybridized against these cDNAs each showed a single band, indicating that each protein is encoded by a single nuclear gene.

References

Alt, J., Morris, J., Westhoff, P., and Herrmann, R.G. (1984). *Curr. Genet.* **8**, 597.

Andersson, B. (1986). In *Encyclopaedia of plant physiology: photosynthesis III* Vol. 19, (ed. L.A. Staehelin and C.J. Arntzen) p. 447. Springer, Berlin.
Arntzen, C.J. and Pakrasi, H.B. (1986). In *Encyclopaedia of plant physiology: photosynthesis III* Vol. 19 (ed. L.A. Staehelin and C.J. Arntzen) p. 457. Springer-Verlag, Berlin.
Bennoun, P., Diner, B.A., Wollman, F.-A., Schmidt, G., and Chua, N.-H. (1981). In *Photosynthesis III. Structure and molecular organization of the photosynthetic apparatus* (ed. G. Akoyonoglou) p. 839. Balaban, Philadelphia, Penn.
Bookjans, G., Stummann, B.M., Rasmussen, O.F., and Henningsen, K.W. (1986). *Plant mol. Biol.* **6**, 359.
Bryant, D.A. (1986). In *Photosynthetic picoplankton, Canadian bulletin of fisheries and aquatic science No. 214* (ed. T. Platt and W.K.W. Li) p. 423. Canadian Department of Fisheries and Oceans, Ottawa.
Carrillo, N., Seyer, P., Tyagi, A., and Herrmann, R.G. (1986). *Curr. Genet.* **10**, 619.
Chang, C.-H., Tiede, D., Tanz, J., Smith, U., Norris, J., and Schiffer, M. (1986). *FEBS Lett.* **205**, 82.
Courtice, G.R.M., Bowman, C.M., Dyer, T.A., and Gray, J.C. (1985). *Curr. Genet.* **10**, 329.
Deisenhofer, J., Epp, O., Miki, K., Huber, R., and Michel, M. (1985). *Nature, Lond.* **318**, 618.
De Vitry, C., Wollman, F.-A., and Delepelaire, P. (1984). *Biochim. Biophys. Acta* **767**, 415.
Erickson, J.M., Rahire, M., and Rochaix, J.-D. (1984). *EMBO J.* **3**, 2753.
Fish, L.E., Kück, U., and Bogorad, J. (1985). *J. biol. Chem.* **260**, 1413.
Golden, S.S., Brusslan, J., and Haselkorn, R. (1986). *EMBO J.* **5**, 2789.
Herrmann, R.G., Alt, J., Schiller, C., Cramer, W., and Widger, W.R. (1984). *FEBS Lett.* **179**, 239.
Herrmann, R.G., Westhoff, P., Alt, J., Tittgen, J., and Nelson, N. (1985). In *Molecular form and function of the plant genome* NATO ASI Series A: Life Sciences Vol. 83, (ed. L. van Vloten-Doting, G.S.P. Groot, and T.C. Hall) p. 233, Plenum Press, New York.
Hird, S.M., Dyer, T.A., and Gray, J.C. (1986). *FEBS Lett.* **209**, 181.
Høj, P.B. and Møller, B.J. (1986). *J. biol. Chem.* **261**, 14292.
Holschuh, C., Bottomley, W., and Herrmann, R.G. (1984). *Nucleic Acids Res.* **12**, 8819.
Hunziker, D., Abramowicz, D.A., Damoder, R., and Dismukes, G.C. (1987). *Biochim. Biophys. Acta* **890** 6.
Ikeuchi, M. and Inoue, Y. (1987). *FEBS Lett.* **210**, 71.
Ikeuchi, M., Yuasa, M., and Inoue, Y. (1985). *FEBS Lett.* **185**, 316.
Kirsch, W., Seyer, P., and Herrmann, R.G. (1986). *Curr. Genet.* **10**, 843.
Kück, U. (1985). *Abstracts, 1st Int. Cong. Plant Mol. Biol.*, p. 125.
Lehmbeck, J., Rasmussen, O.F., Bookjans, G.B., Jepsen, B.R., Stummann, B.M., and Henningsen, K.W. (1986). *Plant mol. Biol.* **7**, 3.
Lambert, D.H., Bryant, D.A., Stirewalt, V.L., Dubbs, J.M., Stevens, E., and Porter, R.D. (1985). *J. Bact.* **164**, 659.
Ljungberg, U., Åkerlund, H.-E., and Andersson, B. (1984*a*). *FEBS Lett.* **175**, 255.

Ljungberg, U., Åkerlund, H.-E., Larsson, C., and Andersson, B. (1984*b*). *Biochim. Biophys. Acta* **767**, 145.
Ljungberg, U., Henrysson, T., Rochester, C.P., Åkerlund, H.-E., and Andersson, B. (1986). *Biochim. Biophys. Acta* **849**, 112.
Malkin, R. (1986). *FEBS Lett.* **208**, 343.
Mayfield, S.P., Rahire, M., Frank, G., Zuber, H., and Rochaix, J.-D. (1987). *Proc. Natn. Acad. Sci. USA* **84**, 749.
Metz, J.G., Pakrasi, H.B., Seibert, M., and Arntzen, C.J. (1986). *FEBS Lett.* **205**, 269.
Michel, H. and Deisenhofer, J. (1986). In *Encyclopaedia of plant physiology: photosynthesis III*, Vol. 19 (ed. L.A. Staehelin and C.J. Arntzen) p. 371. Springer-Verlag, Berlin.
Michel, H., Weyer, K.A., Gruenberg, H., and Lottspeich, F. (1985). *EMBO J.* **4**, 1667.
Michel, H., Epp., O., and Deisenhofer, J. (1986*a*). *EMBO J.* **5**, 2445.
Michel, H., Weyer, K.A., Gruenberg, H., Dunger, I., Oesterhelt, D., and Lottspeich, F. (1986*b*). *EMBO J.* **5**, 1149.
Morris, J. and Herrmann, R.G. (1984). *Nucleic Acids Res.* **12**, 2837.
Nanba, O. and Satoh, K. (1987). *Proc. Natn Acad. Sci. USA* **84** 109.
Ohno, T., Satoh, K., and Katoh, S. (1986). *Biochim. Biophys. Acta* **852**, 1.
Oh-oka, H., Tanaka, S., Wada, K., Kuwabara, T., and Murata, N. (1986). *FEBS Lett.* **197**, 63.
Ohyama, K., Fukuzawa, H., Kohchi, T., Shirai, H., Seno, T., Sano, S., Umesono, K., Shiki, Y., Takeuchi, M., Chang, Z., Aota, S., Inokuchi, H. and Ozeki, H. (1986). *Nature, Lond.* **322**, 572.
Rasmussen, O.F., Bookjans, G., Stummann, B.M., and Henningsen, K.W. (1984). *Plant mol. Biol.* **3**, 191.
Renger, G. and Govindjee (1985). *Photosyn. Res.* **6**, 33.
Rochaix, J.-D., Dron, M., Rahire, M., and Malnoë, P. (1984). *Plant mol. Biol.* **3**, 363.
Rutherford, A.W. and Heathcote, P. (1985). *Photosyn. Res.* **6**, 295.
Satoh, K. (1986). *FEBS Lett.* **204**, 357.
Sayre, R.T., Andersson, B., and Bogorad, L. (1986). *Cell* **47**, 601.
Shinozaki, J., *et al.* (1986). *EMBO, J.* **5**, 2043.
Steinmetz, A.A., Castroviejo, M., Sayre, R.T., and Bogorad, L. (1986). *J. biol. Chem.* **261**, 2485.
Stewart, A.C. and Bendall, D.S. (1979). *FEBS Lett.* **107**, 308.
Stewart, A.C., Siczkowski, M., and Ljungberg, U. (1985*a*). *FEBS Lett.* **193**, 175.
Stewart, A.C., Ljungberg, U., Åkerlund, H.-E., and Andersson, B. (1985*b*). *Biochim. Biophys. Acta* **808**, 353.
Takahashi, Y., Takahashi, M., and Satoh, K. (1986). *FEBS Lett.* **208**, 347.
Tang, X.-S and Satoh, K. (1985). *FEBS Lett.* **179**, 60.
Tittgen, J., Hermans, J., Steppuhn, J., Jansen, T., Jansson, C., Andersson, B., Nechustai, R., Nelson, N. and Herrmann, R.G. (1986). *Mol. gen. Genet.* **204**, 258.
Trebst, A. (1986). *Z. Naturf.* **41c**, 240.
Vater, J., Salnikow, J. and Jansson, C. (1986). *FEBS Lett.* **203**, 230.
Vermaas, W.F.J., Williams, J.G.K., Rutherford, A.W., Mathis, P., and Arntzen,

C.J. (1986). *Proc. Natn Acad. Sci. USA* **83**, 9474.
Westhoff, P., Farchaus, J.W., and Herrmann, R.G. (1986). *Curr. Genet.* **11**, 165.
Westhoff, P., Alt, J., Nelson, N., Bottomley, W., Buhnemann, H., and Herrmann, R.G. (1983). *Plant mol. Biol.* **2**, 95.
Westhoff, P., Jansson, C., Klein-Hitpass, L., Berzborn, R., Larsson, C. and Bartlett, S.G. (1985). *Plant mol. Biol.* **4**, 137.
Widger, W.R., Cramer, W.A., Hermodson, M., and Herrmann, R.G. (1985). *FEBS Lett.* **191**, 186.
Williams, J.C., Steiner, L.A., Feher, G., and Simon, M. (1984). *Proc. Natn Acad. Sci. USA* **81**, 7303.
Wollman, F.-A. (1986). In *Encyclopaedia of plant physiology: photosynthesis III* Vol. 19 (ed. L.A. Staehelin, and C.J. Arntzen) p. 487. Springer-Verlag, Berlin.
Yamagishi, A. and Katoh, S. (1984). *Biochim. Biophys. Acta* **765**, 118.
Youvan, D.C., Bylina, E.J., Alberti, M., Begusch, H., and Hearst, J.E. (1984). *Cell* **37**, 949.

7 Light-induced proton efflux of the cyanobacterium Anabaena variabilis

S. SCHERER

Lehrstuhl für Physiologie und Biochemie der Pflanzen, Universität Konstanz, D-7750 Konstanz, Germany

H. RIEGE

Lehrstuhl für Geomikrobiologie, Universität Oldenburg, D-2900 Oldenburg, Germany

P. BÖGER

Lehrstuhl für Physiologie und Biochemie der Pflanzen, Universität Konstanz, D-7750 Konstanz, Germany

---

Introduction

A light-induced acidification of the medium has been reported for several eukaryotic (Fujii *et al.* 1978; Kura-Hotta and Enami 1981; Spear *et al.* 1969) and prokaryotic (Scholes *et al.* 1969; Scherer and Böger 1984) algae. It is not possible to decide whether a proton efflux or an OH^- influx is observed in these experiments, but despite this uncertainty the term proton efflux is used in this paper. It has been suggested for higher plants (Sze 1984) as well as for *Cyanidium* (Enami and Kura-Hotta 1984) that light-induced proton efflux is due to a unidirectional, proton-translocating ATP-hydrolase, which is driven by photosystem I. In cyanobacteria, light-induced acidification can be demonstrated in species of all taxonomic sections (Scherer and Böger 1984, unpublished observations). Proton translocation due to respiratory electron transport chain associated with the cytoplasmic membrane has been suggested to be responsible for light-induced proton efflux in *Plectonema* (Hawkesford *et al.* 1983).

In this paper we summarize the current state of knowledge and present some data confirming the relationship between light-induced acidification and inorganic carbon uptake in *Anabaena variabilis.*

Kinetics of light-induced acidification

Figure 7.1 shows two typical time courses of light-induced acidification of medium containing *A. variabilis.* In both cases, two kinetically different phases can be seen, the first one starting immediately after illumination, with a rate being at a maximum after 1–3 sec. From well-resolved time course

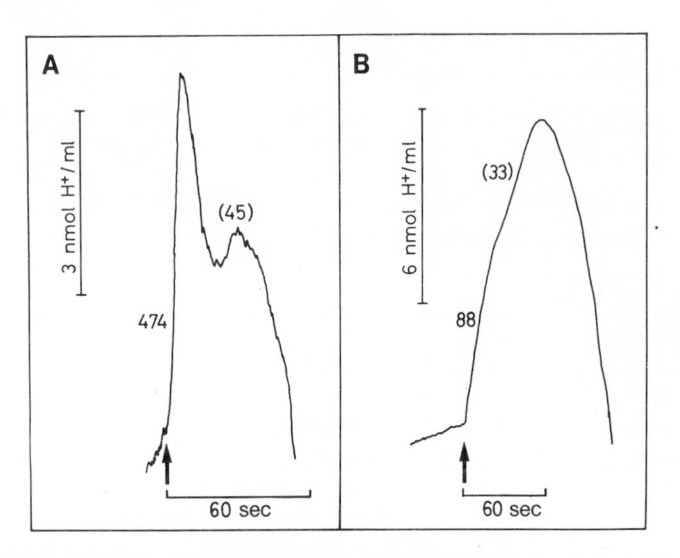

Fig. 7.1. Light-induced acidification of *A. variabilis*. (A) Cells were grown for one day in BG-11 medium at 1.5 per cent CO_2 in air (v/v) and measurements were made in this medium at pH 6.3 and 46 mM NaCl present. (B) Cells were grown for three days in the medium described by Arnon *et al.* (1974), harvested and re-suspended in 75 mM KCL, 75 mM NaCl, 5 mM $MgCl_2$, 3 mM glycylglycine, pH 6.3 (for further experimental details, see Scherer and Böger 1984). The figures represent rates of acidification in μmol H^+ mg^{-1} Chl $hour^{-1}$. Figures in brackets refer to estimated rates (see text).

studies, as shown in Fig. 7.1(A), it can be estimated that phase I ceases after approximately 5–12 sec. Phase II exhibits a maximum rate after 10–20 sec. Direct measurements on young cultures in BG-11 growth medium [Fig. 7.1(A)] showed that these two phases could be resolved clearly, whereas in older cultures [2–4 days, Fig. 7.1(B)] phase II often was seen merely as a shoulder after 10–20 sec. In the latter case, however, the rates can be estimated by assuming that acidification ceases because an alkalization of the medium starts by conversion of HCO_3^- into CO_2 and OH^- (which is utilized by ribulose bisphosphate carboxylase; see Cooper *et al.* 1969; Badger 1980). The latter must be extruded by the cell for pH regulation. The stoichiometry between oxygen evolution and alkalization (O_2:OH^-) in *Anabaena* is a 1:1 ratio at equilibrium at alkaline pH (Miller and Colman 1980; Kaplan 1981; Scherer *et al.*, unpublished observations). By superimposing net alkalization deduced from considering oxygen evolution and acidification, the rate of phase II proton efflux can be estimated, as has been described in detail by Hinrichs *et al.* (1985). The two traces in Fig. 7.1 illustrate the variability of light-induced acidification, depending on growth conditions, age of the culture, and treatment after harvesting the cells. However, under all these conditions, it is evident that two different mechanisms of light-induced acidification are active in *Anabaena*.

Light dependence

In the dark, protons are extruded into the medium by *Anacystis* (Peschek 1983) and *Anabaena* (Scherer *et al.* 1984). There is some dispute as to whether the proton efflux is due to respiratory electron transport or to an ATPase, localized on the cytoplasmic membrane. In *Anabaena*, no conclusive evidence has been presented that respiratory electron transport is localized on the cytoplasmic membrane (Lockau and Pfeffer 1983; Nitschmann and Peschek 1985; Scherer *et al.* 1984). The light-dependence of proton efflux for *Anabaena* is shown in Fig. 7.2. Obviously, phase I is driven by very low light intensities, whereas phase II requires higher light intensities, comparable to those which support oxygen evolution. Furthermore, there is a marked difference with respect to the effect of monochromatic red light (707 nm; Scherer *et al.* 1986): again, phase I was stimulated at very low light intensities,

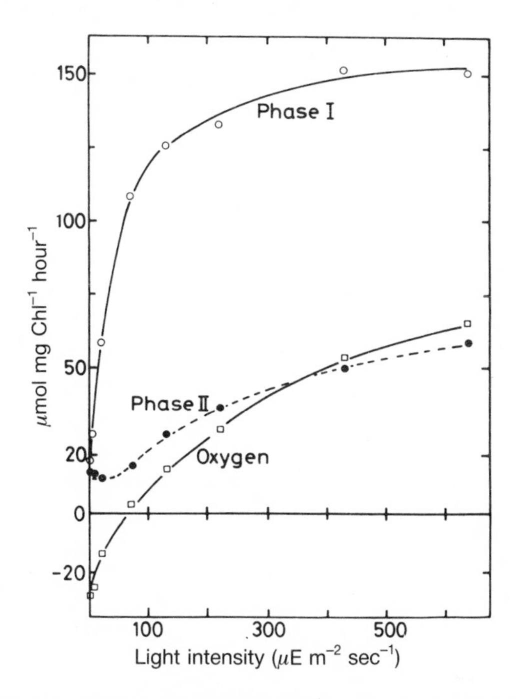

Fig. 7.2. Light-dependence of acidification and oxygen evolution of *A. variabilis.* The concentration of cells in the reaction chamber was 30 μg chlorophyll ml^{-1}. Illumination was by red light, wavelength 600 nm ○●: proton efflux; □ oxygen exchange. (Taken from Hinrichs *et al.* 1985, used by permission)

whereas phase II did not respond at all to this wavelength. This is indicative of photosystem I being active in driving phase I, and both photosystem I and II inducing phase II of proton efflux. An action spectrum corroborated the conclusion that two different mechanisms of proton efflux exist in *Anabaena*.

Influence of Na^+

Light-induced acidification is strictly dependent on the presence of Na^+ (Fig. 7.3); Cl^- and K^+ had no effect, but the provision of either NaCl or Na_2SO_4 induced acidification after a dark-light transition. The Na^+ concentration was found to be optimal at 30–50 mM (Scherer *et al.* 1988). Interestingly, the uptake of dissolved inorganic carbon in *Anabaena* also showed a strikingly similar dependence on Na^+ with an optimum concentration of 40 mM (Kaplan *et al.* 1984; Reinhold *et al.* 1984). This points to a connection between light-induced acidification and active carbon intake. However, Na^+ may be involved also in the pH regulation of cyanobacteria. A Na^+/H^+-antiporter has been suggested to be active in coccoid cyanobacteria (Blumwald *et al.* 1984; Miller *et al.* 1984; Erber *et al.* 1986), but Reed *et al.* (1985)

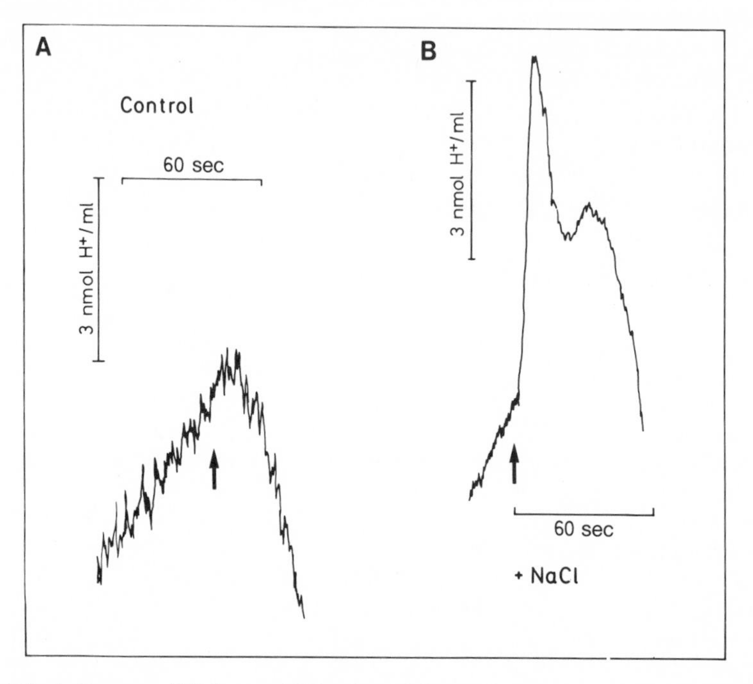

Fig. 7.3. Influence of Na^+ on acidification of the medium by *A. variabilis*. (A) Control without NaCl. (B) 46 mM NaCl added. Measurement was at pH 6.5, with cells grown in BG-11 medium and assayed immediately after harvesting.

provided evidence that in *Synechocystis* Na^+ fluxes may be interpreted as being due to a K^+/Na^+-antiporter. Hitherto, however, mechanisms and bioenergetics of ion fluxes across cyanobacterial membranes have not been thoroughly studied.

Action of inhibitors

The two phases of acidification can also be distinguished by different sensitivity towards inhibitors (see Table 7.1; also Hinrichs *et al.* 1985). As expected, carbonylcyanide-*m*-chlorophenylhydrazone (CCCP) abolished light-induced acidification as well as photosynthetic oxygen evolution completely, whereas 3-(3,4-dichlorophenyl)-1,1-dimethylurea (DCMU) inhibited both phases, although to different degrees. This suggests that proton efflux depends in some way on electron transport and/or on photophosphorylation. Electron acceptors accessible to photosystem I in intact cells, such as metronidazole and methylviologen, which do not inhibit electron transport directly but abolish CO_2 fixation in the Calvin cycle, inhibited both oxygen evolution and phase II but not phase I. The same result was obtained with cyanide, inhibiting CO_2 fixation (Wishnick and Land 1969) and respiratory electron transport. Phase I was somewhat stimulated whereas phase II was inhibited, though to a lesser extent than oxygen evolution or respiration. This result is important with respect to the suggestion that light-induced acidification is due to a respiratory electron transport chain on the cytoplasmic membrane, using reducing equivalents originating from photosynthetic electron transport (Hawkesford *et al.* 1983). This possibility can be excluded for phase I (see Table 7.1) because of the fast kinetics and high rate of phase I acidification as well as its insensitivity to cyanide. Appa-

Table 7.1 Action of inhibitors on acidification and O_2 evolution.

Inhibitor	Phase I (Light-induced proton efflux)	Phase II (Light-induced proton efflux)	Oxygen evolution
CCCP (10 μM)	0	0	0
DCMU (10 μM)	40	22	0
Methylviologen (1 mM)	95	7	0
Metronidazole (4 mM)	100	43	67
Cyanide (1 mM)	100–140	40–60	10
Mercaptoethanol (30 μM)	100	50	80
Vanadate (1 mM)	55–100	15–60	100
Nitrofen (10 μM)	100	50	100
Acetazolamide (1 mM)	45	100	100

Rates of phase I were typically in the range between 200 and 400 μmol mg chl^{-1} h^{-1}; phase II was 70–140 μmol mg chl^{-1} h^{-1}. The rates of oxygen evolution varied between 100–150 μmol mg chl^{-1} h^{-1}. The data are given as % of the corresponding controls; the table represents typical results. Measurements were performed in the medium described in Fig. 7.1(B).

rently, also, with phase II there is no evidence for the operation of an electron transport chain; there was no inhibition by 0.1 mM KCN, which decreases respiratory electron transport to about 30 per cent of the control. Mercaptoethanol inhibited phase II but not phase I, which may indicate a regulatory role of -SH groups in light-induced acidification of phase II.

Most interestingly, vanadate inhibited light-induced acidification, although phase I was less sensitive than phase II. Vanadate is a typical inhibitor of the uni-directional, ion-translocating ATPases localized on the cytoplasmic membranes of plants, yeasts, and fungi (e.g. Macara 1980; Goffeau and Slayman 1981). Furthermore, nitrofen, (2,4-dichlorophenyl-4-nitrophenyl ether), an inhibitor of ATP synthesis (Lambert *et al.* 1979; Huchzermeyer 1982), markedly inhibited phase II but not phase I. We conclude that phase II is ATP-dependent and may be due to a uni-directional, proton-translocating ATP-hydrolase, while phase I does not depend on ATP. A vanadate-sensitive ATPase has been postulated to mediate oxygen-dependent proton efflux in *A. variabilis* in the dark (Scherer *et al.* 1984), an observation confirmed by Nitschmann and Peschek (1985). The inhibition of phase I by acetazolamide (Table 7.1) again points to an involvement of carbon uptake with this acidification, since this inhibitor is known to affect carbonic anhydrase (see Poincelot 1979).

Effect of dissolved inorganic carbon

The cells used in our experiments were grown at high CO_2 concentrations (1.5 per cent CO_2 in air, v/v). It has been shown that both high-CO_2 as well as low-CO_2-grown *A. variabilis* take up carbon by active transport, although low-CO_2-grown cells exhibit a tenfold higher v_{max} (Kaplan *et al.* 1980). Active carbon accumulation is light-dependent and involves a primary electrogenic pump (Kaplan *et al.* 1982), but the mechanism is still unclear (for reviews, see Kaplan 1985; Ogawa *et al.* 1985; Aizawa and Miyachi 1986). Although HCO_3^- and CO_2 can be used by the cells, HCO_3^- seems to be the only carbon species which may cross the cytoplasmic membrane of *A. variabilis* (Volokita *et al.* 1984; but see Shiraiwa and Miyachi 1985).

The fast phase I of light-induced acidification depends on the CO_2 concentration during growth. Cells grown under air at pH 8 to 9 do not exhibit a fast proton efflux, but only a slower one (Scherer *et al* 1988), which could be inhibited completely by nitrofen (data not shown). Whether this slower proton efflux of air-grown cells is identical with phase II of CO_2-grown cells (cf. Fig. 7.1) remains to be demonstrated.

These results and our findings reported above regarding Na^+ dependence and sensitivity towards an inhibitor of carbonic anhydrase suggest that the first phase of acidification depends upon the concentration of dissolved inorganic carbon in the medium. Indeed, phase I of proton efflux was stimulated by the addition of $NaHCO_3$ (Fig. 7.4). A K_m of 100–200 μM for

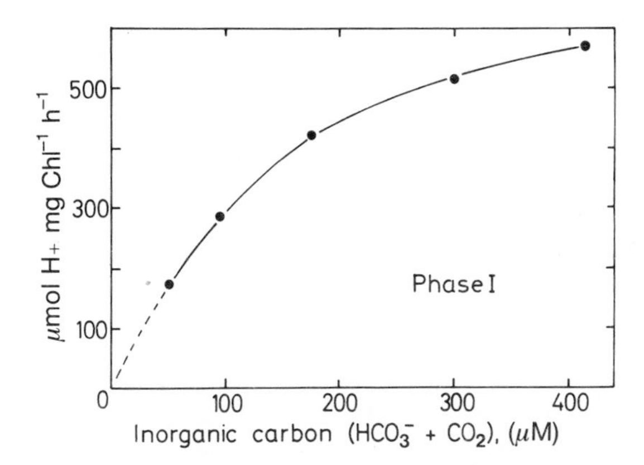

Fig. 7.4. Stimulation of light-induced acidification by $NaHCO_3$ in the presence of saturating concentrations of NaCl (75 mM) in the same medium as for Fig. 7.1(B). Light-induced acidification due to phase I is shown.

HCO_3^- + CO_2 was estimated from these experiments, which matches approximately the apparent K_m of 150 μM for light-induced carbon uptake reported by Kaplan and co-workers (1980) for *Anabaena*. This apparent K_m, however, was found to be variable, depending upon growth state and pre-treatment of the cells before the experiment.

Conclusively, light-induced acidification by phase I may be due to the transport of CO_2 across the cytoplasmic membrane. To be transported across the membrane, CO_2 must be converted to HCO_3^- by the cabonic anhydrase-like activity of the carbonate transporter, as has been suggested by Volokita *et al.* (1984). In this reaction, one proton per CO_2 molecule is released outside the cell, measurable as acidification in our experiments. Interestingly, active uptake of inorganic carbon is sensitive to carbonic anhydrase inhibitors when CO_2 is supplied to the cells, but insensitive when HCO_3^- is given (see Volokita *et al.* 1984). At pH 8 during growth in air, no light-induced acidification due to phase I could be seen (Scherer *et al.* 1988). At this pH the equilibrium of $CO_2 + H_2 \rightleftharpoons H^+ + HCO_3^-$ is completely shifted to the right-hand side; little CO_2 is available and thus no protons can be released.

Conclusion

Based on the available data, we suggest that light-induced acidification of phase II is due to a *net* proton translocation across the cytoplasmic membrane via on ATPase, whereas the first phase reflects the production of HCO_3^- and protons from CO_2 and H_2O at the outside of the cytoplasmic membrane. Therefore our results support the model of active carbon uptake suggested by Volokita *et al.* (1984); namely that HCO_3^-, and not CO_2, is transported across the cytoplasmic membrane of *A. variabilis* cells when grown at high CO_2 concentration.

Acknowledgement

This study was supported by the Deutsche Forschungsgemeinschaft.

References

Aizawa, K. and Miyachi, S. (1986). *FEMS Microbiol. Rev.* **39**, 215.
Arnon, D., McSwain, D., Tsujimuto, H., and Wada, K. (1974). *Biochim. Biophys. Acta* **357**, 231.
Badger, M.R. (1980). *Arch. Biochem. Biophys.* **201**, 247.
Blumwald, E., Wolosin, J.M., and Packer, L. (1984). *Biochim. biophys. Res. Commun.* **122**, 452.
Cooper, T.G., Filmer, M., Wishnick, M., and Lane, M.D. (1969). *J. biol. Chem.* **244**, 1081.
Enami, J. and Kura-Hotta, M. (1984). *Plant Cell Physiol.* **25**, 1101.
Erber, W.W.A., Nitschmann, W.P., Muchl, R., and Peschek, G.A. (1986). *Arch. Biochem. Biophys.* **247**, 28.
Fujii, S., Shimmen, T., and Tazawa, M. (1978). *Plant Cell Physiol.* **19**, 573.
Goffeau, A. and Slayman, C.W. (1981). *Biochim. Biophys. Acta* **639**, 197.
Hawkesford, M.J., Rowell, P., and Stewart, W.D.P. (1983). In *Photosynthetic procaryotes: cell differentiation and function* (ed. G.C. Papageorgiou and L. Packer) p. 199. Elsevier, Amsterdam.
Hinrichs, I., Scherer, S., and Böger, P. (1985). *Physiol. Veg.* **23**, 717.
Huchzermeyer, B. (1982). *Z. Naturf. Sect. C Biosci.* **37**, 787.
Kaplan, A. (1981). *Plant Physiol.* **67**, 201.
Kaplan, A. (1985). In *Inorganic carbon uptake by aquatic photosynthetic organisms* (ed. W.J. Lucas, and J.A. Berry) p. 325. American Society of Plant Physiologists, Rockville, Md.
Kaplan, A., Badger, M.R., and Berry, J.A. (1980). *Planta* **149**, 219.
Kaplan, A., Zenvirth, D., Reinhold, L., and Berry, J.A. (1982). *Plant Physiol.* **69**, 978.
Kaplan, A., Volokita, M., Zenvirth, D., and Reinhold, L. (1984). *FEBS Lett.* **176**, 166.
Kura-Hotta, M. and Enami, J. (1981). *Plant Cell Physiol.* **22**, 1175.
Lambert, R., Kunert, K.J., and Böger, P. (1979). *Pestic. Biochem. Physiol.* **11**, 267.
Lockau, W. and Pfeffer, S. (1983). *Biochim. Biophys. Acta* **733**, 124.
Macara, I.G. (1980). *Trends Biochem. Sci.* **5**, 92.
Miller, A.G. and Colman, B. (1980). *J. Bact.* **143**, 1253.
Miller, A.G., Turpin, D.H., and Canvin, D.T. (1984). *J. Bacteriol.* **159**, 100.
Nitschmann, W.H. and Peschek, G.A. (1985). *Arch. Microbiol.* **141**, 330.
Ogawa, T., Omata, T., Miyano, A. and Inoue, Y. (1985). In *Inorganic carbon uptake by aquatic photosynthetic organisms* (ed. W.J. Lucas, and J.A. Berry)

p. 287. American Society Plant Physiologists, Rockville, Md.
Peschek, G.A. (1983). *J. Bact.* **153**, 539.
Poincelot, R.P. (1979). In *Encyclopedia plant physiology, vol. 6. Photosynthesis II* (ed. M. Gibbs and E. Latzko), p. 230. Springer, Berlin.
Reed, R.H., Rirchardson, D.L., and Stewart, W.D.P. (1985). *Biochim. Biophys. Acta* **814**, 347.
Reinhold, L., Volokita, M., Zenvirth, D. and Kaplan, A. (1984). *Plant Physiol.* **76**, 1090.
Scherer, S. and Böger, P. (1984). *FEMS Microbiol. Lett.* **22**, 215.
Scherer, S., Stürzl, E. and Böger, P. (1984). *J. Bact.* **158**, 609.
Scherer, S., Hinrichs, I., and Böger, P. (1986). *Plant Physiol.* **81**, 939.
Scherer, S., Riege, H., and Böger, P. (1988). *Plant Physiol.* **86**, 769.
Scholes, P., Mitchell, P. and Moyle, J. (1969). *Eur. J. Biochem.* **8**, 450.
Shiraiwa, Y. and Miyachi, S. (1985). *Plant Cell Physiol.* **26**, 109.
Spear, D.G., Barr, J.K., and Barr, C.E. (1969). *J. gen. Physiol.* **43**, 397.
Sze, H. (1984). *Physiol. Plant* **61**, 683.
Volokita, M., Zenvirth, D., Kaplan, A., and Reinhold, L. (1984). *Plant Physiol.* **76**, 599.
Wishnick, M. and Land, M.D. (1969). *J. biol. Chem.* **244**, 55.

8 Dark respiration in cyanobacteria

H.C.P. Matthijs

Biologisch Laboratorium, Vrije Universiteit Amsterdam, De Boelelaan 1087, 1081 HV Amsterdam, The Netherlands

H.J. Lubberding

Laboratorium voor Microbiologie, Faculteit Wiskunde en Natuurwetenschappen, Katholieke Universiteit Nijmegen, Toernooiveld, 6525 ED Nijmegen, The Netherlands

Introduction

This paper deals with the physiological importance of dark respiratory electron transfer. The interaction of electron transfer in the cytochrome *bf* (cyt *bf*) complex in the light and in the dark and its possible mutual regulation is discussed. Recent data on the functional localization of the respiratory chain in either the thylakoids or the cell membrane are also reviewed.

For an excellent encyclopedic treatise with a wealth of information on respiration in cyanobacteria the reader is also referred to two recent reviews on this subject (Peschek 1984, 1987).

Physiological importance of dark respiration

Cyanobacteria are generally considered to be the oldest oxygenic organisms. Their early photoautotrophic efforts created the appropriate atmosphere for oxygen-dependent life. Simultaneously, chemoheterotrophic organisms were supported by cyanobacteria through the storage of potential substrates in biomass. However, being exposed to alternating day and night, even the earliest cyanobacteria must have been equipped with an ATP-generating system for survival in the dark. Bacteria in general possess a wide variety of catabolic energy transducing systems, which do not necessarily depend on oxygen as the terminal electron acceptor, and such diversity, both in respiratory pathways and in substrate-level phosphorylation options, may in principle also be represented in cyanobacteria. Consequently, the generalized schemes presented in this review do not exclude other possibilities in individual strains that have yet to be studied. Nevertheless, this review focuses in general terms on oxygen-dependent dark respiration.

Given their pioneering role in an early world without oxygen it is most remarkable that oxygen deprivation in the dark causes an immediate decrease in the intracellular ATP concentration of the great majority of cyanobacteria (Biggins 1969; Ihlenfeldt and Gibson 1975; Lubberding and Schroten 1984). The general lack of spontaneous restoration of this ATP level is indicative of the absence of fermentation in cyanobacteria, although at least one exception to this 'rule' has been reported (Oren and Shilo 1979). Supply of a pulse of oxygen to non-fermenting cyanobacteria, incubated anaerobically in the dark, immediately gave rise to phosphorylation of the intracellular nucleotide pools (Nitschmann and Peschek 1986). On the other hand, the intracellular ATP concentration has been shown to remain low under continuing anaerobiosis, or if either uncouplers or respiratory electron transfer inhibitors were added before the oxygen pulse (Lubberding and Schroten 1984). These observations clearly demonstrate a potential for oxidative phosphorylation. Nevertheless, a transition from the light to the dark under aerobic conditions results in an immediate drop in the intracellular ATP concentration. This change on switching from a light to a dark mode may indicate a regulatory mechanism, and a number of these will be discussed in this review. After a short while (<5 min) the ATP concentration may either re-equilibrate at its value in the light (Pelroy and Bassham 1972; Ihlenfeldt and Gibson 1975) or it may remain decreased (Bornefeld and Simonis 1974; Kallas and Castenholz 1982; Lubberding and Schroten 1984). This apparent difference in behaviour may be related to the type of strain, to the growth conditions, or to the method of estimation (Lubberding and Schroten 1984), and at this point it should be noted that neither the intracellular ATP concentration as such nor *in vitro* estimation of enzymatic and respiratory activities offer reliable means for monitoring the ATP-generating capacity of cyanobacteria. Determination of activity merely reflects the potential availability of a given catabolic pathway; its actual use *in vivo* would be determined by environmental conditions.

The maintenance energy requirements of cyanobacteria are relatively low (Van Liere *et al.* 1979). However, for growth in the dark a greater demand for ATP is imposed and, to fulfil this, oxidative phosphorylation is apparently the universal choice in cyanobacteria (Biggins 1969; Pelroy and Bassham 1973; Nitschmann and Peschek 1982; Matthijs *et al.* 1984*a*). To sustain this dark aerobic respiration, accumulated reserves of intracellular carbohydrate are consumed preferentially. Externally supplied potential substrates such as glucose or fructose are not used by photoautotrophically growing cyanobacteria unless they are forced to do so in response to starvation or after inhibition of photosystem II activity by addition of 3-3, 4-dichlorophenyl-1, 1,-dimethylurea (DCMU) to a growing culture in the light (Rippka 1972; Hirano *et al.* 1980; Beauclerk and Smith 1978; Raboy and Padan 1978). About half of the cyanobacterial species examined exhibit this type of photoheterotropic growth, via induction in the cell membrane of a transport system for the externally added sugar (Rippka 1972). On the other hand, chemoheterotrophic growth in the dark at the expense of external sugars is very rare amongst cyanobacteria and, if possible at all, the growth rates are very slow (Rippka 1972; Smith 1982). However, in marked contrast to this it has been shown in continuous culture experiments that growth rates in the dark are equal to the rates in the light provided that the internal polyglucose store is used in oxygen-dependent respiration (Foy and Smith 1980; Post *et al.* 1986).

In summary, it may be stated that cyanobacteria have predominantly adapted to two modes of ATP synthesis; i.e. photophosphorylation in the light and oxidative phosphorylation in the dark. In the light, internal carbohydrate reserves are stored which can be used subsequently in the dark. In the nature of things, net carbon dioxide fixation and oxygen evolution in the light exceed the reverse processes in the dark, and the internal carbohydrate store therefore serves as an energy buffer. Although dark respiration plays only a secondary role in overall cyanobacterial energy metabolism, its function in dark oxidative ATP synthesis (and concomitant anabolic conversion of the carbohydrate store which allows for biosynthesis and cell growth at nighttime), as well as its role in the scavenging of oxygen as a prerequisite for nitrogen fixation in heterocysts (Houchins and Hind 1982; Ernst *et al.* 1984), are considered essential for the excellent survival strategies of cyanobacteria. Furthermore, the regulation, or intracellular localization, of respiratory activity may be of vital importance in order to allow such seemingly incompatible functions as oxygenic photosynthesis and nitrogen fixation (or hydrogen evolution) to occur even within a single prokaryotic cell (Gallon 1981; Hawkesford *et al.* 1982; Stal and Krumbein 1985; Mitsui *et al.* 1986).

Models for cyanobacterial energy metabolism

Figure 8.1 depicts a simplified model of the main pathways involved in the energy metabolism of cyanobacteria. Photosystem II activity yields oxygen and passes electrons to the cyt *bf* complex. This complex has been demonstrated to be common to photosynthetic and respiratory electron transfer in all of the cyanobacterial species that have been studied in this respect (Hirano *et al.* 1980; Krinner *et al.* 1982; Peschek 1983; Matthijs *et al.* 1984*b*; Sandmann and Malkin 1984).

The cyt *bf* (*bc*) complex may thus also receive electrons via dehydrogenation of NADPH or NADH, the intermediate electron carriers in the catabolic oxidation of glucose (upper left in Fig. 8.1). After passage through the cyt *bf* complex, the electrons may either proceed to photosystem I or, potentially, also to a terminal oxidase such as cyt aa_3 (cyt *c* oxidase). The former pathway is the likely choice in the light, the latter in the dark (Binder 1982; Matthijs *et al.* 1984*a*; Peschek 1984, 1987; Myers 1986).

Following passage through photosystem I the electrons are used for carbon

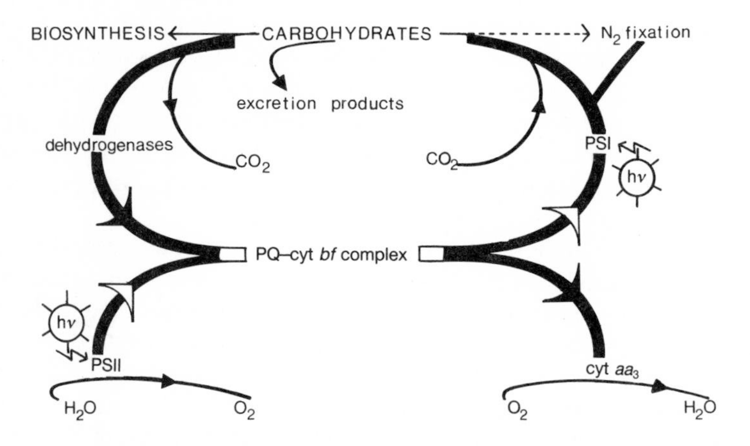

Figure 8.1 Generalized scheme of pathways of energy metabolism in cyanobacteria. PQ = plastoquinone; hv = light.

dioxide fixation rather than for the energetically very expensive process of nitrogen fixation, which occurs only if unavoidable (Zevenboom and Mur 1980) and then only in strains that can be induced for nitrogenase synthesis, in individual cases either in heterocysts (Wolk 1982) or in vegetative cells (Gallon 1981; Hawkesford *et al.* 1982). However, the energy store in carbohydrates may be more than a static deposit, given the reports of carbon dioxide evolution from this stock in the light in a number of cyanobacteria (Scherer and Böger 1982). Net cyanobacterial carbohydrate metabolism has been extensively studied in continuous culture employing different lengths of light and dark periods (Foy and Smith 1980; Post *et al.* 1985). The importance of metabolic control around the carbohydrate pool is illustrated by the fact that growth efficiency can be improved by imposition of a suitable light/dark cycle, the optimum light period being no longer than that allowed for by the (physical) storage capacity of the carbohydrate granules. Active growth during the subsequent dark period lasts as long as polyglucose is available and may be regulated by means of changes in the rate of carbohydrate conversion (Foy and Smith 1980; Post *et al.* 1985, 1986). The growth strategy of an organism is to maximize its doubling rate, and balancing of light and dark carbohydrate metabolism is the likely method by which cyanobacteria in steady state continuous culture cope with light and dark periods of different duration. Catabolic carbohydrate conversion in the dark usually proceeds via the oxidative pentose phosphate cycle (Smith 1982), and glucose-6-phosphate dehydrogenase (GPDH) has been pin-pointed as a possible control site because of its intrinsic regulatory potential; for example, its binding sites for NAD(P)H, ATP, ADP and ribulose bisphosphate (Lubberding and Bot 1984). Interestingly, the enzymes of the ferredoxin-thioredoxin system (FTS), as described for chloroplasts (Buchanan 1980; Cseke and Buchanan 1986), have also been detected in cyanobacteria (Anderson *et al.* 1982; Ip *et al.* 1984; Darling *et al.* 1986; Rowell *et al.* 1988 this volume). Thioredoxins are widely distributed in the animal, plant, and bacterial kingdoms, and these small proteins may exert their regulatory functions via reversible reduction and oxidation of -SH groups in target enzymes. For example, thioredoxins have been shown to change the activity of GPDH in cyanobacteria (Anderson *et al.* 1982). The enzyme is switched off in the light by reduced thioredoxins, these being reduced, in turn, via ferredoxin–thioredoxin reductase (formerly ferralterin) (Cseke and Buchanan 1986; Droux *et al.* 1987). The FTS has multiple functions in higher plant chloroplasts (Buchanan 1980; Cseke and Buchanan 1986), and, among these, its effects on coupling factor ATPase, which results in a switch-off mechanism in the dark for chloroplasts (Cseke and Buchanan 1986; Strotmann 1986), would be harmful in cyanobacterial oxidative phosphorylation. However, as far as we know, this pronounced regulatory effect of thioredoxin appears to be absent in cyanobacteria (Lubberding *et al.* 1981; Hicks and Yocum 1986*b*).

The specific activity of enzymes can also be changed by relatively increased (inducible) *de novo* synthesis, as has been demonstrated for GPDH and 6-phosphogluconate dehydrogenase (Lubberding and Bot 1984), though, in contrast to the soluble enzymes, not much is known about the regulation of the *in vivo* activity of membrane-bound respiratory enzymes. For example, a question of importance which, to our knowledge, has not yet been answered in detail is whether the activities of membrane-bound components of respiratory electron transfer are modulated through *de novo* synthesis. These components include phylloquinone, which has specifically been associated with respiration rather then plastoquinone (Peschek 1980, 1987), the dehydrogenases for NADH and NADPH, soluble *c*-type cytochromes, and cyt aa_3. Interestingly, the salt-induced increase in the rate of dark respiration (Paschinger 1977; Fry *et al.* 1986) has recently been linked with a substantial increase in the activity of cyt aa_3 (Molitor *et al.* 1986). However, it is doubtful whether *de novo* synthesis is involved with the similar changes in activity of cyt aa_3 that are associated with adaptations to other environmental conditions such as different light/dark cycles, photoheterotrophy, and chemoheterotrophy (Matthijs *et al.*, unpublished results with *Plectonema boryanum*).

In a number of studies the suggestion has been made that membrane-bound respiratory electron transfer in cyanobacteria proceeds on the thylakoid membranes (Hirano *et al.* 1980; Matthijs *et al.* 1984*a*, 1984*b*, 1985; Peschek 1984, 1987); the other possible placement would be the cell membrane. The role of the cell membrane in respiration will be discussed later. For reasons of presentation the respiratory electron transfer chain has been shown separately from that of photosynthesis in Fig. 8.2. In this diagram

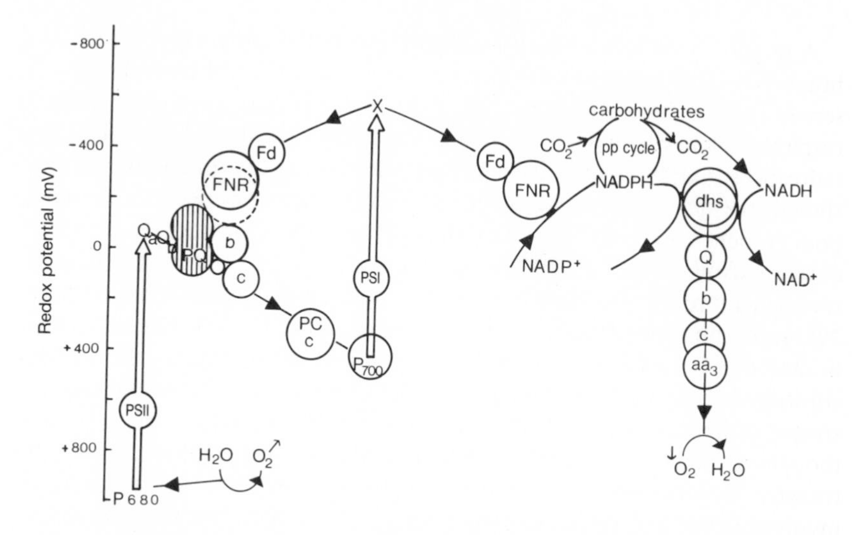

Figure 8.2 A schematic representation of components involved in photosynthetic electron transfer (left) and dark respiratory electron transfer (right) according to redox potential. Abbreviations: aa_3 = cyt aa_3; *b, c* = *b*- and *c*-type cytochromes; dh's = dehydrogenases; Fd = ferredoxin; FNR = ferredoxin: $NADP^+$ oxidoreductase; PC = plastocyanin; pp cycle = pentose phosphate cycle; PQ = plastoquinone; PSI, II = photosystems I, II. Q_a and Q_b = primary and secondary acceptors of PSII.

photosynthesis and respiration have been interconnected via the carbohydrate pool. The reductive pentose phosphate cycle in the light and the oxidative pentose phosphate cycle in the dark are important in carbohydrate metabolism in the cytoplasm. The Embden–Meyerhof pathway may also be used in carbohydrate breakdown in the dark; this process yields NADH instead of NADPH (Houchins and Hind 1982; Schrautemeier and Böhme 1984). An inspection of the standard redox potential of the components of the respiratory chain and their counterparts in photosynthesis, and also taking the central role of the cyt *bc (bf)* complex into account, envisages embedment of the respiratory chain in the thylakoid membranes. In consequence the dehydrogenases for NADH and NADPH have been associated with the thylakoids both physically and mechanistically (Peschek 1983, 1984, 1987; Binder 1982; Houchins and Hind 1982; Sandmann and Malkin 1983*a*, 1983*b*, 1984; Matthijs *et al.* 1984*b*; Schrautemeier and Böhme 1984; Viljoen *et al.* 1985) and this holds true for cyt aa_3 as well (Peschek 1981, 1984, 1987; Peschek *et al.* 1981; Binder 1982; Houchins and Hind 1982). This "all in one membrane" theory suggests two sites for mutual regulation of electron transfer in photosynthesis and respiration, namely before and after the cyt *bf (bc)* complex as depicted in Fig. 8.1.

A possible mechanism for a change in the functional activity of a membrane-bound component of respiration under metabolic control is now presented. Figure 8.2 depicts the organization of the components involved in respiratory and photosynthetic electron transfer according to their standard redox potential. In the Z-scheme of photosynthetic electron transfer, which is shown, the PQ (plastoquinone) pool (cross-hatched) is shown as a diffuse pool which could extend to the site of FNR (ferredoxin: $NADP^+$ oxidoreductase). Recent experiments with higher plant chloroplasts have revealed a binding protein for FNR (Coughlan *et al.* 1985; Ceccarelli *et al.* 1985) and, furthermore, that FNR can be resolved into a soluble and a membrane-bound fraction (Carrillo *et al.* 1980; Matthijs *et al.* 1986, 1987). Reconstitution into completely depleted thylakoid membranes has been demonstrated (Carrillo *et al.* 1980, Matthijs *et al.* 1986). FNR in its soluble form is thought to function in $NADP^+$ reduction in linear photosynthetic electron transfer, whereas FNR in its membrane-bound form (dashed line) may be involved in the transfer of electrons to the PQ pool or to cyt *b* during the process of cyclic electron transfer (Shahak *et al.* 1981; Carrillo and Vallejos 1983). The on/off (dashed/solid) position of FNR in higher plants has been related to light-induced changes in the energy state of the thykaloid membranes, energization favouring the on (= dashed) form. Preliminary experiments in isolated heterocysts of *Anabaena* 7120 have also indicated the existence of a bound pool of FNR on cyanobacterial thylakoid membranes (Matthijs *et al.*, unpublished observations). It is tempting to speculate that a similar on/off conformation of FNR also exists in vegetative cells of cyanobacteria. Apart from a function in mutual fine-tuning of linear and cyclic photosynthetic electron transfer in chloroplasts (Shahak *et al.* 1981) the on/off switching of FNR may then also be involved in the regulation of NADPH-dependent respiration via FNR. In consequence, we propose that the membrane-attached FNR may indeed constitute NADPH dehydrogenase (Gonzalez de la Vara and Gomèz-Lojero 1985). In the dark, membrane-bound FNR is slowly released from higher plant thylakoid membranes due to de-energization (Carrillo and Vallejos 1983). However, in cyanobacteria the cyt *bc*-complex is also used in dark respiration (Hirano *et al.* 1980; Peschek 1983, 1984, 1987; Matthijs *et al.* 1984*b*); thus FNR will not be released as long as respiration keeps the thylakoid membranes energized.

This concept of two functionally distinct pools of FNR in cyanobacteria may also explain the contradictory observations on either NADH or NADPH being the more potent *in vivo* electron donor. NADH dehydrogenase is a membrane-bound complex, in green algal chloroplasts (Godde 1982) and in cyanobacteria (Alpes *et al.* 1985; Sandmann and Malkin 1983*b*; Sandmann 1985; Viljoen *et al.* 1985). Hence, if FNR is released from the thylakoid membrane in its soluble form, NADH may appear to be the predominant electron donor in dark respiration (Houchins and Hind 1982;

Binder *et al.* 1984; Schrautemeier and Böhme 1984). On the other hand, with more of the FNR functionally bound to the membrane, NADPH will be a better electron donor (Biggins 1969; Matthijs *et al.* 1984*b*). In conclusion, further studies of the on/off switch mechanism of FNR in higher plant chloroplasts (Carrillo and Vallejos 1983) can be relevant for a better understanding of the regulation of cyanobacterial respiration, and the more so if a binding protein for FNR could be detected in these organisms.

Concerning regulation at the site subsequent to the cyt *bc* complex, inhibition of respiration in illuminated cyanobacteria at the level of oxygen uptake has been known for a long time (Brown and Webster 1953; Scherer and Böger 1982; Myers 1986). Electrons leaving the cyt *bc (bf)* complex (Figs. 8.1 and 8.2) reduce plastocyanin or a soluble *c*-type cytochrome (Sandmann and Böger 1980; Lockau 1981; Hawkesford *et al.* 1983*a*) and are transferred either to PSI in the light or to cyt aa_3 in the dark (Sandmann and Malkin 1984). These two acceptor systems therefore directly compete for electrons, as has been found in a photoacoustic study with *Anabaena* heterocysts. At low light intensities PSI is inhibited strongly due to electron drain via cyt aa_3 whilst at higher light intensities electron transfer via cyt aa_3 ceases (Matthijs and Carpentier unpublished results). Although competition between PSI and cyt aa_3 may be nothing but a matter of affinity for the electron donor at hand, one other feature of cyanobacteria should not be overlooked: not only the thylakoids but also the cell membrane can take part in respiration. Matters of cell architecture will be related to cyanobacterial energy metabolism in the following section.

Cell architecture and respiration

The all in one concept, i.e. photosynthetic and respiratory electron transfer both in the thylakoid membranes, is in many respects too simple. It does not match with a series of observations. For example, almost complete destruction of the thylakoid membranes in bleaching experiments completely suppressed photosynthesis but did not inhibit respiration by more than 40 per cent (Peschek and Schmetterer 1978). Different break-points in the Arrhenius plots of electron transfer activity vs. temperature further demonstrated that photosynthesis and respiration were likely, at least partially, to use different membrane areas (Scherer *et al.* 1981; Peschek *et al.* 1982*b*). Electron microscopy studies with reducible dyes showed dark electron transfer activity in the thylakoid membranes, but in some cases stain was also deposited in the cell membrane (Peschek *et al.* 1981; Peschek 1984). Careful examination of differently prepared cell-free membranes, in which the thylakoid and cell membranes were either heterogeneously mixed in vesicles or separated in distinct vesicles, also suggested that respiratory electron transfer did not totally coincide with photosynthetic electron transfer (Matthijs *et al.* 1984*b*). Similar conclusions were arrived at in experiments with intact spheroplasts (Peschek *et al.* 1982*a*). Furthermore, the interior of cyanobacteria clearly has to communicate with the outside in, for example, uptake of inorganic nutrients. To this end the cell membrane constitutes a selective barrier which needs to be energized either by ATPase enzymes alone or in combination with respiratory electron transfer (Lockau and Pfeffer 1982; Hawkesford *et al.* 1983*b*; Matthijs *et al.* 1984*a*, 1985; Scherer *et al.* 1984; Peschek 1984, 1987; Peschek *et al.* 1986). Several options for the localization of the components of the respiratory chain(s) of cyanobacteria have been described in the literature (Binder 1982; Houchins and Hind 1982; Matthijs *et al.* 1984*a*; Peschek 1984, 1987), and Fig. 8.3 depicts these. The options are as follows: (a) photosynthetic and respiratory electron transfer (PS and R, respectively) may be located in the same area of the thylakoid membranes; (b) as in (a) but located in different areas; (c) respiration is

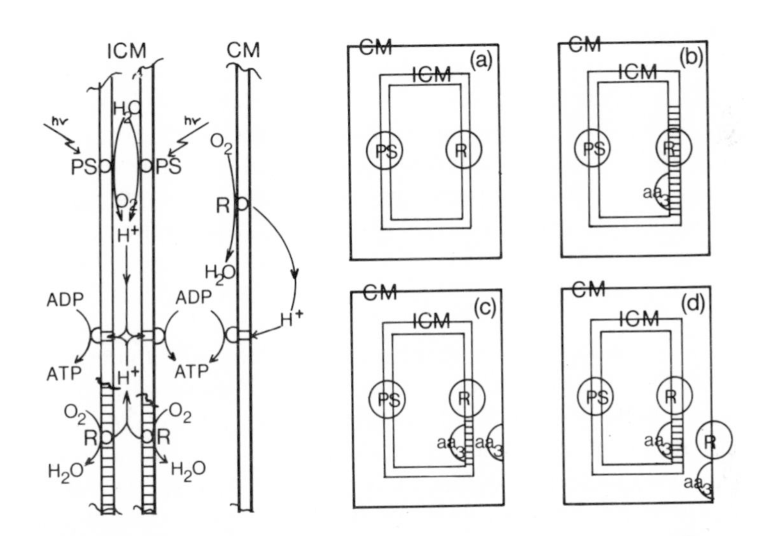

Figure 8.3 Energy transducing membranes in cynobacteria. Two types of membrane, with their potential energy generating options, are depicted on the left; these are the ICM (intracytoplasmic or thylakoid membranes) and the CM (cytoplasmic or cell membrane). R represents respiratory electron transfer, and PS photosynthetic electron transfer. The boxes on the right depict the various options for the localization of R and PS. (a) R and PS are both in the ICM and are mixed up (unlikely); (b) R and PS are in the ICM but cyt aa_3 occupies a separate defined area (Binder 1982; Houchins and Hind 1982; Scherer *et al.* 1981, 1984); (c) PS and R, except for cyt aa_3, are located in the same area of the ICM, while cyt aa_3 is in a different area [as in (b)] or in the cell membrane (Matthijs *et al.* 1984*a*); (d) both the ICM and the CM bear a complete respiratory system including a Q_{bc}-type complex and cyt aa_3 (Peschek 1987).

initiated at the thylakoid membranes, but cyt aa_3 is spread over the two membranes or is present on the cell membrane only; and (d) the thylakoid and cell membranes each bear a complete respiratory chain. Simply because of the fact that chlorophyll is lacking in the cell membrane (Omata and Murata 1984*a*), the possibility that PS activity occurs in the cell membrane is excluded. The option presented in Fig. 8.3(a) is not very likely on the basis of the observed differences in Arrhenius activation energy (see above). On the other hand, the option depicted in Fig. 3(b) is widely supported and all but a few of the published experimental results presented thus far are consistent with this model. The options presented in Fig. 3(c) and (d) need some reservation in their acceptance in that the absence or presence of cyt aa_3 in the cell membrane is still widely disputed (Matthijs *et al.* 1984*a*, 1984*c*; Peschek 1987). The confusion is not only based on results obtained with different cyanobacteria, which might only illustrate the diversity between species, but more confusingly also on experiments with the same strain (Omata and Murata 1984*b*, 1985; Molitor and Peschek 1986; Trnka and Peschek 1986). However, growth conditions apparently play an important role in the activity and intracellular localization of the respiratory enzymes, and changes in respiratory activity in consequence of different light/dark cycles or salt stress have been described (Paschinger 1977; Foy and Smith 1980; Post *et al.* 1985; Fry *et al.* 1986). In addition, changes in growth conditions may also affect the architecture of the energy-transducing membranes. For example, some physical connection between the ICM and CM, and between the inner-thylakoid and periplasmic space, cannot be totally excluded (Jost, 1965). Nevertheless, R and PS activity are likely to reside in different membrane areas.

Recently, a breakthrough in research has been reached with the successful separation of the thylakoid and cell membranes (Murata *et al.* 1981; Molitor and Peschek 1986; Peschek 1987); furthermore, a number of *in vitro* detection methods have been developed for quantitative and qualitative assay of cyt aa_3 in cyanobacteria. Among these are estimation of cyt *c* oxidation activity (Kienzl and Peschek 1982), redox difference spectroscopy (Peschek 1981; Houchins and Hind 1984; Matthijs *et al.* 1984*c*), electron paramagnetic resonance spectroscopy (Matthijs *et al.* 1984*c*; Fry *et al.* 1985), and, most recently, gel electrophoresis and immuno-detection (Trnka and Peschek 1986). This wealth of techniques, in addition to methods for detection of other respiratory chain components, such as spectroscopy for the cyt *bc* complex, (Stewart and Bendall 1980) and detection of iron-sulphur centres for NADH dehydrogenase (Viljoen *et al.* 1985), has enabled detailed studies on the options depicted in Fig. 8.3 for the localization of R activity.

The left-hand side of Fig. 8.3 shows a representation of the intracellular thylakoid membranes (ICM) and the cell membrane (CM); the principal membrane-bound bioenergetic functions are also depicted. PS or R activity results in H^+ translocation from the cytoplasm into the inner thylakoid space (Matthijs *et al.* 1985). The resulting proton gradient can be used for ATP synthesis, either in photophosphorylation or in dark oxidative phosphorylation (Matthijs *et al.* 1981, 1984*a*; Frei *et al.* 1984). Although all coupling factor ATPases are quite similar, those of cyanobacteria are more closely related to those of chloroplasts than to those of other bacteria and mitochondria (Lubberding *et al.* 1981; Hicks *et al.* 1986;, Hicks and Yocum 1986*a*, 1986*b*). However, it remains to be proven whether, in cyanobacteria, the same coupling factor ATPase catalyses both photophosphorylation and oxidative phosphorylation. The CM has also been shown to have a H^+-translocating function; a special ATP hydrolase, which is distinct from the reversible coupling factor ATPase, has been identified (Lockau and Pfeffer 1982), with polarity such that protons are extruded from the cytoplasm to the periplasmic space, i.e. to the cell exterior (Hawkesford *et al.* 1983*b*;, Scherer *et al.* 1984; Nitschmann and Peschek 1985; Peschek *et al.* 1985). Consequently, proton movements caused by electron transfer at both the ICM and the CM are directed away from the cytoplasm, and ATP synthesis may occur in principle at either the ICM or CM, or on both together. For example, ATP synthesis at the CM has been elegantly shown in a Jagendorf-type proton pulse experiment (Muchl and Peschek 1983), whilst ATP synthesis, both in photophosphorylation and in oxidative phosphorylation, has been shown in cell-free preparations (Binder *et al.* 1984, Matthijs *et al.* 1981, 1984*a*). Although the efficiency of oxidative phosphorylation was rather low in these preparations, re-addition of a solubilized protein (very likely a soluble *c*-type cytochrome) gave improved P/O values and this effect has been explained in terms of the separation of respiratory activity between the ICM (bearing the dehydrogenases and the cyt *bf* complex) and the CM (with cyt aa_3)(Matthijs *et al.* 1984*a*). The involvement of the CM in energy transduction via cyt aa_3 and coupling factor ATPase has been demonstrated in numerous studies in Peschek's laboratory (Erber *et al.* 1986, Molitor *et al.* 1986, Nitschmann 1986; Nitschmann and Peschek 1985; Peschek 1987), but the actual occurence of cyt aa_3 in the CM remains subject to dispute (cf. legend to Fig. 8.3). Furthermore, to prove the option shown in Fig. 8.3(d) it is necessary to determine which *b*- and *c*-type cytochromes would take part in a respiratory chain located in a CM. Experiments performed to answer this question, such as reduced minus oxidized difference spectroscopy, demonstrated low contents of unidentified *c*- and *b*-type cytochromes in the CM which apparently were different from the easily detectable *b*- and *c*-type cytochromes in the ICM fraction (Omata and Murata 1984*b*; Matthijs *et al.* 1984*c*). However, a clear answer about the identity, *in vivo* activity, physiological importance, and relative occurrence of respiratory activity in the cell membrane awaits further studies with a greater number of cyanobacteria, preferentially grown under a set of defined conditions. To relate the different options in Fig.8.3 to the physiological needs of cyanobacteria opens an interesting field for future research.

Note added in proof

Although dark fermentation has been amply mentioned in the present paper, recent studies have indicated that different types of fermentative pathways can be found in cyanobacteria, for example in *Oscillatoria* and *Gloeothece* species (J.R. van der Oost and L.J. Stal personal communication.)

Acknowledgement

The authors are indebted to T. Smit and J. Braakman-de Vriend for typing the manuscript, and to R. Kraayenhof and K.J. Hellingwerf for stimulating discussions.

References

Alpes, I., Schrautemeier, B., Scherer, S., and Böger, P. (1985). *FEMS Microbiol. Lett.* **26**, 147.

Anderson, L.E., Ashton, A.R., Mohamed, A.H., and Scheibe, R. (1982). *BioScience* **32**, 103.

Beauclerk, A.A.D. and Smith, A.J. (1978). *Eur. J. Biochem.* **82**, 187.

Biggins, J. (1969). *J. Bact.* **95**, 570.

Binder, A. (1982). *J. Bioenerg. Biomembr.* **14**, 272.

Binder, A., Hauser, R., and Krogmann, O. (1984). *Biochim. Biophys. Acta* **765**, 241.

Bornefeld, T. and Simonis, W. (1974). *Planta* **115**, 309.

Brown, A.H. and Webster, G.C. (1953). *Am. J. Bot.* **40**, 753.

Buchanan, B.B. (1980). *A. Rev. Plant Physiol.* **31**, 341.

Carrillo, N. and Vallejos. R.H. (1983). *Trends biochem. Sci.* **8**, 52.

Carrillo, N., Lucero. H.A., and Vallejos, R.H. (1980). *Plant Physiol.* **65**, 295.

Ceccarelli, E.A., Chan, R.L., and Vallejos, R.H. (1985). *FEBS Lett.* **190**, 165.

Coughlan, S.J., Matthijs, H.C.P., and Hind, G. (1985). *J. Biol. Chem.* **260**, 14891.

Cseke, C. and Buchanan, B.B. (1986). *Biochim. Biophys. Acta* **853**, 43.

Darling, A.J. Rowell, P., and Stewart, W.D.P. (1986). *Biochim. Biophys. Acta* **850**, 116.

Droux, M., Jacquot, J.P., Miginac-Maslow, M., Gadal, P., Huet, J.C., Crawford, N.A., Yee, B.C., and Buchanan, B.B. (1987). *Arch. Biochim. Biophys.* **252**, 426.

Erber, W.W.A., Nitschmann, W.H., Muchl, R., and Peschek, G.A. (1986). *Arch. Biochim. Biophys.* **247**, 28.

Ernst, A., Kirschenlohr, H., Diez, J., and Böger, P. (1984). *Arch. Microbiol.* **140**, 120.
Foy, R.H. and Smith, R.V. (1980). *Br. phycol. J.* **15**, 139.
Frei, R., Binder, A., and Bachofen, R. (1984). *Biochim. Biophys. Acta* **765**, 247.
Fry, I.V., Peschek, G.A., Huflejt, M., Peschek, G.A., and Packer, L. (1985). *Biochim. Biophys. Res. Commun.* **129**, 109.
Fry, I.V., Huflejt, M., Erber, W.W.A., Peschek, G.A., and Packer, L. (1986). *Arch. Biochim. Biophys.* **244**, 686.
Gallon, J.R. (1981). *Trends Biochem. Sci.* **6**, 19.
Godde, D. (1982). *Arch. Microbiol.* **131**, 147.
Gonzalez de la Vara, L. and Gomèz-Lojero, C. (1985). *Photosyn. Res.* **8**, 65.
Hawkesford, M.J., Houchins, J.P., and Hind, G (1983*a*). *FEBS Lett.* **159**, 262.
Hawkesford, M.J., Rowell, P., and Stewart, W.D.P. (1983*b*). In *Photosynthetic prokaryotes/Cell differentiation and function* (ed. G.C. Papageorgiou and L.Packer) p. 199. Elsevier, New York.
Hawkesford, M.J., Reed, R.H., Rowell, P., and Stewart, W.D.P. (1982). *Eur. J. Biochem.* **127**, 63.
Hicks, D.B. and Yocum, C.F. (1986*a*). *Arch. Biochem. Biophys.* **245**, 220.
Hicks, D.B. and Yocum, C.F. (1986*b*). *Arch. Biochem. Biophys.* **245**, 230.
Hicks. D.B., Nelson, N., and Yocum, C.F. (1986). *Biochim. Biophys. Acta* **851**, 217.
Hirano, M., Satoh, K., and Katoh, S. (1980). *Photosyn. Res.* **1**, 149.
Houchins, J.P. and Hind, G. (1982). *Biochim. Biophys. Acta* **682**, 86.
Houchins, J.P. and Hind, G. (1984). *Plant Physiol.* **76**, 456.
Ihlenfeldt, M.J.A. and Gibson, J. (1975). *Arch. Microbiol.* **102**, 13.
Ip, S.M., Rowell, P., Aitken, A., and Stewart, W.D.P. (1984). *Eur. J. Biochem.* **141**, 497.
Jost, M. (1965). *Arch Mikrobiol.* **50**, 211.
Kallas, T. and Castenholz, R.W. (1982). *J. Bact.* **149**, 229.
Kienzl, P.F. and Peschek. G.A. (1982). *Plant Physiol.* **69**, 580.
Krinner, M., Hauska, G., Hurt, E., and Lockau, W. (1982). *Biochim. Biophys. Acta* **681**, 110.
Lockau, W. (1981). *Arch. Microbiol.* **128**, 336.
Lockau, W. and Pfeffer, S. (1982). *Z. Naturf.* **370**, 658.
Lubberding, H.J. and Bot, P.V.M. (1984). *Arch. Microbiol.* **137**, 115.
Lubberding, H.J. and Schroten, W. (1984). *FEMS Microbiol. Lett.* **22**, 93.
Lubberding, H.J., Offerijns, F., Vel, W.A.C., and De Vries, P.J.R. (1981). In *Photosynthesis II,* (ed, G. Akoyunoglou) p. 779. Balaban, Philadelphia, PA.
Matthijs, H.C.P., Coughlan, S.J., and Hind, G. (1986). *J. Biol. Chem.* **261**, 12154.
Matthijs, H.C.P., Scholts, M.J.C., and Schreurs, H. (1981). In *Photosynthesis II* (ed. G. Akoyunoglou) p. 269, Balaban, Philadelphia, PA.
Matthijs, H.C.P., Van Steenbergen. J.M., and Kraayenhof, R. (1985). *Photosyn. Res.* **7**, 59.
Matthijs, H.C.P., Ludérus, E.M.E., Scholts, M.J.C., and Kraayenhof, R. (1984*a*). *Biochim. Biophys. Acta* **766**, 38.
Matthijs, H.C.P., Ludérus, E.M.E., Löffler, H.J., Scholts, M.J.C.,

and Kraayenhof, R. (1984*b*). *Biochim. Biophys. Acta* **766**, 29.
Matthijs, H.C.P., Moore, D., Coughlan, S.J., and Hind, G. (1987), *Photosyn. Res.* **12**, 273.
Matthijs, H.C.P., Van Hoek, A.N., Löffler, H.J.M., and Kraayenhof, R. (1984*c*). In *Advances in Photosynthesis Research*, Vol. II (ed. C. Sybesma), p. 643. Nijhoff/Junk, Den Haag.
Mitsui, A., Kumazawa, S., Takahashi, A., Ikemoto, H., Cao, S., and Arai, T. (1986). *Nature, Lond.* **323**, 720
Molitor, V. and Peschek, G.A. (1986). *FEBS Lett.* **195**, 145.
Molitor, V., Erber, W., and Peschek, G.A. (1986). *FEBS Lett.* **204**, 251.
Muchl, R. and Peschek, G.A. (1983). *FEBS Lett.* **164**, 116.
Murata, N., Sato, N., Omata, T., and Kuwabara, T. (1981). *Plant Cell Physiol.* **22**, 855.
Myers, J. (1986). *Photosyn. Res.* **9**, 135.
Nitschmann, W.H. and Peschek, G.A. (1982). *FEBS Lett.* **139**, 77.
Nitschmann, W.H. and Peschek, G.A. (1985). *Arch. Microbiol.* **141**, 330.
Nitschmann, W.H. and Peschek, G.A. (1986). *J. Bact.* **168**, 1205.
Omata, T. and Murata, N. (1984*a*). *Arch. Microbiol.* **139**, 113.
Omata, T. and Murata, N. (1984*b*). *Biochim. Biophys. Acta* **766**, 395.
Omata, T. and Murata, N. (1985). *Biochim. Biophys. Acta* **810**, 354.
Oren, A. and Shilo, M. (1979). *Arch. Microbiol.* **122**, 77.
Paschinger, H. (1977). *Arch. Microbiol.* **113**, 285.
Pelroy, R.A. and Bassham, J.A. (1972). *Arch. Microbiol.* **86**, 25.
Pelroy, R.A. and Bassham, J.A. (1973). *J. Bact.* **115**, 937.
Peschek, G.A. (1980). *Biochem. J.* **186**, 515.
Peschek, G.A. (1981). *Biochim. Biophys. Acta* **635**, 470.
Peschek, G.A. (1983). *Biochem. J.* **210**, 269.
Peschek, G.A. (1984). In *Subcellular biochemistry*, Vol. 10 (ed. O.B. Roodyn) p. 85. Plenum Press, New York.
Peschek, G.A. (1987). In *The cyanobacteria/a comprehensive review* (ed. P. Fay and C. Van Baalen) p. 119. Elsevier, Amsterdam.
Peschek, G.A. and Schmetterer, G. (1978). *FEMS Microbiol. Lett.* **3**, 295.
Peschek, G.A., Schmetterer, G., and Wagesreiter, H. (1982*a*). *Arch. Microbiol.* **133**, 222.
Peschek, G.A., Czerny, T., Schmetterer, G., and Nitschmann, W.H. (1985). *Plant Physiol.* **79**, 228.
Peschek, G.A., Hinterstoisser, G., Riedler, M., and Muchl, R. (1986). *Arch. Biochim. Biophys.* **247**, 40.
Peschek, G.A., Muchl, R., Kienzl, P.F., and Schmetterer, G. (1982*b*). *Biochim. Biophys. Acta* **697**, 35.
Peschek, G.A., Schmetterer, G., Lockau, W., and Sleytr, U.B. (1981). In *Photosynthesis II*, (ed. G. Akoyunoglou), p. 707. Balaban, Philadelphia, Penn.
Post, A.F., Loogman, J.G., and Mur, L.R. (1985). *FEMS Microbiol. Ecol.* **31**, 97.
Post, A.F., Loogman, J.G., and Mur, L.R. (1986). *J. gen. Microbiol.* **132**, 2129.
Raboy, B. and Padan, E. (1978). *J. biol. Chem.* **253**, 3287.
Rippka, R. (1972). *Arch. Microbiol.* **87**, 93.
Rowell, P., Darling, A.J., Amla, D.V., and Stewart, W.D.P. (1988). In

Biochemistry of the algae and cyanobacteria (ed. L. J. Rogers and J. R. Gallon) p. 201. Clarendon Press, Oxford.
Sandmann, G. (1985). *Photosyn. Res.* **6**, 261.
Sandmann, G. and Böger, P. (1980). *Plant Sci. Lett.* **17**, 417.
Sandmann, G. and Malkin, R. (1983*a*). *Biochim. Biophys. Acta* **725**, 221.
Sandmann, G. and Malkin, R. (1983*b*). *Arch. Microbiol.* **136**, 49.
Sandmann, G. and Malkin, R. (1984). *Arch. Biochim. Biophys.* **234**, 105.
Scherer, S. and Böger, P. (1982). *Arch. Microbiol.* **132**, 329.
Scherer, S., Stürzl, E. and Böger, P. (1981). *Z. Naturf.* **360**, 1036.
Scherer, S., Stürzl, E., and Böger, P. (1984). *J. Bact.* **158**, 609.
Schrautemeier, B. and Böhme, H. (1984). *FEMS Microbiol. Lett.* **25**, 215.
Shahak, Y., Crowther, D., and Hind, G. (1981). *Biochim. Biophys. Acta* **636**, 234.
Smith, A. J. (1982). In *The biology of cyanobacteria* (ed. N.G. Carr and B.A. Whitton) p. 47. Blackwell Scientific Publications, Oxford.
Stal, L. J. and Krumbein, W.E. (1985). *Arch. Microbiol.* **143**, 67.
Stewart, A.L. and Bendall, D.S. (1980). *Biochem. J.* **188**, 351.
Strotmann, H. (1986). In *Encyclopedia of plant physiology, new series*, Vol. 19 (ed. L.A. Staehelin and C.J. Arntzen) p. 584. Springer-Verlag, Berlin.
Trnka, M. and Peschek, G.A. (1986). *Biochim. biophys. Res. Commun.* **136**, 235.
Van Liere, L., Mur, L.R., Gibson, C.E., and Herdman, M. (1979). *Arch. Microbiol.* **123**, 315.
Viljoen, C.C., Cloete, F., and Scott, W.E. (1985). *Biochim. Biophys. Acta* **827**, 247.
Wolk, C.P. (1982). In *The biology of cyanobacteria* (ed. N.G. Carr and B.A. Whitton) p. 359. Blackwell Scientific Publications, Oxford.
Zevenboom, W. and Mur, L.R. (1980). In *Hypertrophic ecosystems* (ed. J. Barica and L.R. Mur) p. 123. Junk, The Hague.

9 Nitrogen fixation

J.R. GALLON and A.E. CHAPLIN

Department of Biochemistry, School of Biological Sciences, University College of Swansea, Singleton Park, Swansea SA2 8PP, UK

Introduction

The ability to fix N_2 (diazotrophy) is confined to prokaryotes. Although over the years there have been reports that the process may occur in certain eukaryotes, including green algae (e.g. Wann 1921), these have not been supported by subsequent investigations (e.g. Bristol and Page 1923). The establishment of a diazotrophic association between the green alga *Chlamydomonas* and the N_2-fixing bacterium *Azotobacter* (Gyurjan *et al.* 1986) illustrates one problem in demonstrating categorically whether or not an alga might fix N_2, and a recent, but unconfirmed, report of microaerobic N_2 fixation by a thermophilic green alga (Yamada and Sakaguchi 1980) may be a consequence of symbiotic association with a diazotroph, though no such organism was detected. In this connection, an N_2-fixing symbiosis involving a marine diatom, *Rhizosolenia*, and a diazotrophic cyanobacterium, *Calothrix*, is known (Mague *et al.* 1974), as are several loose associations of N_2-fixing cyanobacteria with various marine algae (see Stewart *et al.* 1980).

The only prokaryotes that possess a higher plant-type photosynthesis, and could therefore be categorized as algae, are the cyanobacteria (formerly blue-green algae) and the prochlorophytes. To date, the only report of N_2 fixation by a prochlorophyte is that of Paerl (1984), who found nitrogenase activity in an association involving an ascidian (sea-squirt) and the prochlorophyte *Prochloron*. However, because *Prochloron* has not yet been obtained in free-living culture, no detailed information is available. Recently, a free-living prochlorophyte, *Prochlorothrix hollandica*, has been isolated from a shallow lake in the Netherlands. This organism can be grown easily under laboratory conditions but, despite considerable effort to demonstrate nitrogenase activity, it does not appear to fix N_2 (Burger-Wiersma *et al.* 1986; Burger-Wiersma, personal communication).

In this review, therefore, attention will be confined to the cyanobacteria as the only O_2-evolving phototrophs unequivocally demonstrated to have N_2-fixing representatives.

Studies on N_2 fixation by cyanobacteria go back almost 100 years, to the late 1880s, a time when diazotrophy was receiving considerable attention. Indeed, the suggestion that cyanobacteria might fix N_2 (Frank 1889; Prantl

1889) came at about the same time that Hellriegel and Wilforth (1888) and Beijerinck (1888) demonstrated the involvement of root nodule bacteria in N_2 fixation by legumes. Cyanobacteria were, therefore, among the earliest organisms suspected to fix N_2. The early work on cyanobacterial N_2 fixation was performed with impure cultures and with methods considerably more crude than those available today, and, although the observations of these workers have subsequently been verified, there was, at the time, considerable disagreement about whether or not cyanobacteria could fix N_2. Indeed, conclusive evidence in support of the early work was not available until the studies of Drewes (1928) and Fogg (1942). In an article such as this, it is unfortunately not possible to detail the gradual development of our understanding of N_2 fixation from these early studies, but, in Table 9.1, a summary of this work is presented as a series of milestones in our progress from 1889 to the present day. It should be emphasized that these milestones are entirely subjective and there is no intention to belittle the contributions of those many researchers whose names are omitted. Furthermore, some areas, such as studies on the supply of reductant and ATP for cyanobacterial N_2 fixation and the mechanisms whereby cyanobacteria avoid, or limit, contact between nitrogenase and O_2, do not allow themselves to be measured in terms of specific milestones. Here evidence has been, and is, accumulating as a result of a continuous research effort involving many groups, but it is not possible to assign precise dates and specific observations to these areas.

The review by Stewart (1974*a*) gives an excellent historical account of progress in cyanobacterial N_2 fixation up to that date. However, generally reviews take as their starting point a previous review, with the consequence that it can be very difficult for a reader to obtain, from a single source, a comprehensive background to the field. As a compromise, therefore, Table 9.2 shows a list of reviews concerning diazotrophic cyanobacteria from the earliest days up to this present article. The list is not exhaustive, but it is hoped that it will make it easier for the interested reader to obtain a thorough background to the current state of research into cyanobacterial N_2 fixation.

In this review it is intended to delineate areas of current interest and to summarize the present state of knowledge concerning cyanobacterial N_2 fixation. Where a particular area has been reviewed recently, it will receive only a brief mention in this paper. In addition, Scherer *et al.* (1988), Matthijs and Lubberding (1988), Smith (1988*b*). Rowell *et al.* (1988), and Kerby and Stewart (1988), all in this volume, cover topics closely related to N_2 fixation.

Cyanobacteria are not unique in fixing N_2. Diazotrophy is found among many diverse groups of bacteria, both symbiotic and free-living. However, two factors distinguish cyanobacteria from most other diazotrophs. First, cyanobacteria are capable of oxygenic (plant-type) photosynthesis. Whilst this, in theory, allows the ATP and reductant needed for N_2 fixation to be generated directly by photosynthesis, it creates a problem in that

Table 9.1 Some milestones in research into cyanobacterial N_2 fixation.

Year	Author(s)	Observation
1889	Frank; Prantl	First suggestions of N_2 fixation by free-living cyanobacteria
1895	Ward	First suggestion of N_2 fixation in a cyanobacterial symbiosis
1928	Drewes	First demonstration of N_2 fixation by bacteria-free cultures of cyanobacteria
1942	Fogg	Use of purified gases to demonstrate unequivocally that certain cyanobacteria fix N_2
1942, 1943	Burris *et al.*	Incorporation of $^{15}N_2$ by cyanobacteria
1951	Fogg	First conclusive demonstration of N_2 fixation by a cyanobacterium not of the order Nostocales
1955	Bond and Scott	First unequivocal demonstration of N_2 fixation in a symbiotic system involving cyanobacteria
1960	Schneider *et al.*	N_2 fixation in cell-free systems
1961	Dugdale *et al.*	First unequivocal demonstration of N_2 fixation in a non-heterocystous cyanobacterium
1967, 1968	Stewart *et al.*	Application of acetylene reduction technique to cyanobacteria
1968	Fay *et al.*	The heterocyst as the site of N_2 fixation
1969	Wyatt and Silvey	First conclusive demonstration of aerobic N_2 fixation in laboratory cultures of a non-heterocystous cyanobacterium
1970	Stewart and Lex	Microaerobic N_2 fixation by a non-heterocystous cyanobacterium
1971a	Wolk and Wojciuch	Isolation of heterocysts with significant nitrogenase activity
1975	Stewart and Singh	*Nif* gene transfer in a cyanobacterium
1976	Wolk *et al.*	Incorporation of $^{13}N_2$ by cyanobacteria
1979	Hallenbeck *et al.*	Purification of cyanobacterial nitrogenase
1980	Mazur *et al.*	Identification of *nif* genes in cyanobacteria
1981	Mullineaux *et al.*	Temporal separation of photosynthesis and N_2 fixation in a non-heterocystous cyanobacterium
1985	Golden *et al.*	Rearrangement of *nif* genes during heterocyst differentiation
1985	Wolk *et al.*	Gene transfer between *E. coli* and cyanobacteria

Table 9.2 A selection of reviews relevant to cyanobacterial N_2 fixation.

Author(s)	Date	Subject of review
Fogg and Wolfe	1954	Nitrogen metabolism
Allen	1956	Photosynthetic N_2 fixation
Fogg	1956	Comparative physiology and biochemistry
Fogg	1962	N_2 fixation
Fogg and Stewart	1965	N_2 fixation
Holm-Hansen	1968	Ecology, physiology, and biochemistry
Stewart	1970	N_2 fixation
Stewart	1971	Physiology of N_2 fixation
Allen	1972	N_2 fixation
Fay	1973	Heterocysts
Stewart	1973a	N_2 fixation
Stewart	1973*b*	N_2 fixation by phototrophs
Fogg	1974	N_2 fixation
Millbank	1974	N_2 fixation in symbiotic systems
Stewart	1974*a*	N_2 fixation
Stewart	1974*b*	Nitrogenase
Stanier	1974	Relations between photosynthesis and N_2 fixation
Stewart and Tel-Or	1975	N_2 fixation in heterocysts
Stewart	1976	Nitrogenase
Stewart	1977*a*	N_2 fixation: general review
Stewart	1977*b*	General biology, including N_2 fixation
Stewart *et al.*	1977	Physiology and ecology
Haselkorn	1978	Heterocysts
Stewart	1978	Symbiotic N_2 fixation
Gallon	1980	N_2 fixation by phototrophs
Haselkorn *et al.*	1980	Heterocyst differentiation and N_2 fixation
Stewart	1980	Symbiotic N_2 fixation
Stewart *et al.*	1980	Methods for studying N_2 fixation
Trehan and Sinha	1980	Genetics of N_2 fixation
Wolk	1981	N_2 fixation by heterocysts
Bothe	1982	N_2 fixation: general review
Stewart *et al.*	1982	N_2 fixation and its regulation
Kallas *et al.*	1983	Aerobic N_2 fixation in non-heterocystous species
Stewart *et al.*	1983	Symbioses involving cyanobacteria
Bothe *et al.*	1984	N_2 fixation: general review
Houchins	1984	Nitrogenase in context of hydrogen metabolism
Thomas and Apte	1984	Na^+ and N_2 fixation
Houchins	1985	Electron transfer to nitrogenase
Venkataraman	1985	Genetics of N_2 fixation
Haselkorn	1986	Organization of the *nif* genes
Haselkorn *et al.*	1986	Organization of the *nif* genes
Stewart and Rowell	1986	N_2 fixation by phototrophs
Gallon and Chaplin	1988	N_2 fixation in non-heterocystous cyanobacteria

cyanobacterial N_2 fixation, an O_2-sensitive process, occurs alongside photosynthetic O_2 evolution. Second, in many cyanobacteria, the ability to fix N_2 is accompanied by differentiation of vegetative cells into heterocysts. Although cellular differentiation is observed in other diazotrophs, for example in the bacteroids of legume root nodules, and in the vesicles of *Frankia* (the symbiont of the non-legumes), cyanobacterial differentiation occurs in free-living culture under well-defined conditions.

Not all cyanobacteria are diazotrophs. Originally, N_2 fixation was thought to be confined to the Nostocales (equivalent to section IV in the classification of Rippka *et al.* 1979), and later to heterocystous cyanobacteria only (Table 9.1). However, it is now known that many genera of non-heterocystous cyanobacteria contain strains that can fix N_2, though only a few of these do so under aerobic conditions. Lists of cyanobacterial genera that have diazotrophic representatives are available (see Rippka *et al.* 1979; Stewart and Rowell 1986; Gallon and Chaplin 1988).

N_2 fixation by cyanobacteria represents only one area within the general field of N_2 fixation, and many of our ideas on the process in cyanobacteria have been influenced by findings with other diazotrophs. Some discussion of recent developments with other organisms is therefore included in this review. However, for a more detailed resumé of the field of N_2 fixation in general, see the recent books by Postgate (1987) and Gallon and Chaplin (1987).

Nitrogenases

There are two enzyme systems that catalyze N_2 fixation. The conventional, molybdenum-based system is well documented in a variety of diazotrophs, including cyanobacteria, but the 'alternative' vanadium-based system has been known only since 1980 and has been studied only in *Azotobacter* spp. (Bishop 1986; Hales *et al.* 1986; Robson *et al.* 1986; Smith *et al.* 1987).

In *Azotobacter chroococcum* and *Azotobacter vinelandii*, vanadium-based N_2 fixation is expressed only under conditions of molybdenum deficiency. In the presence of even 0.1 μM molybdenum, synthesis of the vanadium system is repressed and all measurable N_2 fixation can be attributed to the molybdenum system. Table 9.3 shows a comparison of the molybdenum system with the vanadium system from *A. chroococcum* (Robson *et al.* 1986; Smith *et al.* 1987). The properties of the vanadium enzyme from *A. vinelandii* are similar but, perhaps, not identical (Hales *et al.* 1986). Although there are many similarities between the two N_2-fixing systems, compared with molybdenum nitrogenase, vanadium nitrogenase is less efficient at carrying out N_2 fixation relative to H_2 evolution and is also less efficient at reducing acetylene. Consequently, use of the conventional acetylene reduction assay for N_2 fixation can underestimate rates of N_2

Table 9.3 A comparison of molybdenum nitrogenase and vanadium nitrogenase.

Property	Mo-nitrogenase	V-nitrogenase
	Fe protein (*Protein 2*)	*Fe protein* (*Protein 2**)
M_r	57 000–72 000	63 000
Subunit structure	α_2	α_2
Gene	*nif H*	*nif H**
Metal content	4Fe:4S	4Fe:4S
$t_{\frac{1}{2}}$ in air	45 sec	–
	MoFe protein (*Protein 1*)	*VFe-protein* (*Protein 1**)
M_r	200 000–240 000	210 000
Subunit structure	$\alpha_2\beta_2$	$\alpha_2\beta_2$ (γ?)
Subunit M_r and gene	α = 50 000 (*nif D*)	α = 50 000 (not identified yet)
	β = 60 000 (*nif K*)	β = 55 000 (*nif K**)
		γ = 6 000 (ORF adjacent to *nif H**)
Metal content	2Mo:24–32Fe:24–30S	2V:20Fe:20S
$t_{\frac{1}{2}}$ in air	8 min	40 sec
Reaction	$N_2 + 8H^+ + 8e^- \rightarrow 2NH_3 + H_2$	$N_2 + 12H^+ + 12e^- \rightarrow 2NH_3 + 3H_2$
Relative activity as substrate (N_2:C_2H_2:$2H^+$)	1:3.8:0.9	1:0.6:17

ORF = open reading frame, a long stretch of DNA triplet codons uninterrupted by a translational stop codon, i.e. a putative gene. The data for V-nitrogenase are those for the enzyme from *A. chroococcum* (Smith *et al.* 1987); those for the Mo-nitrogenase are from a variety of organisms.

fixation by the vanadium enzyme. Vanadium nitrogenase is also more sensitive to inactivation by O_2 than is the molybdenum enzyme.

Bortels (1930) demonstrated that molybdenum was essential for N_2 fixation in *A. vinelandii*, but six years later he reported that vanadium could substitute for molybdenum (Bortels 1936). Although a role for molybdenum was also proposed in cyanobacterial N_2 fixation (Bortels 1940), neither Allen and Arnon (1955) nor Holm-Hansen (1968) found any evidence that vanadium could replace molybdenum in the growth medium used to culture diazotrophic cyanobacteria. These latter findings were extended by Fay and de Vasconcelos (1974), who reported that vanadium inhibited N_2 fixation by *Anabaena cylindrica*.

In *Azotobacter* spp., molybdenum deficiency could be established by growth in molybdenum-free medium or by addition of tungstate, an inhibitor of molybdenum uptake and/or assimilation in these bacteria (Takahashi and Nason 1957; Keeler and Varner 1957). Strains capable of producing vanadium nitrogenase grew under both these conditions. Although some cyanobacteria were unable to grow in the presence of tungstate, and exhibited all the symptoms of molybdenum deficiency (Ramos and Madueno 1986), others not only tolerated tungstate but showed a tungsten requirement (Tyagi 1974; Singh *et al.* 1978). Furthermore, synthesis of nitrogenase proteins has been observed in certain cyanobacteria growing in tungsten-containing media (Wolk and Wojciuch 1971*b*; Nagatani and Haselkorn 1978; Hallenbeck and Benemann 1980), though these proteins were not necessarily active in N_2 fixation. Since a requirement for tungsten might be explained in terms of trace contamination of tungstate with vanadium (Robson *et al.* 1986), the possibility exists that certain cyanobacteria may, like *Azotobacter* spp., produce a vanadium nitrogenase under molybdenum-deficient conditions.

Inspired by the report of Bishop (1981) that tungsten-tolerant strains of *A. vinelandii* might contain a second nitrogenase system, we initiated a search for tungsten-tolerant N_2-fixing cyanobacteria. In nature, tungsten ores are commonly found in association with tin, so we obtained a variety of samples from the tin mining region of Cornwall, England. From a water channel draining from a tin mine, we isolated *Nostoc* sp. UCSB7 (Biochemistry Department Culture Collection, University College of Swansea), which would grow diazotrophically, and also reduce acetylene, in a defined medium (Gallon *et al.* 1978) supplemented with 1 mM Na_2WO_4 (Table 9.4). In contrast, a variety of other *Nostoc* spp., *Anabaena* spp., and also the unicellular *Gloeothece* sp. ATCC 27152 (American Type Culture

Table 9.4 Acetylene reduction and ammonia production by tungsten-tolerant and tungsten-sensitive strains of *Nostoc*.

Organism	Acetylene reduction (nmol min^{-1} mg protein^{-1})		Ammonia production (nmol min^{-1} mg protein^{-1})	
	$-$ Na_2WO_4	+ 1 mM Na_2WO_4	$-$ Na_2WO_4	+ 1 mM Na_2WO_4
Nostoc sp. UCSB7	1.92	2.19	2.6	2.9
Nostoc muscorum CCAP 1453/23	2.16	0	–	–
Nostoc sp. UCSB18	–	–	3.4	0

Cultures were grown for seven days in nitrogen-free medium (Tözüm and Gallon 1979) with or without added 1 mM Na_2WO_4. Acetylene reduction was measured as described by Tözüm *et al.* (1977), whilst ammonia production was determined by incubation for 6 hours in the presence of L-methionine-D, L-sulphoximine, an inhibitor of ammonia assimilation. Under these conditions, cultures released newly fixed ammonia which could then be measured in the culture medium using the method of Maryan and Vorley (1979).

Collection) would not grow in the presence of 1 mM tungstate. *Nostoc* sp. UCSB7 also excreted ammonia when incubated in tungstate-supplemented medium along with 2 mM L-methionine D, L-sulphoximine, an inhibitor of ammonia assimilation (Table 9.4). We therefore concluded that *Nostoc* sp. UCSB7 was capable of N_2 fixation in the presence of 1 mM tungstate. Incubation of *Nostoc* sp. UCSB7 (but not a tungsten-sensitive strain of *Nostoc*) with $Na_2{}^{185}WO_4$ or $^{55}FeCl_3$ resulted in incorporation of radioactivity into a protein that correspond to the ^{55}Fe-protein, assumed by Mullineaux *et al.* (1983) to be the MoFe-protein of nitrogenase (Fig. 9.1). It is not known whether this protein is a WFe-analogue of the MoFe-protein, nor is it known whether it is catalytically active. However, in the light of other studies (Benemann *et al.* 1973; Nagatani and Brill 1974) it is probably inactive in N_2 fixation.

Nostoc sp. UCSB7 would not grow in nitrogen-free medium from which molybdenum had been removed (Eady and Robson 1984) unless either molybdate or tungstate was added, and it was concluded that this cyanobacterium was merely a good scavenger of molybdenum, even in the presence of tungstate. The ability to fix N_2 when tungstate was added was, we assumed, due to

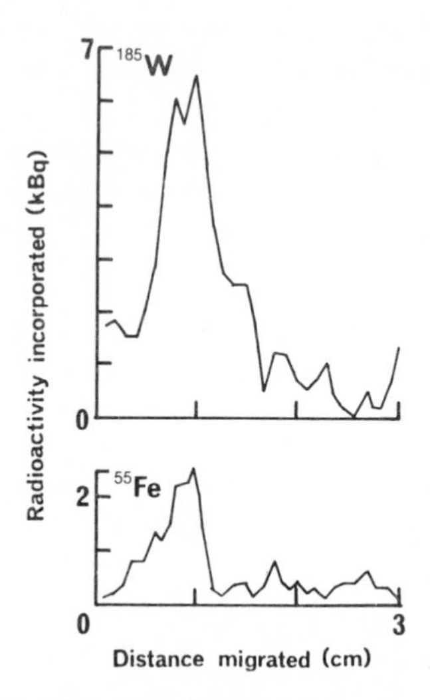

Fig. 9.1. Incorporation of ^{185}W and ^{55}Fe into proteins of *Nostoc* sp. UCSB7. Cultures were grown for seven days in medium (Tözüm and Gallon 1979) containing 1 mM Na_2WO_4, but were washed several times and suspended in tungsten-free medium prior to experimentation. After incubation for 2 hours with 37 kBq ml^{-1} (about 50 μg) of $Na_2{}^{185}WO_4$ or $^{55}FeCl_3$, the cells were collected and broken, and a portion of cell extract, containing about 0.2 mg protein, was subjected to PAGE under non-denaturing conditions as described by Mullineaux *et al.* (1983). Radioactivity was then measured in 1 mm slices of the gel.

contamination of tungstate with traces of molybdate. However, in the light of the observation that tungstate may be contaminated with vanadate (Robson *et al.* 1986), and also because tungsten-resistance frequently correlates with defective molybdenum incorporation, it remains possible that *Nostoc* sp. UCSB7 has a vanadium nitrogenase that is expressed under conditions of molybdenum deficiency. Unfortunately, the original isolate of this organism has been lost, so further work awaits its re-isolation.

In *A. chroococcum*, the vanadium nitrogenase system is encoded in genes separate from the conventional N_2 fixation genes. One of these separate genes, *nif H**, which encodes the Fe-protein of the vanadium system (Table 9.3), is very similar to, though not identical with, *nif H*, the gene that encodes the Fe-protein of the molybdenum system (Smith *et al.* 1987). Similar reiteration of *nif H* was shown by Rice *et al.* (1982) to occur in the cyanobacterium *Anabaena* PCC 7120 (Pasteur Institute Culture Collection). Adjacent to, and following, *nif H** in *A. chroococcum* is an open reading frame (Table 9.3) whose product is a small, ferredoxin-like protein that may interact with the VFe-protein (Smith *et al.* 1987). Perhaps significantly, there is also a small open reading frame adjacent to, though preceding, *nif H** in *Anabaena* 7120 (Haselkorn 1986). These findings would be consistent with an 'alternative' vanadium nitrogenase in this cyanobacterium, though, of course, they do not provide direct evidence for such a system.

Reductant and ATP

In heterocystous cyanobacteria, aerobic N_2 fixation is confined to heterocysts, though, over the years, there have been suggestions that under anaerobic or microaerobic conditions, and, in some species, even under aerobic conditions, nitrogenase activity may also be found in vegetative cells (see Gallon 1980). Recently, however, immuno-cytochemical evidence has shown that the nitrogenase proteins are present only in heterocysts, under both aerobic and microaerobic conditions (Murry *et al.* 1984; Bergman *et al.* 1986). It would be interesting to see whether, in the heterocyst-less mutants described by Rippka and Stanier (1978), nitrogenase is confined to certain cells, whose differentiation into heterocysts is blocked, or is evenly distributed among the vegetative cells as originally suggested by these authors. Despite lacking heterocysts, these mutants are capable of anaerobic N_2 fixation.

The location of nitrogenase in heterocysts of heterocystous cyanobacteria separates N_2 fixation from the photochemical generation of reductant and ATP from water. Heterocysts possess, at best, limited photosystem II activity and, since they also lack ribulose-1, 5-bisphosphate carboxylase, neither evolve O_2 nor fix CO_2 (Stewart *et al.* 1985). Nevertheless, N_2 fixation requires both ATP and a source of reducing power. Although heterocysts have the ability to generate ATP by photosystem I-mediated cyclic photophosphorylation, reductant for N_2 fixation must, ultimately, be dependent upon a supply of carbon compounds from the photosynthesizing vegetative cells.

On the other hand, in non-heterocystous cyanobacteria, photosynthesis can, in theory, directly supply both ATP and reductant, though Gallon (1980) was unable to demonstrate categorically that this was the case in the unicellular cyanobacterium *Gloeothece*. Subsequently, it has been demonstrated that, when grown under alternating 12 hour light and 12 hour darkness, cultures of this organism fixed N_2 almost exclusively in the dark (Mullineaux *et al.* 1981) and could not therefore immediately utilize photosynthetically generated ATP and reductant. This pattern of behaviour is found in many other non-heterocystous cyanobacteria (Pearson *et al.* 1981; Stal and Krumbein 1985*a*; Grobbelaar *et al.* 1986; Mitsui *et al.* 1986; Khamees *et al.* 1987) though not in all such organisms (Saino and Hattori 1978; Bryceson and Fay 1981). Furthermore, in the unicellular cyanobacterium *Gloeothece*, N_2 fixation under constant illumination is supported by respiration rather than photosynthesis, except perhaps at high light intensities (Maryan *et al* 1986*a*). It therefore appears that in most, if not all, diazotrophic cyanobacteria, the provision of reductant and ATP for N_2 fixation is only indirectly linked to photosynthesis, through oxidation of photosynthetically fixed carbon.

Stewart and Rowell (1986) provide a comprehensive account of how reductant and ATP might be provided for cyanobacterial N_2 fixation. Consequently, only a brief summary will be provided here.

Oxidation of fixed carbon can generate reductant as NADH and NADPH, through the activity of enzymes such as glucose-6-phosphate dehydrogenase, glyceraldehyde-3-phosphate dehydrogenase, isocitrate dehydrogenase, and malate dehydrogenase. NAD(P)H may, in turn, reduce ferredoxin (which is probably the immediate electron donor to nitrogenase in most cyanobacteria) directly by the action of ferredoxin: NADP oxidoreductase, or indirectly through the involvement of photosystem I of photosynthesis, or in a reaction involving the electrical component ($\Delta\Psi$) of an energized plasmalemma. Alternatively, or indeed simultaneously, reduced ferredoxin may be generated directly from oxidation of pyruvate by pyruvate: ferredoxin oxidoreductase.

Reductant for N_2 fixation may also be provided by exogenous H_2 or that produced by the nitrogenase reaction itself (Table 9.3). This involves an 'uptake' hydrogenase enzyme. Although H_2-supported N_2 fixation has clearly been demonstrated in the laboratory (Houchins 1984) the physiological role of H_2 as an electron donor to cyanobacterial N_2 fixation is less clear (Daday *et al.* 1985; Jensen *et al.* 1986). Despite detailed studies on the heterocystous cyanobacterium *A. cylindrica*, Smith (1988*a*) was unable to discover any advantage, in terms of growth or N_2 fixation, that might be conferred by possession of uptake hydrogenase. He therefore concluded that if such an advantage exists it must be confined to a restricted range of, as yet unidentified, growth conditions.

The ATP requirement for N_2 fixation may be met through respiratory oxidation of NADPH and NADH, perhaps supplemented under some conditions by cyclic photophosphorylation. The Na^+ requirement for N_2 fixation in certain cyanobacteria may be related to a role for this ion in ATP synthesis. Under Na^+-deficient conditions uptake and utilization of phosphate was decreased (Thomas and Apte 1984).

Regulation of N_2 fixation

Regulation of N_2 fixation may be exerted at the level of nitrogenase activity or nitrogenase synthesis and there has been extensive research into the factors that regulate activity and synthesis of the N_2-fixing enzyme (for example, see Eady 1981; Gallon and Chaplin 1987). However, before considering regulation in diazotrophic cyanobacteria it is worth digressing briefly to summarize what is currently known about regulation of nitrogenase synthesis in *Klebsiella pneumoniae*, the diazotroph about which our knowledge is most complete (Dixon 1984; Cannon *et al.* 1985; also see Haselkorn 1986).

In *K. pneumoniae* the N_2 fixation (*nif*) genes are organized into a regulon

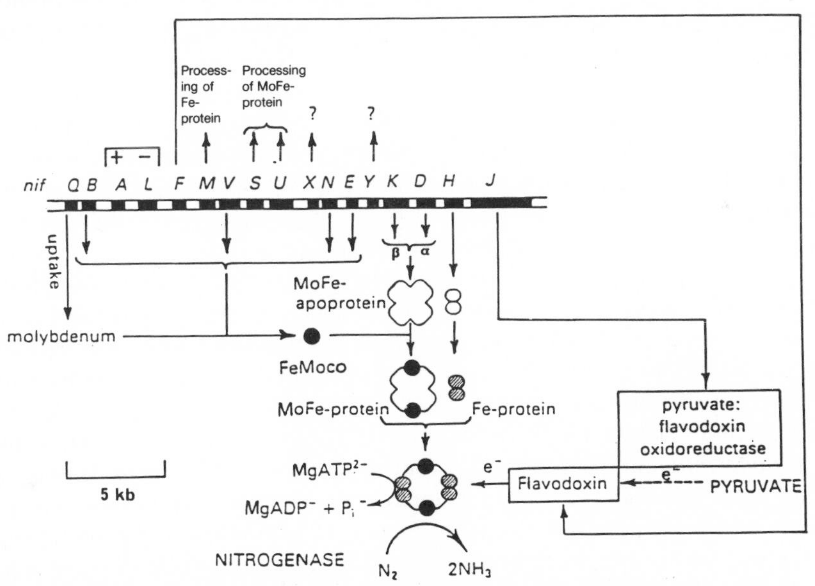

Fig. 9.2. The N_2 fixation (*nif*) regulon of *K. pneumoniae*, showing the roles of the gene products in N_2 fixation (based on Gallon and Chaplin 1987). FeMoco = iron-molybdenum cofactor.

of 17 genes, consisting of eight operons each of which is transcribed into a single, usually polycistronic, mRNA. Although only five of the gene products have been purified and properly characterized, functions have been assigned to all of the genes except for *nif X* and *nif Y* (Fig. 9.2). However, for the purposes of this review it is necessary to consider only the genes *nif H* (which codes for the Fe-protein of *Klebsiella* nitrogenase), *nif D* and *nif K* (which encode, respectively, the α and β subunits of the MoFe-protein) and *nif A* and *nif L* which perform a regulatory function (Fig. 9.3).

Regulation of *nif* gene expression in *K. pneumoniae* has two elements, an external system, designated *ntr*, and an internal system mediated by *nif A* and *nif L*. The *ntr* system responds to conditions of nitrogen starvation by activating genes that enable the organism to utilize 'unusual' nitrogen sources such as arginine, proline, and histidine, as well as N_2 itself, in the last case by switching on the *nif* genes. The interrelations between external and internal regulation of the *nif* genes in *K. pneumoniae*, and the conditions under which nitrogenase synthesis occurs, are summarized in Fig. 9.3.

As in other diazotrophs, regulation of cyanobacterial N_2 fixation involves regulation of both nitrogenase synthesis and nitrogenase activity. Nitrogenase activity may, for example, be regulated by availability of reductant and ATP, which in diazotrophic cyanobacteria are supplied by catabolism of carbon reserves rather than directly by photosynthesis. In many photosynthetic organisms catabolism of fixed carbon does not proceed simultaneously with photosynthetic CO_2 fixation, a consequence of the thioredoxin-mediated modulation of enzyme activities by light (see Stewart and Rowell 1986). Nevertheless, diazotrophic cyanobacteria, including those strains that under alternating light and darkness fix N_2 only in the dark, are capable under continuous illumination of simultaneously fixing CO_2 and N_2, the latter process apparently requiring oxidative breakdown of stored carbon reserves. How this may be achieved is discussed in detail by Stewart and Rowell (1986), so will not be dealt with here.

In general, non-heterocystous cyanobacteria fix N_2 during the dark phase of a cycle of alternating light and darkness, though the marine *Trichodesmium* spp. are a notable exception to this (Saino and Hattori 1978; Bryceson and Fay 1981). However, it is possible that spatial separation of nitrogenase from photosynthesis may occur in these latter organisms (see Gallon and Chaplin 1988), so, in terms of the organization of N_2 fixation, *Trichodesmium* may have more in common with heterocystous species than with other non-heterocystous cyanobacteria. Heterocystous cyanobacteria fix N_2 either exclusively in the light phase of an alternating cycle of light and darkness, or both in the light and dark (Khamees *et al.* 1987).

In the unicellular cyanobacterium *Synechococcus* Miami BG 43511, studied by Mitsui *et al.* (1986), and perhaps also in *Synechococcus* RF1 (Grobbelaar *et al.* 1986), the separation of N_2 fixation in the dark from

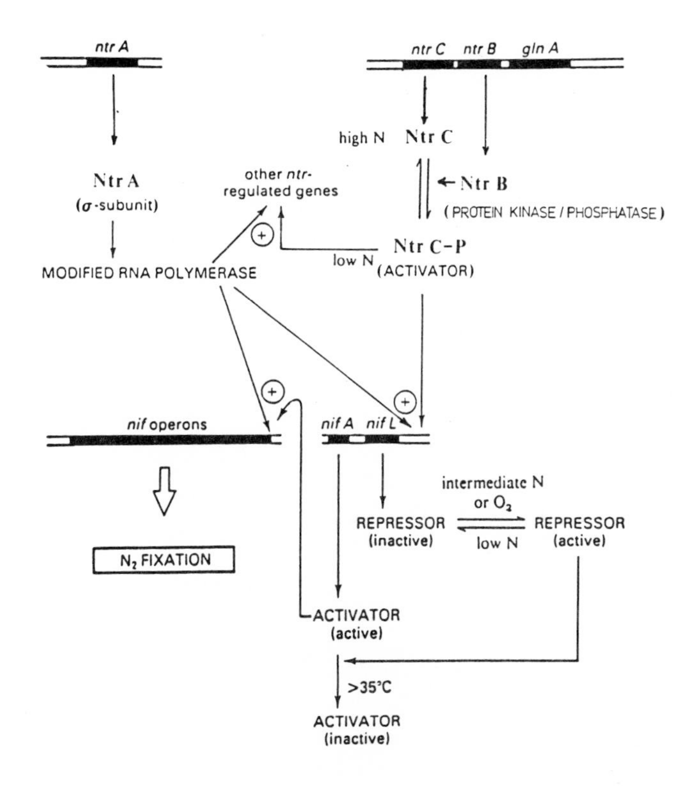

Fig. 9.3. Regulation of *nif* gene expression in *K. pneumoniae*. The *ntr A* gene product (NtrA) is a σ-factor of RNA polymerase which recognizes the promoters of the *nif* and other *ntr*-regulated genes. These promoters have a structure different from that of typical bacterial promoters. NtrA allows RNA polymerase to bind at the *nif* promoters and to initiate transcription there. The *ntr B* gene product (NtrB) is an enzyme that functions both as a protein kinase and as a phosphatase, the substrate of which is NtrC (the *ntrC* gene product). Whether kinase or phosphatase activity predominates depends upon the nitrogen status of the bacterium, and the consequence of this is that, under conditions of nitrogen starvation, NtrC exists in its phosphorylated form (NtrC-P). NtrC-P acts as an activator of, among other operons, *nif LA*. The *nif A* product is an activator of transcription of the other *nif* genes, whilst the *nif L* product, in the presence of either intermediate concentrations of fixed nitrogen (4 mM or greater) or of O_2 (0.1 μM or greater), inactivates the *nif A* product, thereby preventing transcription of the other *nif* genes. For further details see Dixon (1984), Cannon *et al.* (1985), and Haselkorn (1986).

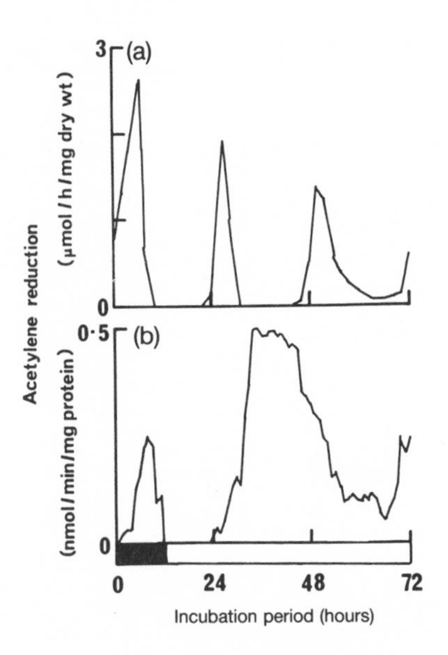

Fig. 9.4. The effect on acetylene reduction (N_2 fixation) of transfer of cultures of (a) *Synechococcus* sp. Miami BG 43511, and (b) *Gloeothece* sp. ATCC 27152 from alternating 12 hour light and 12 hour darkness to continuous illumination. The data in (a) are derived from Mitsui *et al.* (1986). Measurements in (b) were made every hour, but for clarity individual datum points are not marked.

photosynthetic activity in the light may be an endogenous property of the cell cycle, cell division occurring synchronously approximately every 24 hours. Cultures transferred from alternating light and dark to continuous illumination retained their diurnal fluctuations in N_2 fixation [Fig. 9.4(a)]. However, whilst cultures of *Gloeothece*, which have a doubling time of about 60 hours, also showed diurnal fluctuations of N_2 fixation [Fig. 9.5(a)] and photosynthesis under alternating 12 hour light and 12 hour dark, this behaviour was not retained following transfer to constant illumination [Fig. 9.4(b)]. In this cyanobacterium the fluctuations are not endogenous; rather they are imposed at the level of nitrogenase synthesis, probably by the availability of storage glucan.

About 2 hours before cultures enter the dark phase, nitrogenase synthesis commences [Fig. 9.5(b)]. In the dark this synthesis, as well as nitrogenase activity, is supported by ATP and reductant generated by breakdown of glucan that accumulated during the light period [Fig. 9.5(d)]. After 6 hours in the dark, the rate of glucan utilization markedly decreases, perhaps because of depletion of utilizable glucan reserves, though this cannot be stated for

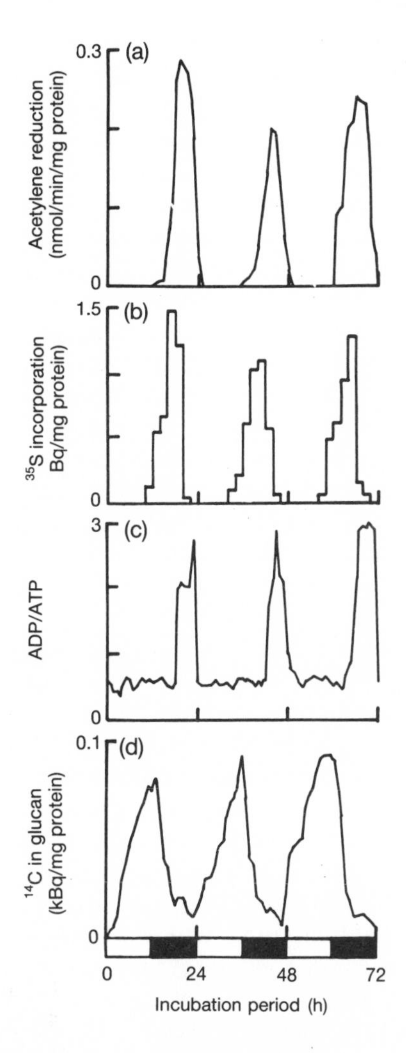

Fig. 9.5. The effect of incubation under alternating 12 hour light and 12 hour darkness on (a) acetylene reduction, (b) nitrogenase synthesis, (c) the intracellular ratio of ADP to ATP, and (d) radioactivity in intracellular glucan, in cultures of *Gloeothece* sp. ATCC 27152. Nitrogenase synthesis was measured as incorporation of radioactivity from $Na_2{}^{35}SO_4$ into the MoFe-protein of nitrogenase using the radioimmunoassay of Maryan *et al.* (1986*b*). The ratio of ADP/ATP was measured by bioluminescence (Spielmann *et al.* 1981), using luciferin and luciferase prepared as described by Kimmich *et al.* (1975). Incorporation of radioactivity from $Na_2{}^{14}CO_3$ into glucan was measured as described by Mullineaux *et al.* (1980).

certain without fractionating *Gloeothece* polysaccharides and measuring storage glucan directly. Coincident with the decrease in glucan utilization the intracellular ratio of ADP to ATP (the D/T ratio) markedly increases [Fig. 9.5(c)]. As the D/T ratio increases, nitrogenase synthesis ceases and activity disappears, mainly as a result of O_2 inactivation of previously synthesized nitrogenase. When cultures re-enter the light phase, the D/T ratio decreases, glucan reserves start to be replenished, and, after a 10 hours lag, nitrogenase synthesis recommences. It is the onset of illumination that, after 10 hours, triggers nitrogenase synthesis; the onset of darkness initiates a series of events (cessation of photosynthesis, glucan utilization, followed by an increase in the D/T ratio) that , a few hours later, interrupts nitrogenase synthesis.

Regulation of nitrogenase activity through the D/T ratio is well documented in other diazotrophs (for example, see Upchurch and Mortenson 1980) and stems from the observation that, whilst $MgATP^{2-}$ is needed for N_2 fixation, $MgADP^-$ is an inhibitor of the process. However, in *Gloeothece* the D/T ratio may also affect nitrogenase synthesis. Although the D/T ratio does, under certain conditions, affect the rate of N_2 fixation (Chaplin *et al.* 1984), nitrogenase activity can still be supported at a D/T ratio close to 3 [Fig. 9.5(a,c)].

In some diazotrophs, nitrogenase activity may be modulated by covalent modification (an up-to-date summary is included in Gallon and Chaplin 1987), but there is no evidence as yet for this in cyanobacteria. Although preparations of the iron-protein of *Gloeothece* nitrogenase often showed low activity (consistent with covalent modification), they could not be activated by incubation with the activating enzyme of *Azospirillum brasilense*, a component of the covalent modification system of that organism (P.S. Maryan, unpublished observation).

The two factors that have been most extensively studied for their regulatory effect on N_2 fixation are O_2 and fixed nitrogen. Nitrogenase is an extremely O_2-sensitive enzyme, even in cyanobacteria, and the fact that these organisms also photoevolve O_2 has attracted particular attention to the methods whereby N_2 fixation in these organisms is maintained in the presence of both photosynthetic and atmospheric O_2. The popularity of this area of research is reflected in the regularity with which this topic appears in reviews on cyanobacterial N_2 fixation (see, most recently, Stewart and Rowell 1986; Gallon and Chaplin 1988), so, for the main part, only a summary will be presented here.

Cyanobacteria employ a variety of methods in order to avoid, or to limit, contact between O_2 and nitrogenase. The relative significance of each individual strategy varies from organism to organism and, furthermore, most diazotrophic cyanobacteria rely upon a combination of methods to sustain an active nitrogenase. The strategies are listed below.

1. Behavioural methods, such as avoidance (as in the microaerobic diazotroph *Plectonema boryanum*) and the clumping of cells together to create a low O_2 tension in the centre of the clumps (e.g. *Trichodesmium*).

2. Spatial separation of N_2 fixation, in the heterocysts, from photosynthesis, in vegetative cells. Heterocysts do not photoevolve O_2 and the rate of diffusion of O_2 into these cells from the atmosphere is such that observed rates of respiration would maintain the dissolved O_2 concentration close to zero (Walsby 1985). The non-heterocystous marine cyanobacterium, *Trichodesmium*, may also effect a spatial separation of N_2 fixation and photosynthesis but, in this case, without any cellular differentiation.

3. Temporal separation of N_2 fixation, in the dark, from photosynthesis, in the light, when cultures are grown under alternating light and darkness. This strategy is found in most non-heterocystous cyanobacteria (Khamees *et al.* 1987).

4. Respiratory consumption of O_2 is observed in both heterocystous and non-heterocystous cyanobacteria. Respiration may be carbon-supported or, in some cyanobacteria, supported by electrons derived from the H_2 produced simultaneously with reduction of N_2 by nitrogenase. Some carbon-supported O_2 consumption may be dependent upon illumination (Smith *et al.* 1986), possibly explaining the increased O_2 sensitivity of N_2 fixation in the dark (Gallon and Chaplin 1988). However, when exposed to atmospheres containing elevated concentrations of O_2, only in the short term was increased respiratory activity effective in ameliorating the effects of O_2 on *Gloeothece* nitrogenase (Gallon and Chaplin 1988).

5. Both non-heterocystous (Tözüm and Gallon 1979) and heterocystous (Tel-Or *et al.* 1986) cyanobacteria contain a variety of enzymes that remove the toxic radicals generated by electron donation to O_2 (see Gallon 1980).

6. Conversion of active nitrogenase to an inactive but O_2-stable form during exposure to elevated concentrations of O_2. Reconversion to the active form occurs immediately following return to more favourable conditions. This mechanism is well documented in the diazotrophic aerobe *Azotobacter* and may also occur in the non-heterocystous, filamentous cyanobacterium *Oscillatoria* sp. strain 23 (Stal and Krumbein 1985*b*) and, perhaps, in heterocystous *Anabaena* spp. (Pienkos *et al.* 1983). In the latter organisms, the O_2-stable form may retain activity (Stewart and Rowell 1986), implying a mechanism different from that in *Azotobacter*.

7. Continuous synthesis of nitrogenase may replace any cyanobacterial enzyme inactivated irreversibly by O_2. In many diazotrophs, synthesis of nitrogenase is repressed by O_2, but in cyanobacteria this repression may be only transient (Pienkos *et al.* 1983; Maryan *et al.* 1986*b*; Gallon and Chaplin 1988). The inhibition of transcription of *nif H*, observed over 45 min in

Anabaena PCC 7120 by Haselkorn *et al.* (1983), may also have been transient rather than permanent.

The mechanism whereby O_2 represses nitrogenase synthesis may, in *K. pneumoniae*, be exerted through the *nif LA* system (Fig. 9.3; Dixon 1984) or through inhibition of DNA topoisomerase II (DNA gyrase), an enzyme that introduces negative superhelices into DNA (Kranz and Haselkorn 1986). [For a definition of positive and negative superhelicity, see Champoux (1978).] Inhibitors of DNA topoisomerase II prevented transcription of the *nif* genes in *K. pneumoniae* and also in the photosynthetic bacterium *Rhodopseudomonas capsulata*, so it appears that *nif* gene transcription is most effective when the genes are in a negatively supercoiled conformation. Because, in some bacteria, anaerobiosis increases DNA topoisomerase II activity and also the degree of negative supercoiling of the chromosome, it is possible that, in at least some diazotrophs, the effect of O_2 on nitrogenase synthesis could be related to decreased topoisomerase II activity (Kranz and Haselkorn 1986). In cyanobacteria, the transient nature of O_2 repression of nitrogenase synthesis could argue against a common (i.e. *nif LA*-mediated) regulation of synthesis by O_2 and fixed nitrogen (which, as in other diazotrophs, permanently represses nitrogenase synthesis). Furthermore, our own studies on *Gloeothece* suggest that the topoisomerase II inhibitors nalidixic acid and novobiocin are less inhibitory to nitrogenase synthesis under microaerobic conditions than under aerobic conditions (Table 9.5). This could be related to increased toposomerase II activity, and to a higher degree of negative superhelicity, at lower O_2 concentrations, but it would be premature, at present, to rule out alternative explanations.

There are reports that the tolerance of cyanobacterial N_2 fixation to O_2 is stimulated by exposure of cultures to high O_2 concentrations (Mackey and Smith 1983; Pienkos *et al.* 1983; Stewart and Rowell 1986) or by nitrogen starvation (He *et al.* 1985). On the other hand, the O_2 sensitivity of the process is increased by incubation of cyanobacterial cultures under microaerobic conditions (Kallas *et al.* 1983; Gallon and Chaplin 1988), by calcium deficiency (Hamadi and Gallon 1981; Gallon and Hamadi 1984), or at elevated temperatures (Pedersen *et al.* 1986).

Nitrogenase synthesis in cyanobacteria, as in other diazotrophs, is repressed by addition of fixed nitrogen as, for example, ammonia, urea, or amino acids. There is currently some uncertainty as to whether ammonia itself acts as the repressor of nitrogenase synthesis (Singh *et al.* 1983; Turpin *et al.* 1984; Mackerras and Smith 1986) or whether the repressor is a product of ammonia assimilation, such as glutamine (Thomas *et al.* 1982; Stewart *et al.* 1985). In part this is the result of a dispute as to the mechanism whereby L-methionine D, L-sulphoximine prevents exogenous ammonia from inhibiting nitrogenase synthesis. Stewart and Rowell (1986) present a detailed dis-

Table 9.5 Effect of nalidixic acid and novobiocin on synthesis of nitrogenase in *Gloeothece* in different gas phases (air and N_2).

Addition	Nitrogenase activity (relative rate)			
	Air		N_2	
	Control (+$CaCl_2$)	Experimental (+EDTA + $CaCl_2$)	Control (+$CaCl_2$)	Experimental (+EDTA + $CaCl_2$)
Nalidixic acid				
10 μM	128	104 (81)	94	103 (110)
50 μM	145	81 (56)	95	90 (95)
100 μM	147	69 (47)	82	74 (90)
200 μM	131	51 (39)	76	65 (86)
Novobiocin				
10 μM	120	100 (83)	95	95 (100)
50 μM	150	78 (52)	109	96 (88)
100 μM	134	48 (36)	84	72 (86)
200 μM	117	26 (22)	82	67 (82)
None	57.7	50.8	92.5	85.0

In experimental cultures, pre-existing nitrogenase was inactivated by exposure to 1 mM EDTA for 20 min in the presence of 5 mM HEPES/NaOH buffer, pH 7.5. At this point, where indicated, cultures were flushed with N_2 for 5 min and sealed. $CaCl_2$ (5 mM) was then added to cell cultures, along with the compound shown, and, after incubation for 6 hours in the light, recovered nitrogenase activity (which depends upon nitrogenase synthesis) was measured as acetylene reduction (Gallon 1980). Control cultures were treated identically except that no EDTA was added; they therefore show the effect of the addition on nitrogenase activity during the period of the assay. The relative activities recorded are those in the presence of nalidixic acid or novobiocin with respect to the activity found in their absence (shown as pmol min^{-1} mg $protein^{-1}$ at the foot of each column). The figures in parentheses show the relative activity after EDTA/$CaCl_2$ treatment, corrected for any effect of the addition on nitrogenase activity found in the appropriate control culture. They are therefore a measure of nitrogenase synthesis. Although these measurements are indirect, previous data (Gallon 1980) correlate well with data from other techniques for measuring nitrogenase synthesis in *Gloeothece* (Mullineaux *et al.* 1983; Maryan *et al.* 1986*b*).

cussion of this controversy; also see Kerby *et al.* (1986) and Rai *et al.* (1986*a*). Addition of ammonia has been reported, under certain conditions, to affect cyanobacterial nitrogenase activity as well as synthesis (Reich *et al.* 1986).

The regulation of nitrogenase synthesis by fixed nitrogen may be exerted in a manner analagous to that in *K. pneumoniae* (Fig. 9.3). However, since studies on cyanobacterial genetics are not very far advanced (see Haselkorn 1986), details are not known. There may be genes analagous to *ntr* in cyanobacteria (Machray and Stewart 1985) but these have not yet been characterized.

Fixed nitrogen in the form of nitrate inhibited N_2 fixation in several cyanobacteria (Gallon 1980). As in *K. pneumoniae* (Hom *et al.* 1980) the effect of nitrate may be exerted through reduction to nitrite rather than to ammonia.

Böhme (1986) reported that nitrite irreversibly inactivated cyanobacterial nitrogenase.

Finally, nitrogenase synthesis in *K. pneumoniae* is inhibited by incubation at temperatures greater than 35 °C (Fig. 9.3). This may also be the case in the unicellular cyanobacterium, *Gloeothece*. Nitrogenase synthesis, measured as incorporation of $Na_2{}^{35}SO_4$ into the MoFe-protein (Maryan *et al.* 1986*b*), was severely inhibited when cultures grown at 22 °C were transferred to 37 °C for 30 min, though synthesis subsequently recovered following return to 22 °C (S. Perry, unpublished observations).

Cellular differentiation

In heterocystous cyanobacteria, the acquisition of nitrogenase activity in response to nitrogen starvation is accompanied by the differentiation of vegetative cells into heterocysts. This process has been studied in some detail (Haselkorn 1978, 1986) and is discussed by R.J. Smith (1988*b*, this volume, Chapter 10). However, in as much as heterocyst differentiation and N_2 fixation are linked in these organisms, it is necessary to make some mention of differentiation here.

All non-heterocystous cyanobacteria so far examined possess the genes *nif H*, *nif D*, and *nif K* as a cluster (Kallas *et al.* 1985), an arrangement similar to that in *K. pneumoniae* (Fig. 9.2). On the other hand, in the DNA of vegetative cells of those heterocystous cyanobacteria that have been examined, the gene *nif K* is separated from the genes *nif D* and *nif H* (Kallas *et al.* 1985; Haselkorn 1986; Haselkorn *et al.* 1986). During the differentiation of vegetative cells into heterocysts the intervening DNA, consisting in *Anabaena* PCC 7120 of about 11 000 base pairs (11 kb), is excised as a circle, resulting in a clustered *nif HDK* operon (Haselkorn 1986; Haselkorn *et al.* 1986). Excision is catalysed by the product of a gene, *xis A*, located within the excised 11 kb region (Fig. 9.6). The actual sites of excision are characterized by an 11-base region of sequence -GGATTACTCCG-, and a proposed mechanism for the rearrangement is described by Haselkorn *et al.* (1986). Because one of the excision sites is located within *nif D*, the carboxy-terminal 43 amino acids of the α-subunit of the MoFe-protein of the heterocyst nitrogenase are encoded on the *nif K* side of the intervening 11 kb segment.

A second rearrangement has recently been found during heterocyst differentiation. This occurs, on the one hand, in the region of *nif S*, a gene involved in *K. pneumoniae* in the processing of the MoFe-protein of nitrogenase, and, on the other, close to *nif B*, a gene involved in synthesis of FeMoco, the iron–molybdenum centre of the MoFe-protein (Fig. 9.3). In this second rearrangement, a segment of DNA of approximately 50 kb is excised, again as a circle, and after this rearrangement an operon with the structure *nif B*:ORF-1:*nif S*:ORF-2 is formed (Fig. 9.6). Excision occurs between ORF-1

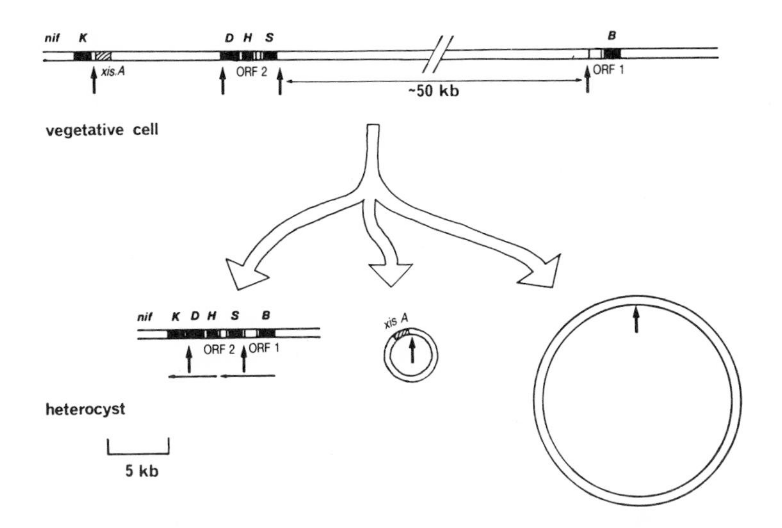

Fig. 9.6. Rearrangements of the *nif* genes in *Anabaena* PCC 7120 during heterocyst differentiation. The vertical arrows show the sites of DNA excision: the horizontal arrows indicate the transcriptional units (operons) formed following rearrangement. ORF = open reading frame (see Table 9.3). [Based on Haselkorn (1986), Haselkorn *et al.* (1986), and Haselkorn (personal communication)].

and *nif S*, at a region with a base sequence -TATTC-AGAA- (Haselkorn *et al.* 1986). This excision is not catalysed by the *xis A* product.

Thus, heterocyst differentiation, and the acquisition of aerotolerant nitrogenase activity, in cyanobacteria such as *Anabaena* PCC 7120 is accompanied by a marked rearrangement of the DNA chromosome of the differentiating cell.

The *nif* gene promoters in *Anabaena* PCC 7120 differ in structure from both *E. coli* promoters and the *nif* promoters of *K. pneumoniae* (Haselkorn *et al.* 1983). In support of this, no expression of the *nif H* gene from *Anabaena variabilis* was found when it was inserted, with its promoter, into *K. pneumoniae* (Hirschberg *et al.* 1985).

Symbiotic systems

Diazotrophic cyanobacteria are found in a variety of symbiotic associations, variously involving, as co-symbiont, algae, fungi, bryophytes, pteridophytes, gymnosperms, an angiosperm, and even animals (Wilkinson and Fay 1979). Detailed reviews are available (Millbank 1974; Stewart 1978; Stewart *et al.* 1980). Over the years, many of these associations have been studied and some common factors have emerged. For example, in symbiotic association the cyanobacterial partner often exhibits increased heterocyst frequency, including the formation of multiple heterocysts, though maximum rates of N_2 fixation may occur in regions where single heterocysts predominate (Lindblad *et al.* 1985). Other changes include decreased photosynthesis, caused by partial inhibition of transcription of genes associated with this process (Nierzwicki-Bauer and Haselkorn 1986), rendering the cyanobiont dependent upon a supply of fixed carbon from the co-symbiont (this is not observed in those symbioses in which the cyanobacterium is also the sole photosynthetic partner), and decreased ammonia utilization, with consequent excretion of fixed nitrogen to the co-symbiont. Decreased ammonia uptake is usually a reflection of decreased activity of the assimilatory enzyme glutamine synthetase, but this is not always the case (Rai *et al.* 1986*b*). The mechanism whereby establishment of the symbiosis induces these various changes in the cyanobiont is not clear, but recent studies suggest that they may be related to the availability of exogenous carbon (Kumar *et al.* 1986; Rozen *et al.* 1986) to microaerobiosis (Rai *et al.* 1986*b*) or to physical effects that may be mimicked by immobilization in polyvinyl foam (Hall *et al.* 1985; D-J Shi, personal communication).

Acknowledgements

Our own work, reported here, was supported by the Royal Society, the Science and Engineering Research Council, and the Natural Environment Research Council. The contributions to this work of Dr H.S. Khamees, Dr S. Perry, Dr T.M.A. Rajab, and Mrs J.H. Thomas are gratefully acknowledged. We also thank Dr B. Bergman, Dr T. Burger-Wiersma, Dr R.R. Eady, Prof. R. Haselkorn, and Dr D-J. Shi for providing us with research findings in advance of publication.

References

Allen, M.B. (1956). *Science Monthly* **83**, 100.
Allen, M.B. (1972). *Photophysiology* **7**, 73.
Allen, M.B. and Arnon, D.I. (1955). *Plant. Physiol.* **30**, 366.
Beijerinck, M.W. (1888). *Bot. Ztg* **46**, 725.
Benemann, J.R., Smith, G.M., Kostel, P.K., and McKenna, C.E. (1973). *FEBS Lett.* **29**, 219.
Bergman, B., Lindblad, P., and Rai, A.N. (1986). *FEMS Microbiol. Lett.* **35**, 75.
Bishop, P.E. (1981). In *Current perspectives in nitrogen fixation* (ed. A.H. Gibson and W.E. Newton) p. 171. Australian Academy of Science, Canberra.
Bishop, P.E. (1986). *Trends biochem. Sci.* **11**, 225.
Böhme, H. (1986). *Arch. Microbiol.* **146**, 99.
Bond, G. and Scott, D.G. (1955). *Ann. Bot.* **19**, 67.
Bortels, H. (1930). *Arch. Microbiol.* **1**, 333.
Bortels, H. (1936). *Zent. Bakteriol. ParasitKde Abt. II* **95**, 193.

Bortels, H. (1940). *Arch. Microbiol.* **11**, 155.
Bothe, H. (1982). In *The biology of the cyanobacteria*, (ed. N.G. Carr and B.A. Whitton), p. 87. Blackwell Scientific Publications, Oxford.
Bothe, H., Nelles, H., Hager, K.P., Papen, H., and Neuer, G. (1984). In *Advances in nitrogen fixation research* (ed. C. Veeger and W.E. Newton) p. 199. Martinus Nijhoff, The Hague.
Bristol, B.M. and Page, H.J. (1923). *Ann. appl. Biol.* **10**, 378.
Bryceson, I. and Fay, P. (1981). *Mar. Biol.* **61**, 159.
Burger-Wiersma, T., Veenhuis, M., Korthals, H.J., Van de Wiel, C.C.M., and Mur, L.R. (1986). *Nature, Lond.* **320**, 262.
Burris, R.H., Eppling, F.J., Wahlin, H.B., and Wilson, P.W. (1942). *Proc. soil Sci. Soc. Am.* **7**, 258.
Burris, R.H., Eppling, F.J., Wahlin, H.B., and Wilson, P.W. (1943). *J.biol.Chem.* **148**, 349.
Cannon, F., Beynon, J., Buchanon-Wollaston, V., Burghoff, R., Cannon, M., Kwatiowski, R., Laver, G. and Rubin, R. (1985). In *Nitrogen fixation research progress* (ed. H.J. Evans, P.J. Bottomley, and W.E. Newton) p. 453. Martinus Nijhoff, The Hague.
Champoux, J.J. (1978). *A. Rev. Biochem.* **47**, 449.
Chaplin, A.E., Rajab, T.M.A., and Gallon, J.R. (1984). In *Advances in nitrogen fixation research* (ed. C. Veeger and W.E. Newton) p. 217. Martinus Nijhoff, The Hague.
Daday, A., Mackerras, A., and Smith, G.D. (1985). *J. gen. Microbiol.* **131**, 231.
Dixon, R.A. (1984). *J. gen. Microbiol.* **130**, 2745.
Drewes, K. (1928). *Zent. Bakt. ParasitKde Abt. II* **76**, 88.
Dugdale, R.C., Menzel, D.W., and Ryther, J.A. (1961). *Deep-Sea Res.* **7**, 297.
Eady, R.R. (1981). In *Current perspectives in nitrogen fixation* (ed. A.H. Gibson and W.E. Newton) p. 172. Australian Academy of Science, Canberra.
Eady, R.R. and Robson, R.L. (1984). *Biochem, J.* **224**, 853.
Fay, P. (1973). In *The biology of blue-green algae* (ed. N.G. Carr and B.A. Whitton) p. 238. Blackwell Scientific Publications, Oxford.
Fay, P. and de Vasconcelos, L. (1974). Arch. Microbiol. **99**, 221.
Fay, P., Stewart, W.D.P., Walsby, A.E., and Fogg, G.E. (1968). *Nature, Lond.* **220**, 810.
Fogg, G.E. (1942). *J. exp. Biol.* **19**, 78.
Fogg, G.E. (1951). *J. exp. Bot.* **2**, 117.
Fogg, G.E. (1956). *Bact. Rev.* **20**, 148.
Fogg, G.E. (1962). In *Physiology and biochemistry of algae* (ed. R.A. Lewin) p. 161. Academic Press, New York.
Fogg, G.E. (1974). In *Algal physiology and biochemistry* (ed. W.D.P. Stewart) p. 560. Blackwell Scientific Publications, Oxford.
Fogg, G.E. and Stewart, W.D.P. (1965). *Sci. Prog., Lond.* **53**, 191.
Fogg, G.E. and Wolfe, M. (1954). In *Autotrophic microorganisms* (ed. B.A. Fry and J.L. Peel) p. 99. Cambridge University Press.
Frank, B. (1889). *Ber. deut. bot. Ges.* **7**, 34.
Gallon, J.R. (1980). In *Nitrogen fixation* (ed. W.D.P. Stewart and J.R. Gallon) p. 197. Academic Press, London.
Gallon, J.R. and Chaplin, A.E. (1987). *An introduction to nitrogen fixation.* Cassell, Eastbourne, Sussex.

Gallon, J.R. and Chaplin, A.E. (1988). In *Phycotalk* Vol. 2 (ed. H.D. Kumar). Print House (India), Lucknow. (in press).
Gallon, J.R. and Hamadi, A.F. (1984). *J. gen Microbiol.* **130**, 495.
Gallon, J.R., Ul-Haque, M., and Chaplin, A.E. (1978). *J. gen. Microbiol.* **106**, 329.
Golden, J.W., Robinson, S.J., and Haselkorn, R. (1985). *Nature, Lond.* **314**, 419.
Grobbelaar, N., Huang, T.C., Lin, H.Y., and Chow, T.J. (1986). *FEMS Microbiol Lett.* **37**, 173.
Gyurjan, I., Nghia, N.H., Toth, G., and Turtoczky, I. (1986). *Biochem. Physiol. Pflanz* **181**, 147.
Hales, B.J., Case, E.E., Morningstar, J.E., Dzeda, M.F., and Mauterer, M.F. (1986). *Biochemistry* **25**, 7251.
Hall, D.O., Affolter, D.A., Brouers, M., Shi, D–J., Yang, L-W., and Rao, K.K. (1985). In *Plant products and the new technology* (ed. K.W. Fuller and J.R. Gallon) p. 161. Oxford University Press.
Hallenbeck, P.C. and Benemann, J.R. (1980). *FEMS Microbiol. Lett* **9**, 121.
Hallenbeck, P.C., Kostel, P.J., and Benemann, J.R. (1979). *Eur. J. Biochem.* **98**, 275.
Hamadi, A.F. and Gallon, J.R. (1981). *J. gen. Microbiol.* **125**, 391.
Haselkorn, R. (1978). *A. Rev. plant Physiol.* **29**, 319.
Haselkorn, R. (1986). *A. Rev. Microbiol.* **40**, 525.
Haselkorn, R., Golden, J.W., Lammers, P.J., and Mulligan, M.E. (1986). *Trends Genet.* **2**, 255.
Haselkorn, R., Rice, D., Curtis, S.E., and Robinson, S.J. (1983). *Ann. Inst. Pasteur, Paris* **134B**, 181.
Haselkorn, R., Mazur, B., Orr, J., Rice, D., Wood, N., and Rippka, R. (1980). In *Nitrogen fixation* (ed. W.E. Newton and W.H. Orme-Johnson) p. 95. Plenum, New York.
He, Z., Du, D., Lin, H., Dai, L., and Xing, W. (1985). *Shuisheng Shengwuxue Jikan* **9**, 324.
Hellriegel, H. and Wilforth, H. (1888). *Z. Ver. Rübenzucker Industrie Deutschen Reichs* (Beilageheft).
Hirschberg, R., *et al.* (1985). *J. Biotechnol.* **2**, 23.
Holm-Hanson, O. (1968). *A. Rev. Microbiol.* **22**, 47.
Hom, S.S.M., Hennecke, H., and Shanmugam, K.T. (1980). *J. gen. Microbiol.* **117**, 169.
Houchins, J.P. (1984). *Biochim. Biophys. Acta* **765**, 227.
Houchins, J.P. (1985). In *Nitrogen fixation and CO_2 metabolism* (ed. P.W. Ludden and J.E. Burris) p. 261. Elsevier, New York.
Jensen, B.B., Cox, R.P., and Burris, R.H. (1986). *Arch. Microbiol.* **145**, 241.
Kallas, T., Coursin, T., and Rippka, R. (1985). *Plant molec. Biol.* **5**, 321.
Kallas, T., Rippka, R., Coursin, T., Rebière, M-C., Tandeau de Marsac, N., and Cohen-Bazire, G. (1983). In *Photosynthetic prokaryotes: cell differentiation and function* (ed. G.C. Papageorgiou and L. Packer) p. 281. Elsevier, New York.
Keeler, R.F. and Varner, J.E. (1957). *Arch. Biochem. Biophys.* **70**, 585.
Kerby, N.W. and Stewart, W.D.P. (1988). In *Biochemistry of the algae and cyanobacteria*, p. 319. Clarendon Press, Oxford.
Kerby, N.W., Rowell, P., and Stewart, W.D.P. (1986). *Arch. Microbiol.* **143**, 353.

Khamees, H., Gallon, J.R., and Chaplin, A.E. (1987). *Br. phycol. J.* **22**, 55.
Kimmich, G.A., Randles, J., and Brand, J.S. (1975). *Anal. Biochem.* **69**, 187.
Kranz, R.G. and Haselkorn, R. (1986). *Proc. natn. Acad. Sci. USA* **83**, 6805.
Kumar, A.P., Perraju, B.T.V.V., and Singh, H.N. (1986). *New Phytol.* **104**, 115.
Lindblad, P., Hallbom, L., and Bergman, B. (1985). *Symbiosis* **1**, 19.
Machray, G.C. and Stewart, W.D.P. (1985). *Proc. R. Soc. Edin.* **85B**, 239.
Mackerras, A.H. and Smith, G.D. (1986). *Biochem. biophys. Res. Commun.* **134**, 835.
Mackey, E.J. and Smith, G.D. (1983). *FEBS Lett.* **156**, 108.
Mague, T.H., Weare, N.M., and Holm-Hansen, O. (1974). *Mar. Biol.* **24**, 109.
Matthijs, H.C.P. and Lubberding, H.J. (1988). In *Biochemistry of the algae and cyanobacteria*, p. 000. Clarendon Press, Oxford.
Maryan, P.S. and Vorley, T. (1979). *Laboratory Practice* **28**, 251.
Maryan, P.S., Eady, R.R., Chaplin, A.E., and Gallon, J.R. (1986*a*). *J. gen. Microbiol.* **132**, 798.
Maryan, P.S., Eady, R.R., Chaplin, A.E., and Gallon, J.R. (1986*b*). *FEMS Microbiol. Lett.* **34**, 251.
Mazur, B.J., Rice, D., and Haselkorn, R. (1980). *Proc. natn. Acad. Sci. USA* **77**, 186.
Millbank, J.W., (1974). In *The biology of nitrogen fixation*, (ed. A. Quispel) p. 238. North-Holland, Amsterdam.
Mitsui, A., Kumazawa, S., Takahashi, A., Ikemoto, H., Cao, S., and Arai, T. (1986). *Nature, Lond.* **323**, 720.
Mullineaux, P.M., Chaplin, A.E., and Gallon, J.R. (1980). *J. gen Microbiol.* **120**, 227.
Mullineaux, P.M., Chaplin, A.E., and Gallon, J.R. (1983). *J. gen. Microbiol.* **129**, 1689.
Mullineaux, P.M., Gallon, J.R., and Chaplin, A.E. (1981). *FEMS Microbiol. Lett.* **10**, 245.
Murry, M.A., Hallenbeck, P.C., and Benemann, J.R. (1984). *Arch. Microbiol.* **137**, 194.
Nagatani, H.H. and Brill, W.J. (1974). *Biochim. Biophys. Acta* **362**, 160.
Nagatani, H.H. and Haselkorn, R. (1978). *J. Bact.* **134**, 597.
Nierzwicki-Bauer, S. and Haselkorn, R. (1986). *EMBO J.* **5**, 29.
Paerl, H. (1984). *Mar. Biol.* **87**, 251.
Pearson, H.W., Malin, G., and Howsley, R. (1981). *Br. phycol. J.* **16**, 139.
Pederson, D.M., Daday, A., and Smith, G.D. (1986). *Biochimie* **68**, 113.
Pienkos, P.T., Bodmer, S., and Tabita, F.R. (1983). *J. Bact.* **153**, 182.
Postgate, J. (1987) *Nitrogen fixation*, (2nd edn). Edward Arnold, London.
Prantl, K. (1889). *Hedwigia* **28**, 135.
Rai, A.N., Singh, D.T., and Singh, H.N. (1986*a*). *Physiologia Pl.* **68**, 320.
Rai, A.N., Lindblad, P., and Bergman, B. (1986*b*). *Planta* **169**, 379.
Ramos, J.L. and Madueno, F. (1986). *FEMS Microbiol. Lett.* **36**, 73.
Reich, S., Almon, H., and Böger, P. (1986). *FEMS Microbiol. Lett.* **34**, 53.
Rice, D., Mazur, B.J., and Haselkorn, R. (1982). *J. biol. Chem.* **257**, 13157.
Rippka, R. and Stanier, R.Y. (1978). *J. gen. Microbiol.* **105**, 83.
Rippka, R., Deruelles, J., Waterbury, J.B., Herdman, M., and Stanier, R.Y. (1979). *J. gen. Microbiol.* **111**, 1.

Robson, R.L., Eady, R.R., Richardson, T.H., Miller, R.W., Hawkins, M., and Postgate, J.R. (1986). *Nature, Lond.* **322**, 388.
Rowell, P., Darling, A.J., Amla, D.V., and Stewart, W.D.P. In *Biochemistry of the algae and cyanobacteria*, p. 201. Clarendon Press, Oxford.
Rozen, A., Arad, H., Schonfeld, M., and Tel-Or, E. (1986). *Arch. Microbiol.* **145**, 187.
Saino, T. and Hattori, A. (1978). *Deep-Sea Res.* **25**, 1259.
Schneider, K.C., Bradbeer, C., Singh, R.N., Wang, L.C., and Burris, R.H. (1960). *Proc. natn. Acad. Sci. USA* **46**, 726.
Scherer, S., Riege, H., and Böger, P. (1988). In *Biochemistry of the algae and cyanobacteria*, p. 121. Clarendon Press, Oxford.
Singh, N.H., Vaishampayan, A., and Singh, R.K. (1978). *Biochem. biophys. Res. Commun.* **81**, 67.
Singh, H.N., Rai, U.N., Rao, V.V., and Bagchi, S.N. (1983). *Biochem. biophys. Res. Commun.* **111**, 180.
Smith, B.E., *et al.* (1987). *Phil. Trans. R. Soc. B.* **317,** 131.
Smith, D.L., Patriquin, D.G., Dijak, M., and Curry, G.M. (1986). *Can. J. Bot.* **64**, 1843.
Smith, G.D. (1988*a*). In *Phycotalk* Vol. 2, (ed. H.D. Kumar). Print House (India), Lucknow. (in press).
Smith, R.J. (1988*b*). In *Biochemistry of the algae and cyanobacteria*, p. 185. Clarendon Press, Oxford.
Spielmann, H., Jacob-Müller, U., and Schulz, P. (1981). *Anal. Biochem.* **113**, 172.
Stal, L.J. and Krumbein, W.E. (1985*a*). *Arch. Microbiol.* **143**, 67.
Stal, L.J. and Krumbein, W.E. (1985*b*). *Arch. Microbiol.* **143**, 72.
Stanier, R.Y. (1974). *Aust. J. exp. Biol. med. Sci.* **52**, 3.
Stewart, W.D.P. (1970). *Plant Soil* **32**, 555.
Stewart, W.D.P. (1971). *Plant Soil Spec. Vol.* p. 377.
Stewart, W.D.P. (1973*a*). In *The biology of blue-green algae* (ed. N.G. Carr and B.A. Whitton) p. 260, Blackwell Scientific Publications, Oxford.
Stewart, W.D.P. (1973*b*). *A. Rev. Microbiol.* **27**, 283.
Stewart, W.D.P. (1974*a*). In *The biology of nitrogen fixation* (ed. A. Quispel) p. 202. North-Holland, Amsterdam.
Stewart, W.D.P. (1974*b*). In *The biology of nitrogen fixation* (ed. A. Quispel) p. 687. North-Holland, Amsterdam.
Stewart, W.D.P. (1976). In *Proc. 1st Int. Symp. on Nitrogen Fixation* Vol. 1. (ed. W.E. Newton and C.J. Nyman) p. 257. Washington State Press, Pullman.
Stewart, W.D.P. (1977*a*). In *A treatise on dinitrogen fixation* Vol. 3 (ed. R.W.F. Hardy and W.D. Silver) p. 63. John Wiley, New York.
Stewart, W.D.P. (1977*b*). *Br. phycol. J.* **12**, 89.
Stewart, W.D.P. (1978). *Endeavour* **2**, 170.
Stewart, W.D.P. (1980). In *Methods for the evaluation of biological nitrogen fixation* (ed. F.J. Bergersen) p. 583, John Wiley, Chichester, Sussex.
Stewart, W.D.P. and Lex, M. (1970). *Arch. Microbiol.* **73**, 250.
Stewart, W.D.P. and Rowell, P. (1986). *Plant Soil* **90**, 167.
Stewart, W.D.P. and Singh, H.N. (1975). *Biochem. biophys. Res. Commun.* **62,** 62.
Stewart, W.D.P. and Tel-Or, E. (1975). *Biochem. Soc. Trans.* **3**, 357.

Stewart, W.D.P., Fitzgerald, G.P., and Burris, R.H. (1967). *Proc. natn. Acad. Sci. USA* **58**, 2071.

Stewart, W.D.P., Fitzgerald, G.P., and Burris, R.H. (1968). *Arch. Microbiol.* **62**, 336.

Stewart, W.D.P., Rowell, P., and Apte, S.K. (1977). In *Recent developments in nitrogen fixation* (ed. W.E. Newton, J.R. Postgate, and C. Rodriguez Barrueco) p. 287. Academic Press, London.

Stewart, W.D.P., Rowell, R., and Rai, A.N. (1980). In *Nitrogen fixation* (ed. W.D.P. Stewart and J.R. Gallon) p. 239. Academic Press, London.

Stewart, W.D.P., Rowell, P., and Rai, A.N. (1983). *Ann. Inst. Pasteur, Paris* **134B**, 205.

Stewart, W.D.P., Rowell, P., Cossar, D., and Kerby, N.W. (1985). In *Nitrogen fixation and CO_2 metabolism* (ed. P.W. Ludden and J.E. Burris) p. 269. Elsevier, New York.

Stewart, W.D.P., Rowell, P., Hawkesford, M., Sampaio, M.J.A.M., and Ernst, A. (1982). *Israel J. Bot.* **31**, 168.

Takahashi, H. and Nason, A. (1957). *Biochem. Biophys. Acta* **23**, 433.

Tel-Or, E., Huflejt, M.E., and Packer, L. (1986). *Arch. biochem. Biophys.* **246**, 396.

Thomas, J. and Apte, S.K. (1984). *J. Biosci.* **6**, 771.

Thomas, J.H., Mullineaux, P.M. Cronshaw, A.D., Chaplin, A.E., and Gallon, J.R. (1982). *J. gen. Microbiol.* **128**, 885.

Tözüm, S.R.D. and Gallon, J.R. (1979). *J. gen. Microbiol.* **111**, 313.

Tözüm, D., Ul-Haque, M.I., Chaplin, A.E., and Gallon, J.R. (1977). *Biochem. Soc. Trans.* **5**, 1482.

Trehan, K. and Sinha, U. (1980). *J. Cytol. Genet.* **15**, 1.

Turpin, D.H., Edie, S.A., and Canvin, D.T. (1984). *Plant Physiol., Lancaster* **74**, 701.

Tyagi, V.V.S. (1974). *Ann. Bot.* **38**, 485.

Upchurch, R.G. and Mortenson, L.E. (1980). *J. Bact.* **143**, 274.

Venkataraman, G.S. (1985). *Curr. Sci.* **54**, 493.

Walsby, A.E. (1985). *Proc. R. Soc.* B **226**, 345.

Wann, F.B. (1921). *Am. J. Bot.* **8**, 1.

Ward, H.M. (1895). *Sci. Prag.* **3**, 251.

Wilkinson, C.R. and Fay, P. (1979). *Nature, Lond.* **279**, 527.

Wolk, C.P. (1981). In *Genetic engineering and symbiotic nitrogen fixation for the conservation of fixed nitrogen* (ed. J.M. Lyons, R.C. Valentine, P.A. Phillips, D.W. Rains, and R.C. Huffaker) p. 315. Plenum, New York.

Wolk, C.P. and Wojciuch, E. (1971*a*). *Planta* **97**, 126.

Wolk, C.P. and Wojciuch, E. (1971*b*). *J. Phycol.* **7**, 339.

Wolk, C.P., Flores, E., Schmetterer, G., Herrero, A., and Elhai, J. (1985). In *Nitrogen fixation research progress* (ed. H.J. Evans, P.J. Bottomley, and W.E. Newton) p. 491. Martinus Nijhoff, The Hague.

Wolk, C.P., Thomas, J., Shaffer, P.W., Austin, S., and Galonsky, A., (1976). *J. biol. Chem.* **251**, 5027.

Wyatt, J.T. and Silvey, J.K.G. (1969). *Science, N.Y.* **165**, 908.

Yamada, T. and Sakaguchi, K. (1980). *Arch. Microbiol.* **124**, 161.

Poster abstracts

Effects of temperature on photosynthesis in the thermo-acidophilic alga *Cyanidium caldarium*

T. W. Ford
Department of Botany, Royal Holloway and Bedford New College, University of London, Egham Hill, Egham, Surrey, UK

Ultrastructurally, the unicellular alga *Cyanidium caldarium* shows features in common with the red algae, especially in the structure of the chloroplast. However, unlike other rhodophytes, it shows a relatively constant growth rate in continuous culture up to temperatures of 50°C, and growth is still possible at 55°C. This ability to grow at high temperatures is reflected in a more thermostable photosynthetic electron transport system and an RUBP-carboxylase with enhanced thermal stability. The upper temperature limit for life is lower for photoautotrophic than for heterotrophic micro-organisms, suggesting that photosynthesis may be the thermally limiting metabolic activity. When the culture temperature of *Cyanidium* is raised to 60°C, growth ceases. The thermal stability of photosynthetic electron transport coincides with this maximum growth temperature, whilst RUBP-carboxylase is still active *in vivo* at higher temperatures. It is suggested that one or more components of the photosynthetic electron transport chain, rather than enzymes such as RUBP-carboxylase, dictate the growth temperature maximum for this alga. However, this component is unlikely to be the pigment complexes which are stable, *in vivo* up to 60°C. The mobile electron carrier plastoquinone seems a more suitable candidate since the system also shows homeoviscous adaptation in response to growth temperature, probably mediated via substitution of high melting point lipids in the thylakoid membranes, hence stabilizing the electron transport chain.

The use of *Scenedesmus obliquus* to probe the mode of action of nitrodiphenyl ether herbicides

J.R. Bowyer, B.J. Smith, and J. Howard
Department of Biochemistry, Royal Holloway and Bedford New College, Egham Hill, Egham, Surrey, UK
P. Camilleri
Sittingbourne Research Centre, Sittingbourne ME9 8AG, Kent, UK

Nitrodiphenyl ether (DPE) herbicides cause rapid light- and oxygen dependent membrane lipid peroxidation in higher plant leaf tissue. These effects are similar to those of paraquat, but, unlike the case of paraquat, inhibition of photosynthetic electron transport, either by mutation leading to loss of photosystem (PS) I or II (barley) or by addition of diuron, does not provide resistance to DPE herbicides. In contrast, in the alga *Scenedesmus obliquus* diuron protects against both paraquat and DPEs. We are able to show that mutants of *S. obliquus* lacking PSI or PSII are also resistant to DPE herbicides, that the role of photosynthetic electron transport is not to reduce the DPE to an anion radical, and that the probable role of photosynthetic electron transport is to maintain a sufficiently high oxygen concentration to support lipid peroxidation and chlorophyll bleaching.

Electron transport proteins from *Porphyra umbilicalis*—one flavodoxin or two?

N. T. Price, A. J. Smith, and L. J. Rogers
Department of Biochemistry, University College of Wales, Aberystwyth, Dyfed SY23 3DD, UK

In the red macroalga *Porphyra umbilicalis* the predominant low potential redox protein, a well-characterized ferredoxin, is accompanied by a small amount of flavodoxin. This was purified from cell extracts by exploiting ammonium sulphate fractionation, ion-exchange chromatography on DEAE-cellulose, and gel-filtration on Sephadex G-100, in succession. Although PAGE of the purified flavodoxin on disc gels suggested its homogeneity, application of SDS-PAGE on slab gels revealed two components differing slightly in molecular weight. The ratio of the larger M_r flavodoxin to the smaller species was about 5:1 throughout the purification, and in batches of alga processed on different occasions. The two proteins were separated by preparative PAGE on slab gels and recovered individually by electo-elution. Both possessed the characteristic flavodoxin absorption spectrum, and had one FMN per molecule. A molecular weight of 21 000 for the larger flavodoxin was determined from meniscus depletion sedimentation equilibrium studies, and was consistent with estimates from SDS-PAGE and amino acid composition analysis. The apoprotein produced by trichloroacetic acid precipitation gave several components of lower mobility on PAGE; these would not re-associate with FMN. However, apoprotein prepared in the presence of dithiothreitol gave a single species which, though of substantially lower mobility on PAGE, could be shown by sedimentation equilibrium studies to be monomeric. Removal of FMN therefore induces an extreme conformational change in the apoprotein; nevertheless, this apoprotein would re-associate with FMN to give a holoprotein with electrophoretic characteristics indistinguishable from those of the native flavodoxin. Parallel studies with the minor flavodoxin were not feasible because of the small amounts which were obtained. Preliminary investigations of the relationship of the two flavodoxins by mapping of staphylococcal protease peptides suggests they are related. However, the smaller flavodoxin appears not to be a product of proteolysis *in vitro* since inclusion of protease inhibitors during the extraction procedure does not influence the ratio in which the two proteins are obtained.

Amino acid sequence of the ferredoxin from *Synechococcus* 6301 (*Anacystis nidulans*)

K. Wada, R. Masui, and H. Matsubara
Department of Biology, Osaka University, Osaka 560, Japan
L.J. Rogers
Department of Biochemistry, University College of Wales, Aberystwyth, Dyfed SY23 3DD, UK

The ferredoxin from *Synechococcus* PCC 6301 (*Anacystis nidulans*) was isolated from 60 g of acetone-dried powder obtained from 280 g wet wt. photoautotrophically grown cells. Purification of the ferredoxin by our established methods gave 150 mg of a 2Fe–2S ferredoxin; only one soluble ferredoxin was apparently present. The ferredoxin possessed absorption maxima at 276, 282 (shoulder), 331, 423, and 464 nm; the A_{331}/A_{276} ratio was 0.75. The midpoint redox potential determined by potentiometric titration was about −380 mV.

The amino acid sequence determination for the carboxymethylferredoxin was based on Edman degradation of peptides obtained by enzymic digestion. The sequence thus derived for *Synechococcus* 6301 ferredoxin was:

```
         10        20        30        40        50
ATYKVTLVNAAEGLNTTIDVADDTYILDAAEEQGIDLPYSCRAGACSTCA

51       60        70        80        90
GKVVSGTVDQSDQSFLDDDQIAAGFVLTCVAYPTSDVTIETHKEEDLY
```

The sequence differs markedly compared to the ferredoxins from other unicellular cyanobacteria, with 25 differences even when compared to the ferredoxin sequence for another *Synechococcus* sp., but is surprisingly similar to those of the ferredoxins from the filamentous cyanobacteria *Aphanizomenon flos-aquae, Chlorogloeopsis fritschii* and *Mastigocladus laminosus* (13–15 differences, with deletions counted as one).

The coding sequence of a ferredoxin gene from *Synechococcus* PCC 7942 (*A. nidulans* R2) has been reported recently (Van der Plas *et al.* 1986).

Neglecting an N-terminal methionine, probably removed by post-translational processing, the amino acid sequences correspond. Studies of genetic relationships (Wilmotte and Stam 1984) have supported the close resemblance of these two *Synechococcus* strains, and this observation is now underlined at the biochemical level in the identical sequences of their ferredoxins.

Van der Plas, J., deGroot, R.P., Woortman, M.R., Weisbeck, P.J., and van Arkel, G.A. (1986). *Nucleic Acids Res.* **14**, 7804.
Wilmotte, A.M.R. and Stam, W.T. (1984). *J. gen. Microbiol.* **130**, 2737.

Cytochrome oxidase in cyanobacterial membrane preparations: topography, activity, identity

G.A. Peschek, V. Molitor, M. Trnka, M. Wastyn, and W. Erber
Biophysical Chemistry Group, Institute of Physical Chemistry, University of Vienna, Währingerstrasse 42, A-1090 Vienna, Austria

Plasma and thylakoid membranes (CM and ICM, respectively) were separated from lysozyme-treated and French-pressed *Anacystis nidulans, Plectonema boryanum, Synechocystis* 6714, *Nostoc* Mac, and *Anabaena* ATCC 29413 (both vegetative cells and heterocysts) by discontinuous sucrose density gradient centrifugation (Omata and Murata 1983; Molitor *et al.* 1987). The rate of oxidation of horse heart cytochrome *c* by the individual membrane fractions was tested at different steps of purification. The activity was inhibited more than 95 per cent by, for example, 1.2 μM KCN, 0.1 mM Na_2S, 1.0 mM NaN_3, 100 mM NaCl or KCl, and CO (Molitor *et al.* 1987; Trnka and Peschek 1986). CM was identified by labelling of intact cells with ^{35}S-diazobenzenesulfonate or fluorescamine prior to membrane isolation (Molitor *et al.* 1986, 1987). Substantial cytochrome oxidase activity was found for the CM from *Anacystis* and *Plectonema*, with intermediate activity in the CM of *Nostoc*, and negligible activity in the CM of *Synechocystis* and *Anabaena*. A late (instead of early) 'logarithmic' growth phase, and the presence of 0.4 M NaCl (but not KCl) in the growth medium, resulted in markedly enhanced levels of cytochrome oxidase in the CM (but not ICM) of *Anacystis* (and *Synechocystis*). Bioenergetically, this complements the increased activity of the amiloride-sensitive Na^+/H^+-antiport by intact cells grown with NaCl (Molitor *et al.* 1986).

Unequivocally positive immunological cross-reaction was obtained between antibodies against subunits I and II and the holoenzyme of *Paracoccus denitrificans* aa_3-type cytochrome oxidase, and Western blots after SDS-PAGE of the membranes from *Anacystis* (Molitor *et al.* 1987; Trnka and Peschek 1986), *Plectonema*, and *Synechocystis*, though each produced a distinct pattern.

Molitor, V., Erber, W., and Peschek, G.A. (1986). *FEBS Lett.* **204**, 251.
Molitor, V., Trnka, M., and Peschek, G.A. (1987). *Curr. Microbiol.* **4**, 263.
Omata, T. and Murata, N. (1983). *Plant Cell Physiol.* **24**, 1101.
Trnka, M. and Peschek, G.A. (1986) *Biochem Biophys. Res. Commun.* **136**, 235.

Nitrogen fixation by cyanobacteria in situ

A.E. Chaplin, M.S.H. Griffiths, and J.R. Gallon
Department of Biochemistry, University College of Swansea, Swansea SA2 8PP, UK

When pure cultures of diazotrophic cyanobacteria were maintained on a diurnal cycle of light and darkness, three patterns of N_2 fixation were observed, different organisms fixing N_2 mainly in the dark, mainly in the light, or both in the light and in the dark. We examined whether these patterns are maintained under field conditions.

Measurements taken hourly at the entrance to a limestone cave, at sites dominated by the heterocystous *Nostoc* sp., showed acetylene reduction only during the day. This contrasts with the pattern of pure cultures of the *Nostoc* isolated from this site which reduced acetylene in both the light and the dark. Failure of this *Nostoc* sp. to fix N_2 in the dark *in situ* may be a consequence of the lower temperature recorded at night.

At the rear of the same cave were sites dominated by the unicellular *Gloeothece* sp. At these sites, acetylene reduction could be detected only at night, a pattern identical to that of unialgal cultures of the *Gloeothece* isolated from this site.

At Landimore Marsh, a salt-marsh containing loose mats of the non-heterocystous, filamentous *Oscillatoria* sp., N_2 fixation could be detected only at night. This pattern is identical to that of pure cultures of *Oscillatoria* sp. maintained under laboratory conditions.

Thus cyanobacterial N_2 fixation *in situ* is not confined to the light period, particularly at sites dominated by non-heterocystous species.

The diurnal pattern of nitrogen fixation by *Gloeothece*

S. Perry, J.R. Gallon, and A.E. Chaplin
Department of Biochemistry, University College of Swansea, Swansea SA2 8PP, UK

Gloeothece is a unicellular cyanobacterium capable of evolving O_2 by photosynthesis and, simultaneously, fixing N_2. However, when grown under an alternating cycle of light and darkness, *Gloeothece* fixes N_2 predominantly in the dark thereby effecting a temporal separation of the two processes (Mullineaux *et al.* 1981). Similar diurnal variations in N_2 fixation have been reported in *Synechococcus* sp., and Mitsui *et al.* (1986) have proposed that here this pattern is imposed by the cell cycle. However, we propose that in *Gloeothece* the pattern is imposed by metabolic events unrelated to the cell cycle.

In *Gloeothece*, diurnal variations in the rate of N_2 fixation almost parallel variations in the rate of nitrogenase synthesis, which is detectable only during that part of the light–dark cycle during which nitrogenase activity is increasing or about to increase (i.e. the first 6–8 hours of darkness). During this period, the energy needed for nitrogenase synthesis and activity is provided by the oxidation of glucan synthesized in the preceding light period. Cessation of nitrogenase synthesis after 8 hours darkness coincides with depletion of glucan reserves and a sharp rise in the ADP/ATP ratio. Subsequent restoration of nitrogenase synthesis requires a lag period of 12–16 hours including at least 4 hours of light. These events are repeated every 24 hours and do not relate to the doubling time of *Gloeothece* (55–70 hours).

Mitsui, A., Kumazawa, S., Takahashi, A., Ikemoto, H., Cao, S., and Arai, T. (1986). *Nature, Lond.* **323**, 720.

Mullineaux, P. M., Gallon, J. R., and Chaplin, A. E. (1981). *FEMS Microbiol. Lett.* **10**, 245.

Diurnal fluctuations in the concentration of nitrogen-containing compounds in cultures of *Gloeothece*

J.H. Thomas, S. Perry, J.R. Gallon, and A.E. Chaplin
Department of Biochemistry, University College of Swansea, Swansea SA2 8PP, UK

Gloeothece, sp., a unicellular cyanobacterium, is capable of evolving O_2 by photosynthesis and, simultaneously, fixing N_2. When grown under a cycle of alternating light and darkness, *Gloeothece* effects a temporal separation of the two processes. Thus N_2 fixation, typically zero at the start of the dark period, rises to reach a maximum rate some 6–8 hours after the start of the dark period and subsequently declines, becoming zero at or shortly after the start of the next light period.

The rate of protein synthesis is some 80 per cent lower in the dark than in the light. On the other hand, nitrogenase synthesis occurs only in the dark. If protein synthesis decreases in the dark, what is the fate of the ammonia produced by nitrogen fixation during the dark period and what, in the absence of N_2 fixation, provides the nitrogen required for protein synthesis in the light?

Of the nitrogen-containing compounds examined, including cyanophycin, phycocyanin, total free amino acids, or total free nucleotides, although some fluctuations were observed, none fluctuated in sympathy with fluctuations in the rate of N_2 fixation. The only nitrogen-containing compound known to be synthesized extensively during the period of maximum nitrogenase activity is nitrogenase. An intriguing possibility is that nitrogenase itself acts as the major store of nitrogen fixed in the dark period.

Nitrogen fixation by *Oscillatoria* sp. UCSB8 under alternating light and darkness

M.A. Hashem, J.R. Gallon, and A.E. Chaplin
Department of Biochemistry, University College of Swansea, Swansea SA2 8PP, UK

Oscillatoria sp. UCSB8 is a non-heterocystous, filamentous cyanobacterium capable of aerobic N_2 fixation. It can grow photoheterotrophically in the light in the presence of glucose and 20 μM 3-(3,4-dichlorophenyl)-1,1-dimethyl urea, a specific inhibitor of photosystem II. However, the doubling time of the organism under photoheterotrophic conditions (15 days) was longer than that observed under photoautotrophic conditions (10 days).

When grown under a cycle of alternating light and darkness, cultures of this organism fixed nitrogen mainly in the dark period. However, marked differences were observed in this pattern of nitrogen fixation when cultures grown autotrophically were compared with those grown photoheterotrophically. In autotrophically grown cultures acetylene reduction activity, typically zero at the start of the dark period, became measurable after 2–3 hours of darkness, reached a peak after 6–7 hours, and then declined steadily, becoming negligible soon after the end of the dark period. No measurable rates of activity occurred in the light period. Conversely, in photoheterotrophically grown cultures, N_2 fixation occurred both in the dark and in the light but at a lower rate in the light than in the dark. Significant rates of activity were observed throughout each 24 hour period. The specific activity (1.6–18.9 nmol C_2H_4 min^{-1} mg $protein^{-1}$) of photoheterotrophic cultures was far greater than that of autotrophic cultures (0.06–1.2 nmol C_2H_4 min^{-1} mg $protein^{-1}$) throughout the period of observation.

Part 4: Regulation

10 Calcium-Mediated regulation in the cyanobacteria?

R.J. SMITH

Department of Biological Sciences, University of Lancaster, Bailrigg, Lancaster LA1 4YQ, UK

Calcium-mediated regulation

The role of Ca^{2+} in contraction and neuromuscular transduction was first described a century ago (Ringer 1882; Locke 1894). During the last two decades the recognition of Ca^{2+} as a second messenger which effects the regulation of diverse events in both animal and plant cells has been well substantiated (Means *et al.* 1982; Klee and Vanaman 1982; Case 1980; Rasmussen and Goodman 1977; Klee *et al.* 1980); a process assisted by the discovery of calmodulin (Cheung 1980*a*; Cormier 1983). In general terms, cells maintain a low intracellular concentration of Ca^{2+} (Ashley 1983; Ashley and Campbell 1979). Calcium efflux against the electrochemical gradient requires energy and employs Ca^{2+}- and ATP-dependent membrane transport systems (Carafoli *et al.* 1986). In most bacteria that have been investigated, H^+/Ca^+-antiporter mechanisms are present (Hasan and Rosen 1979), though a Na^+-antiporter is found in halobacteria (Ando *et al.* 1981). The influx of Ca^{2+} down the electrochemical gradient, which is maintained by efflux, is a passive event. The passage of Ca^{2+} across the membrane utilizes Ca^{2+}-specific channels which differ in their ability to respond to signals such as membrane depolarization and hormonal triggers (Reuter 1986).

Cytoplasm has a Ca^{2+} binding capacity in excess of the normal Ca^{2+} content (Hodgkin and Keynes 1957) and therefore the propagation of a Ca^{2+} flux through the cytoplasm is limited by this buffering capacity (Rose and Loewenstein 1975). Mechanisms have evolved in eukaryotes, such as that based on inositol triphosphate (Berridge 1986), which are capable of propagating Ca^{2+}-mediated regulation. The concept of free and bound Ca^{2+} is further extended in eukaryotes by the ability of intracellular inclusions and membranes to bind free Ca^{2+}. This additional binding may assist in buffering the cytoplasmic concentration of Ca^{2+} and also in the control of Ca^{2+}-mediated events through the distal triggering of Ca^{2+} release (Muallem *et al.* 1985; Volpe *et al.* 1985).

The number of processes which are thought to be under Ca^{2+}-mediated regulation in animal (Marme and Dieter 1983) and plant (Dieter 1985) cells is steadily increasing. Such processes ultimately depend upon proteins, for instance calmodulin (Cheung 1980*b*; Klee *et al.* 1986), which are capable of specific binding to Ca^{2+}. These proteins promote, through allosteric interaction, not only an amplification of the regulatory signal, but also a more prolonged response through the activation of protein kinases and phosphorylases. However, the mechanism dependent upon calmodulin is complex and interactive and in animal cells it is interwoven with cyclic nucleotide-mediated regulation. Although the Ca^{2+}-binding domains of several proteins have received much attention (Williams 1986) this interest has not produced any substantial evidence that proteins containing such domains are present in prokaryotes. This lack of evidence appears to have engendered the belief that Ca^{2+}-dependent regulation is not a feature of the prokaryotic organism (Cormier 1983). On the other hand, Ca^{2+}-binding proteins have recently been shown to be present in extracts of *Escherechia coli* (Harman *et al.* 1985). Though they may possess Ca^{2+} binding sites which differ from those present in eukaryotes, these proteins offer a means by which a Ca^{2+}-mediated regulatory signal could be detected in a prokaryote.

Calcium and cyanobacteria

Although reference to Ca^{2+} occurs only infrequently in the literature describing cyanobacteria, there have recently been published a few reports of Ca^{2+}-dependent processes. Prominent amongst these are those which note a Ca^{2+} requirement within photosystems (PS) I and II. For example, the *in vitro* PSII activity of extracts derived from *Anacystis nidulans*, a unicellular cyanobacterium, is dependent upon the concentration of Ca^{2+} present in the extraction buffers (England and Evans 1983) and upon the concentration of Ca^{2+} in the medium upon which the culture was grown (Roberts and Evans, this volume p 233). Extended observations of this phenomenon have prompted speculation about a role for Ca^{2+} in the regulation of PSII (Evans *et al.* 1983). A Ca^{2+} requirement *in vivo* is suggested by the decreased activity of PSII in *A. nidulans* cultures depleted of Ca^{2+} (Becker and Brand 1985). Calcium-mediated activation of O_2 evolution has also been observed in extracts of *Phormidium luridum*. Addition of Ca^{2+} allowed a four fold increase in the number of active photosynthetic units, though Mg^{2+}, which promoted a doubling of turnover rate, was also required to account for an overall eightfold increase in O_2 evolution (Piccioni and Mauzerall 1978). The Ca^{2+} requirement for O_2 evolution is thought to be involved in electron transport between component Z and P680 on the water-splitting side of PSII (Evans *et al.* 1983; Satoh and Katoh 1985).

A Ca^{2+}-dependent ATPase is associated with a particle named AF_1, which resembles the CF_1 particle of higher plant chloroplasts. The AF_1 particle has been isolated from the thermophilic cyanobacterium *Mastigocladus laminosus* and reconstituted into the organism's thylakoid membranes (Wolf *et al.* 1981). A similar latent Ca^{2+}-ATPase has been isolated from the alkalinophile *Spirulina platensis*. This preparation provides high rates of phosphorylation following reconstitution into thylakoid membranes and, like Ca^{2+}-dependent proteins in eukaryotes (Klee *et al.* 1986), may be activated by limited digestion with trypsin (Owers-Nahri *et al.* 1979).

A Ca^{2+} requirement has been associated with cyanobacterial N_2 fixation. Eyster (1972) described Ca^{2+} as an essential requirement for the growth of *Nostoc muscorum* and distinguished two critical concentrations of Ca^{2+} in the medium; one was required for growth in the presence of combined nitrogen (7.5×10^{-8} M) and another, two orders of magnitude greater (7.5×10^{-6} M), was required for growth on N_2. The latter requirement may be similar to that found in *Gloeothece* sp. (previously known as *Gloeocapsa* sp.) by Hamadi and Gallon (1981; also see Gallon and Hamadi 1984). They proposed, from observations of the effect of Ca^{2+} chelating agents upon nitrogenase activity in the presence and absence of O_2, that Ca^{2+} may be involved in a process by which nitrogenase is protected from inactivation by O_2.

The trichomes of most filamentous cyanobacteria are motile (Castenholz 1973). Although there is general agreement that external Ca^{2+} is an essential factor in this motility (Castenholz 1967; Halfen and Castenholz 1971), the precise role of the element is unclear. The contractile function may perhaps be attributed to the parallel arrays of fibres which are present in the peptidoglycan layer of the cell envelopes of *Oscillatoria princeps* (Tahmida Khan and Godward 1977) and *S. platensis* (Van Eykelenburg 1977). Several agents which were shown to stimulate the motility of *Spirulina subsala* in the presence of Ca^{2+} and Na^+ failed to do so with Ca^{2+} alone. Thus increased influx of Ca^{2+} did not itself appear to stimulate motility. These observations, together with the presence of the fibres in the cell wall, led Abeliovitch and Gan (1982) to suggest that the site of action of Ca^{2+} is extracellular. However, the effects of the calcium channel antogonists, ruthenium red and lanthanum, on *Phormidium uncinatum* (Haeder 1982) would suggest otherwise. The work of Murvanidze and Glagolev (1983) may offer a rationalization of these conflicting opinions. They have observed a wave of membrane depolarization which is transmitted along filaments of *P. uncinatum* on switching from light to dark. This depolarization, along with the photophobic reversions in trichome movement, were found to be Ca^{2+}-dependent (Murvanidze 1981) and would suggest that a change in transmembrane potential, as well as Ca^{2+} influx, is required as a signal for phototaxis. It is plausible that Na^+ as well as Ca^{2+} is required for the depolarization event.

An increase in the rate of uptake of inorganic phosphate has been reported on addition of micromolar concentrations of Ca^{2+} to cyanobacterial cultures. This increase has been attributed to the deposition of polyphosphate granules as a pH-dependent complex with divalent cations (Falkner *et al.* 1980). However, the effect of trifluoroperazine (Kerson *et al.* 1984), an inhibitor of Ca^{2+} binding by proteins, suggests a more specific interaction which deserves further investigation. The sporulation of akinetes in cultures of *Anabaena cylindrica* Lemm containing low concentrations of inorganic phosphate exhibits Ca^{2+} dependency (Wolk 1965). In these conditions a variety of Ca^{2+} salts prevented formation of abnormal cells, while addition of calcium glucuronate, but not other Ca^{2+} salts, enhanced akinete formation. The more recent discovery of the dependency of phosphate uptake upon Ca^{2+} necessitates further assessment of the question, but does not eliminate the possibility that a Ca^{2+} requirement may be involved in some specific aspect of akinete formation.

Although the demonstration of a Ca^{2+} requirement does not necessarily infer Ca^{2+}-mediated regulation of the process, it suggests the possibility. Indeed, any regulatory mechanism is likely to be first detected and described as a particular requirement. A number of cyanobacterial processes have been shown to have a Ca^{2+} requirement, but, in order that these requirements may be considered as Ca^{2+}-mediated regulatory processes, the effects due to Ca^{2+} must be shown to be present within the physiological range of the intracellular concentration of Ca^{2+}. Furthermore, it must be shown that the intracellular Ca^{2+} concentration (or perhaps the free Ca^{2+} concentration) does indeed vary within this range in response to physiological conditions.

A passive Ca^{2+} influx such as that associated with Ca^{2+} channels has been observed with inverted vesicles of plasmalemma prepared from *Anabaena variabilis* (Lockau and Pfeffer 1983). These authors' main concern, however, was the use of improved membrane fractionation techniques to locate and characterize a Ca^{2+}-ATPase in the plasmalemma fraction. This activity may represent the Ca^{2+} pump, which excludes Ca^{2+} from the cyanobacterial cell. *Streptococcus faecalis* is the only other prokaryote in which an $MgATP^{2-}$-dependent membrane transport of Ca^{2+} has been detected (Kobayashi *et al.* 1978).

The Ca^{2+} influx associated with trichome motility suggests that, like most other organisms, the cyanobacteria maintain a low intracellular concentration of Ca^{2+}, and a Ca^{2+} gradient across the cell membrane which facilitates Ca^{2+} influx. The influx may involve gated channels since it is blocked by known inhibitors of eukaryotic Ca^{2+} channels, such as lanthanum and ruthenium red. In addition, the neurotransmitter acetylcholine, which promotes Ca^{2+}-channel-mediated influx in certain eukaryotic tissues, has been shown to lower the Ca^{2+} concentration threshold of trichome tip movement and to increase the number of responsive trichomes when applied to cultures of *S. platensis*. The latter effect is dependent upon the presence of Ca^{2+} (Abeliovitch and Gan 1982). *In toto*, the effects of these Ca^{2+} channel agonists and antagonists suggest that a considerable conservation of the Ca^{2+} channel structure may have occurred despite the wide evolutionary divide. This inference would also be supported by the action of abscisic acid in plants, animals, and cyanobacteria (Huddart *et al.* 1986).

If 'increased permeability' may be interpreted as accumulation, then decreased energy production through either inhibition of PSI, PSII, or oxidative phosphorylation in darkened cultures leads to accumulation of Ca^{2+} (Abeliovitch and Gan 1982). Presumably this results from energy limitation of Ca^{2+} efflux. In cultures of *A. nidulans* the loss of $^{45}Ca^{2+}$ from illuminated cells was more rapid and established a lower steady state intracellular concentration than that observed in darkened cultures (Becker and Brand 1985). However, the assumption that this effect is also due to energy limitation conflicts with the observed maintenance of intracellular ATP concentrations on light step-down procedures (Akinyanju and Smith 1982). Alternative explanations, such as a regulated mechanism, have yet to be dismissed.

Ca^{2+}, heterocyst frequency, and nitrogenase activity

Prompted by studies on the effects of the phytohormone, abscisic acid, on cyanobacteria, recent work in our laboratory has provided evidence that the cyanobacterial processes which may be influenced by Ca^{2+} extend to heterocyst differentiation and N_2 fixation. The heterocyst is a thick-walled, specialized cell which assists N_2 fixation by harbouring the oxygen-labile nitrogenase complex within a microanaerobic environment (Carr 1979; Adams and Carr 1981; Wolk 1983). The long-standing suggestion that the characteristic thickened cell walls of the heterocyst provide a permeability barrier to O_2 has recently been supported by experimental evidence (Walsby 1985). This relative impermeability and the absence of PSII activity (Bradley and Carr 1977) would allow the maintenance of a low concentration of dissolved O_2 within the heterocyst. However, the availability to the heterocyst of the required energy and carbohydrate (reductant) under different physiological conditions (Stewart 1980) and their regulation with respect to the requirements of N_2 fixation (Wolk 1978) are unclear.

Cytodifferentiation of the heterocyst and induction of N_2-fixing activity in filamentous cyanobacteria occurs following removal of exogenous sources of combined nitrogen (Carr 1979; Wolk 1983). The differentiation of the smaller product of a vegetative cell division into a heterocyst proceeds sequentially through considerable morphological changes (Bradley and Carr 1977), including membrane rearrangment, cell wall formation, and cell pore deposition. These changes require substantial protein synthesis (Fleming and Haselkorn 1974) and realignment of the transcriptional specificity (Lynn *et al.* 1986). Both the molecular and morphological progression of differentiation proceeds via commitment steps, which are defined by the ability or otherwise of partially differentiated cells to revert to the vegetative condition (Bradley and Carr 1976, 1977).

The mechanisms initiating and controlling this progressive differentiation are unknown, though current opinion ascribes the establishment and maintenance of the regular spacing of heterocysts along the filament to an unidentified inhibitor and to a gradient of inhibition which is generated by the heterocyst and passes through adjacent vegatative cells (Wilcox *et al.* 1973). An assumption of this theory, that the inhibitor would reflect the state of nitrogen metabolism within the filament, is in agreement with the effects on heterocyst spacing of various inhibitory agents, notably methionine sulphoximine (Wolk *et al.* 1976). However, the pattern is established prior to the formation of mature, N_2-fixing heterocysts and also in nitrogen-free cultures incubated under argon (Wilcox 1970; Neilson *et al.* 1971), indicating that neither heterocyst maturation nor N_2 fixation itself is necessary for the operation of the regulatory process.

Unlike the heterotrophic bacteria, the intracellular amino-acid pools of cyanobacteria are not rapidly depleted on removal of exogenous sources of combined nitrogen (Ownby *et al.* 1979). The accessory pigment phycocyanin appears to act as an amino acid reservoir (Foulds and Carr 1977) which is mobilized by specific and regulated proteolysis (Gupta and Carr 1981). This would suggest that the mechanism governing heterocyst cytodifferentiation responds to a signal which is more refined than gross nitrogen deficiency.

The inhibitor model of the regulation of heterocyst differentiation has much to commend it, and even though the nature of the inhibitor molecule(s) and the mode of action have yet to be defined there is considerable circumstantial evidence in support of the theory. However, the model does not exclude the possibility that other regulatory mechanisms may contribute to the modulation of heterocyst frequency and nitrogenase activity either directly or indirectly. Encouraged by our observations of the effects of abscisic acid on heterocyst frequency we have used a variety of techniques to accumulate evidence for a Ca^{2+}-mediated effect.

In cultures of *Nostoc* 6720 (Rippka *et al.* 1979), caused to form heterocysts by removal of supplemented nitrate, heterocyst frequency was found to vary directly with the concentration of Ca^{2+} present in the growth medium. Over the range 10^{-5}–10^{-4} M Ca^{2+} the frequency increased from 5 per cent to 9 per cent and continued to rise to 11–12 per cent as the concentration of Ca^{2+} was increased to 10^{-3} M. Above 2×10^{-3} M the heterocyst frequency was rapidly decreased, but this decrease was associated with an inhibition of growth which may be caused by phosphate limitation resulting from precipitation of calcium phosphate salts. The acetylene reduction (N_2-fixing) capability of cultures of *Nostoc* 6720 also varied with respect to the prevailing concentration of Ca^{2+} and decreased by around 20 per cent over the range 10^{-5} to 10^{-4} M Ca^{2+}, remained constant between 10^{-4} to 10^{-3} M Ca^{2+}, and increased in

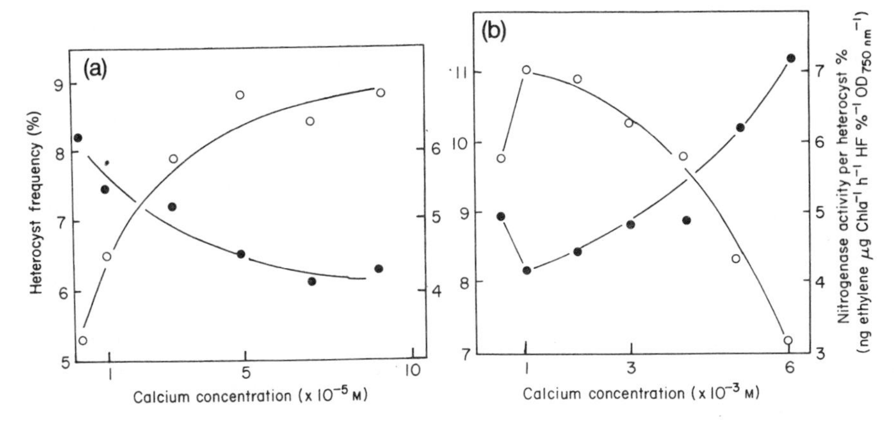

Fig. 10.1. The effect of the concentration of Ca^{2+} in the growth medium on the heterocyst frequency (○) and the nitrogenase activity per heterocyst % (●) in cultures of *Nostoc* 6720 caused to differentiate heterocysts by removal of nitrate. (a) $Ca^{2+} \times 10^{-5}$ M; (b) $Ca^{2+} \times 10^{-3}$ M.

growth-limited cultures exposed to greater than 2×10^{-3} M Ca^{2+}. The rate of acetylene reduction per heterocyst was significantly decreased at concentrations of Ca^{2+} over the range of 10^{-5}–10^{-4} M because of the increase in the heterocyst frequency (Fig. 10.1). The effect of the external concentration of Ca^{2+} on both heterocyst frequency and acetylene reduction was light-dependent. Over the range 10^{-5}–10^{-4} M Ca^{2+}, neither the increase in heterocyst frequency nor the decrease in acetylene reduction was observed in cultures incubated under irradiances exceeding 30 μE m^2 s^{-1}.

Compound A23187 (calimycin) has been described as a Ca^{2+} ionophore (Campbell 1983), but is also known to affect the permeability of membranes to other cations (Hinds and Vincenzi 1985). Compound A23187 increased the heterocyst frequency in cultures of *Nostoc* 6720 in a dose-dependent manner, increasing the frequency by 50 per cent in comparison to untreated controls at an optimum ionophore concentration of 10^{-7} M. A decline in the heterocyst frequency occurred in excess of the optimum concentration, which may result from detrimental effects of the ionophore on the permeability and functional integrity of the membrane. In the presence of A23187, the increase in heterocyst frequency with respect to the concentration of Ca^{2+} in the growth medium exceeded that observed with Ca^{2+} alone (Fig. 10.2). The interaction between the concentration of Ca^{2+} and the ionophore was found to be significant ($p > 0.001$, analysis of variance) in both experiments. In the presence of A23187 the acetylene reduction capability of the culture was inhibited when the concentration of Ca^{2+} added to the medium was less than 10^{-4} M, but was increased in excess of that found in untreated controls at

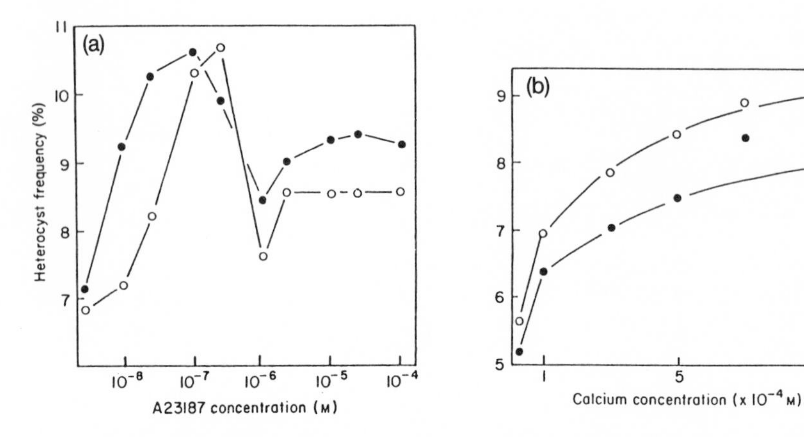

Fig. 10.2. (a) The effect of the ionophore A23187, with 10^{-3} M Ca^{2+} (○) and 10^{-4} M Ca^{2+} (●), on the heterocyst frequency of cultures of *Nostoc* 6720. (b) The effect of Ca^{2+} on the heterocyst frequency of cultures of *Nostoc* 6720 in the presence (○) and absence (●) of 5×10^{-7} M A23187.

concentrations of Ca^{2+} greater than 10^{-3} M (Fig. 10.3).

A similar response of both heterocyst frequency and acetylene reduction capability was present in cultures treated with abscisic acid (ABA) (Fig.

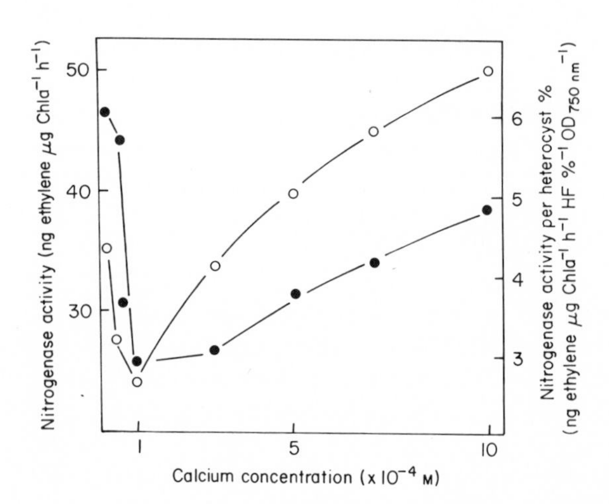

Fig. 10.3. The effect of Ca^{2+} on the nitrogenase activity (○) and the nitrogenase activity per heterocyst (%) (●) in cultures of *Nostoc* 6720 incubated in the presence of 5×10^{-7} M A23187.

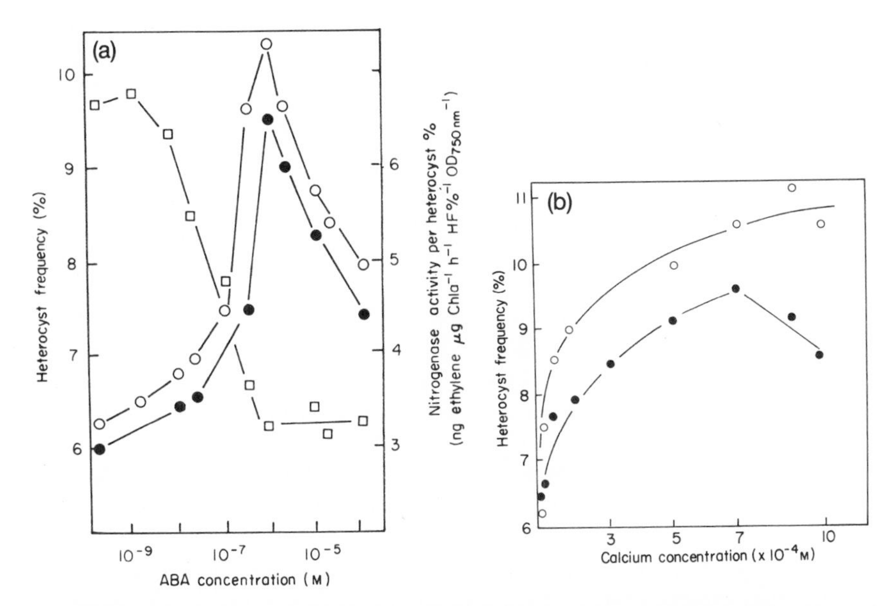

Fig. 10.4. (a) The effect of abscisic acid (ABA) on heterocyst frequency (○, ●) and nitrogenase activity per heterocyst (%) (□) in cultures of *Nostoc* 6720. Cultures were incubated in the presence of 10^{-4} M Ca^{2+} (open symbols) or 10^{-5} M Ca^{2+} (closed symbols). (b) The effect of Ca^{2+} on the heterocyst frequency of cultures of *Nostoc* 6720 in the presence (○) and absence (●) of ABA.

10.4). De Silva *et al.* (1985*a*, 1985*b*) have shown that the action of ABA on the inhibition of stomatal opening is dependent on Ca^{2+} and is inhibited by lanthanum (La^{3+}) and the calcium channel antagonists, verapamil and nifedipine. Further studies have shown that ABA also stimulated K^+-induced rat muscle contraction, which, coupled with its effect on the prokaryotic *Nostoc* 6720 (Fig. 10.4), raises the question of whether ABA is a universal Ca^{2+} agonist (Huddart *et al.* 1986). The effects of ABA on heterocyst frequency, like those of A23187 (Fig. 10.2), showed a significant interaction between the phytohormone and the extracellular concentration of Ca^{2+}. Also the heterocyst frequency declined when the optimum concentration of ABA (10^{-6} M) was exceeded (Fig. 10.4). A negative response to ABA concentration is observed in acetylcholine-induced contractures of rat urinary bladder smooth muscle (Huddart and Langton, personal communication) and may result from some impairment of membrane function. The similar effects of A23187 and ABA presumably devolve from their common ability to potentiate the entry of Ca^{2+} into the cyanobacterial cell. Both also behaved similarly, in that as the incident irradiance of treated cultures was increased the effects of A23187 and ABA on heterocyst frequency and N_2 fixation were decreased.

Lanthanum is a competitive inhibitor of Ca^{2+}-dependent processes (Van Breemar *et al.* 1972) and has been used as a Ca^{2+} channel antagonist (Hodgson *et al.* 1972). The dose response of heterocyst frequency to La^{3+} concentrations between 10^{-12} and 10^{-4} M is one of steady decline (Fig. 10.5). However, although La^{3+} is a competitive inhibitor it is notable that no significant difference in this dose response was observed at competing concentrations of Ca^{2+} which differed by an order of magnitude (10^{-5} versus 10^{-4} M). Furthermore, La^{3+} decreased the heterocyst frequency over a range of concentrations of Ca^{2+} with no evidence of alleviation of the inhibitor's action at the higher concentrations [Fig. 10.5(b)]. The acetylene reduction capability of cultures treated with La^{3+} was enhanced relative to an untreated control at concentrations up to 10^{-9} M La^{3+}. At concentrations which exceeded this optimum, acetylene reduction declined [Fig. 10.5(a)].

Lanthanum becomes a more effective inhibitor of heterocyst differentiation as the irradiance of cultures is increased. After extended incubation of cultures in the presence of La^{3+}, cell division ceased, cells and filaments were disrupted, and a progressive loss of chlorophyll was observed. All these effects appeared more rapidly as the irradiance of the culture was increased.

The enhancement of acetylene reduction capability in cultures possessing increased intracellular concentrations of Ca^{2+} (e.g. A23187- and ABA-treated cultures in the presence of concentrations $>10^{-3}$ M Ca^{2+}) could be attributed to the protection of nitrogenese from O_2 by a mechanism akin to

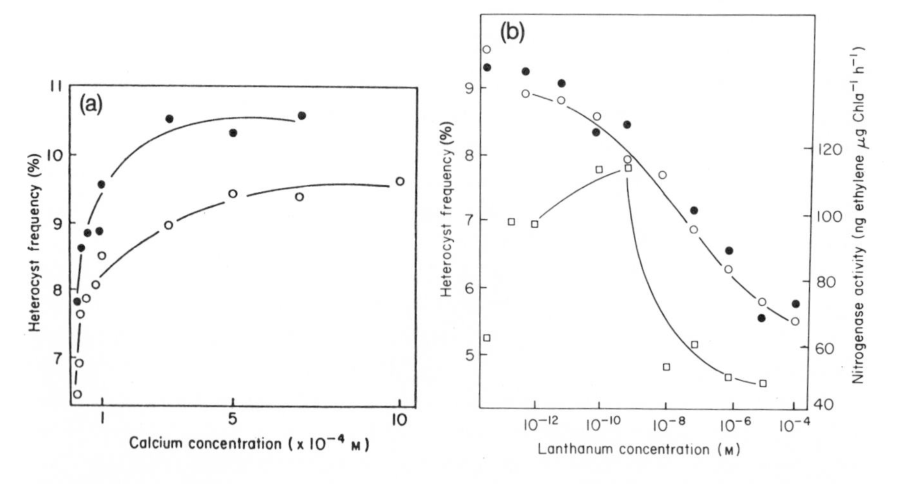

Fig. 10.5. (a) The effect of La^{3+} on the heterocyst frequency (○, ●) and nitrogenase activity (□) in cultures of *Nostoc* 6720. Cultures were incubated in the presence of 10^{-4} M Ca^{2+} (open symbols) or 10^{-3} M Ca^{2+} (closed symbols). (b) The effect of Ca^{2+} on the heterocyst frequency of cultures of *Nostoc* 6720 in the presence (○) and absence (●) of 10^{-4} M La^{3+}.

that indicated in *Gloeothece (Gloeocapsa)* sp. (Gallon and Hamadi 1984). Other explanations, such as increased reductant and carbohydrate from enhanced photosynthesis (Stewart 1980), may also be valid. On the other hand, the enhanced acetylene reduction observed in Ca^{2+}-depleted [$< 10^{-4}$ M Ca^{2+} in the presence of ABA or A23187; Fig. 10.1(b) and Fig. 10.3] or lanthanum-treated cultures (Fig. 10.5) runs contrary to these explanations and implies that a second Ca^{2+}-mediated mechanism might be present (Smith *et al.* 1987*a*, 1987*b*).

The effects upon heterocyst frequency and acetylene reduction reported here are attributed to changes in the intracellular Ca^{2+} content. However, the effects of the treatments upon the cellular contents of other ions (e.g. Mg^{2+}) have not been assessed. Nevertheless, all four treatments produce consistent results and the interactions between A23187, ABA, La^{3+}, and the external concentration of Ca^{2+} are significant, indicating that the observed effects are Ca^{2+}-mediated. Other observations of the effects of Ca^{2+} channel antagonists and inhibitors of the binding of Ca^{2+} to proteins support this conclusion (Smith *et al.* 1987*b*, unpublished data).

The intracellular concentration of Ca^{2+} in cyanobacteria

The four treatments described above were chosen for their ability to promote changes in the intracellular concentration of Ca^{2+} in the cyanobacteria, based on the assumption that they would have the same mode of action as that demonstrated on other cell types. A technique based on $^{45}Ca^{2+}$ labelling was developed in order to test these assumptions. In pulse-labelled cultures the association of $^{45}Ca^{2+}$ with cyanobacterial cells was found to be biphasic. The two components were distinguished by brief washing with EGTA at 4°C. The EGTA-removable fraction (REM), which is also composed of two components, persists in heat-killed and sonicated cultures. The EGTA-resistant fraction (RES) is absent after these treatments. On this basis the RES fraction is taken as that due to intracellular accumulation of Ca^{2+} and the REM fraction as that due to external Ca^{2+} binding.

After incubation of a culture with $^{45}Ca^{2+}$ for at least one mean generation time it may be assumed that the radioisotope is in equilibrium with cellular Ca^{2+} and therefore provides a relative estimate of the Ca^{2+} content of the culture. In cultures containing 10^{-4} M Ca^{2+}, and incubated in the presence of ABA or A23187, the RES fraction was significantly increased in comparison to untreated controls (Smith *et al.* 1987*a*, 1987*b*). These results are in accordance with the assumed mode of action of these Ca^{2+} agonists. In cultures incubated with decreased concentrations of Ca^{2+} in the growth medium over the range 10^{-5}–10^{-4} M the Ca^{2+} content of the cells was significantly decreased. Thus the effects of the concentrations of Ca^{2+}, ABA, and A23187 on heterocyst frequency are correlated with their effects on the intracellular concentration of Ca^{2+}. However, the concentration of Ca^{2+} in La^{3+}-treated cultures was found to increase between two' and threefold. Since La^{3+} decreased the heterocyst frequency a decrease in the intracellular Ca^{2+} content would have been expected. Although often assumed otherwise, La^{3+} may enter cells (Dunbar 1982). The presence of lanthanum inside the cell could disrupt Ca^{2+}/protein interactions and lead to the observed effects on Ca^{2+} uptake and external binding. Evidence that La^{3+} induces the expression of a gene encoding a Ca^{2+}-binding protein and causes accumulation of Ca^{2+} by inhibiting the normal regulation of Ca^{2+} influx (Jackson *et al.*, this volume p 233) supports this notion.

Although the experiments described here suggest that Ca^{2+} may modulate both heterocyst frequency and nitrogenase activity in *Nostoc* 6720, they are based on artificial means of manipulating the intracellular concentration of Ca^{2+} and thus provide no information concerning the physiological role of the mechanism involved. However, it is notable that all the effects described are dependent upon culture irradiance, such that both illumination and culture density affect the response. Work in progress (Smith and Wilkins, this volume p 234) has demonstrated that the intracellular concentration of Ca^{2+} in cultures of *Nostoc* 6720 varies inversely with the incident irradiance at illuminations below 30 $\mu E\ m^{-2}\ s^{-1}$. In agreement with the decreased rate of efflux of Ca^{2+} from darkened *A. nidulans* cultures (Becker and Brand 1985) the efflux of Ca^{2+} from *Nostoc* 6720 cultures was also found to be light-dependent. Furthermore, whereas heterocyst frequency in *Nostoc* 6720 is positively correlated with the external concentration of Ca^{2+}, it shows negative correlation with incident irradiance, at low intensities, similar to that found in *Anabeana cylindrica* (Fogg 1949).

These effects on heterocyst frequency and acetylene reduction in *Nostoc* 6720 add to the list of cyanobacterial processes which may undergo Ca^{2+}-mediated regulation. They also demonstrate that the range of intracellular Ca^{2+} concentrations which may be obtained by artificial means in growing cultures is sufficient to promote such regulation.

Concluding remarks

The literature supports the concept that cyanobacteria maintain a low intracellular concentration and electrochemical gradient of Ca^{2+}, despite passive Ca^{2+} influx, by employing an energy-dependent efflux. There are many examples recorded in which the artificial manipulation of this equilibrium results in changes in the metabolic and physiological responses of cyanobacterial cultures, indicating that these responses are within the range of intracellular concentrations of Ca^{2+} that are accommodated by cyanobacteria. Furthermore, the demonstration that the heterocyst frequency and the intracellular content of Ca^{2+} in *Nostoc* 6720 are inversely related to the incident irradiance provides a *raison d'etre* for Ca^{2+}-mediated regulation in the form of a response to an environmental variable. Thus the capability is present in cyanobacteria for both transient regulatory phenomena, such as the triggering of motility by Ca^{2+} influx, and long-term regulation, such as the increase in heterocyst frequency in *A. cylindrica* at low light intensities.

The demonstration of a negative correlation between the intracellular content of Ca^{2+} and incident irradiance also suggests a Ca^{2+}-mediated response attuned to high cell density and decreased illumination. This opens the way to a number of interesting speculations. For example, the activation of photosynthetic active centres by Ca^{2+} suggests a mechanism by which cyanobacteria may enhance photosynthesis to offset decreased photon flux; the enhancement of N_2 fixation by Ca^{2+}, through protection of the O_2-labile nitrogenase, could act against the effects of O_2 which might accumulate in the heterocyst due to the limitations on reductant and energy supply under low irradiances. Although too little is known at present to support realistically such speculation, these possibilities offer interesting avenues for future research. In summary one may conclude that there is much to be said in support of the hypothesis for Ca^{2+}-mediated regulation of physiological processes in the cyanobacteria, and as yet no substantial evidence against the concept. Further research is required to investigate this promising hypothesis.

Acknowledgements

I thank the *New Phytologist* for allowing the reproduction of figures. This work was financed in part by the Lancaster University Research Fund. I thank my colleagues Professors T.A. Mansfield P.J. Lea and Drs H. Huddart and A.M. Hetherington for their assistance and interest. I am also indebted to A. Wilkins, S. Hobson, I. Ellis, G. Jackson, and S. Temple for their diligent assistance.

References

Abeliovich, A. and Gan, J. (1982). In *Photosynthetic prokaryotes. Cell differentiation and function* (ed. G.C. Papageorgiou and L. Packer) p. 353. Elsevier, New York.

Adams, D.G. and Carr, N.G. (1981). *Crit. Rev. Microbiol.* **9**, 45.

Akinyanju, J.A. and Smith, R.J. (1982). *J. Bacteriol.* **149**, 68.

Ando, A., Yabuki, M., and Kusaka, I. (1981). *Biochem. Biophys. Acta* **640**, 179.

Ashley, C.C. (1983). In *Calcium in biology* (ed. T.G. Spiro) p. 109. John Wiley, New York.

Ashley, C.C. and Campbell, A.K. (1979). *The detection and measurement of free Ca^{2+}* in cells. Elsevier, New York.

Becker, D.W. and Brand, J.J. (1985). *Plant Physiol.* **79**, 552.

Berridge, M.J. (1986) In *Calcium and the cell*, (D. Evered and J. Whelan), p. 39. John Wiley, Chichester, Sussex.
Bradley, S. and Carr, N.G. (1976). *J. gen. Microbiol.* **96**, 175.
Bradley, S. and Carr, N.G. (1977). *J. gen. Microbiol.* **101**, 291.
Campbell, A.K. (1983). *Intracellular calcium*. John Wiley, Chichester, Sussex.
Carafoli, E., Zurini, M., and Benaim, G. (1986). In *Calcium and the cell*, (D. Evered and J. Whelan), p. 58. John Wiley, Chichester, Sussex.
Carr, N.G. (1979). In *Developmental biology of prokaryotes*, (ed. E. Parish), p. 168. Blackwell Scientific Publications, Oxford.
Case, R.M. (ed.) (1980). *Cell calcium*, Vol. 2. Churchill Livingstone, New York.
Castenholz, R.W. (1967). *Nature, Lond.* **215**, 1285.
Castenholz, R.W. (1973). In *The biology of blue-green algae* (ed. N.G. Carr and B.A. Whitton) p. 320. Blackwell Scientific Publications, Oxford.
Cheung, W.Y. (1980*a*). *Calcium and cell function I: calmodulin*. Academic Press, New York.
Cheung, W.Y. (1980*b*). *Science, Washington* **207**, 19.
Cormier, M.J. (1983). In *Calcium in biology* (ed. T.G. Spiro) p. 55. John Wiley, New York.
De Silva D.L.R., Hetherington, A.M., and Mansfield, T.A. (1985*a*). *New Phytol.* **100**, 478.
De Silva D.L.R., Cox, R.C., Hetherington, A.M., and Mansfield, T.A. (1985*b*). *New Phytologist* **101**, 555.
Dunbar, S.J. (1982). *Comp. Biochem. Physiol.* **72**, 199.
Dieter, P. (1985). *Plant Cell and Environ.* **7**, 371.
England, R.R. and Evans, E.H. (1983). *Biochem. J.* **210**, 473.
Evans, E.H., England, R.R., and Manwaring, J. (1983). In *Photosynthetic prokaryotes: cell differentiation and function* (ed. G.C. Papageorgiou and L. Packer) p. 175. Elsevier, New York.
Eyster, C. (1972). In *Taxonomy and biology of blue green algae* (ed. T.V. Desikachary) p. 508. University of Madras Press, Bangladore.
Falkner, G., Horner, F., and Simonis, W. (1980). *Planta* **149**, 138.
Fleming, H. and Haselkorn, R. (1974). *Cell* **3**, 159.
Fogg, G.E. (1949). *Ann. Bot.* **13**, 241.
Foulds, T.J. and Carr, N.G. (1977). *FEMS Microbiol. Lett.* **2**, 117.
Gallon, J.R. and Hamadi, A.F. (1984). *J. gen. Microbiol.* **125**, 391.
Gupta, M. and Carr, N.G. (1981). *J. gen. Microbiol.* **125**, 17.
Haeder, D.P. (1982). *Arch. Microbiol.* **131**, 77.
Halfen, L.N. and Castenholz, R.W. (1971). *J. Phycol.* **7**, 258.
Hamadi, A.F. and Gallon J.R. (1981). *J. gen. Microbiol.* **125**, 391.
Harman, A.C., Prasher, D., and Cormier, M.J. (1985). *Biochem biophys. Res Commun.* **127**, 31.
Hasan, S.M. and Rosen, B.P. (1979). *J. Bact.* **140**, 745.
Hinds, T.R. and Vincenzi, F.F. (1985). *Cell Calcium* **6**, 265.
Hodgkin, A.L. and Keynes, R.D. (1957). *J. Physiology, Lond.* **138**, 253.
Hodgson, B.J., Kidwai, A., and Daniel, E.E. (1972). *J. Physiol. Pharmacol.* **50**, 730.
Huddart, H., Smith, R.J., Langton, P.D., Hetherington, A.M., and Mansfield, T.A. (1986). *New Phytol.* **104**, 161.
Kerson, G.W., Miernyk, J.A., and Budd, K. (1984). *Plant Physiol.* **75**, 222.

Klee, C.B. and Vanaman, C.B. (1982). *Adv. Protein Chem.* **35**, 213.
Klee, C.B., Crouch, T.H., and Richman, P.G. (1980). *A. Rev. Biochem.* **49**, 489.
Klee, C.B., Newton, D.L., Wei-Chao, N., and Haiech, J. (1986). In *Calcium and the cell*, (D. Evered and J. Whelan), p. 162. John Wiley, Chichester, Sussex.
Kobayashi, H., Brunt, J.V., and Horald, F.M. (1978). *J. biol. Chem.* **253**, 2085.
Lockau, W. and Pfeffer, S. (1983). *Biochim. Biophys. Acta* **733**, 123.
Locke, F.S. (1894). *Zent Physiol.* **8**, 166.
Lynn, M.E., Bantle, J.A., and Ownby, J.D. (1986). *J. Bact.* **167**, 940.
Marme, D. and Dieter, P. (1983). In *Calcium and cell function*, Vol. 4 (ed. W.Y. Cheung) p. 264. Academic Press, New York.
Means, A.R., Tash, J.R., and Chafauleas, J.G. (1982). *Physiol. Rev.* **62**, 1.
Muallem, S., Schoeffield, M., Pandol, S., and Sachs, G. (1985). *Proc. natn. Acad Sci. USA* **82**, 4433.
Murvanidze, G.V. (1981). *Soobshch. Akad. Nauk. Gruz. SSR* **104**, 173.
Murvanidze, G.V. and Glagolev, A.N. (1983). *Biofizika* **28**, 838.
Neilson, A., Rippka, R., and Kumsawa, R. (1971). *Arch. Microbiol.* **76**, 139.
Owers-Nahri, L., Robinson, S.J., Selvius De Roo, C., and Yocum, C.F. (1979). *Biochem. biophys. Res. Commun.* **90**, 1025.
Ownby, J.D., Shannahan, M., and Hood, E. (1979). *J. gen. Microbiol.* **110**, 255.
Piccioni, R.G. and Mauzerall, D.C. (1978). *Biochim. Biophys. Acta* **504**, 384.
Rasmussen, H. and Goodman D.B.P. (1977). *Physiol. Rev.* **57**, 422.
Reuter, H. (1986). In *Calcium and the cell*, (D. Evered and J. Whelan), p. 5. John Wiley, Chichester, Sussex.
Ringer, S. (1882). *J. Physiol.* **3**, 380.
Rippka, R., Deruelles, J., Waterbury, J.R., Herdman, M., and Stainer, R.Y. (1979). *J. gen. Microbiol.* **111**, 1.
Rose, B. and Loewenstein, W.R. (1975). *Science, Washington* **190**, 1204.
Satoh, K. and Katoh, S. (1985). *FEBS. Lett.* **190**, 199.
Smith, R.J., Hobson, S., and Ellis, I. (1987*a*). *New Phytol.* **105**, 531.
Smith, R.J., Hobson, S., and Ellis, I. (1987*b*). *New Phytol.* **105**, 543.
Stewart, W.D.P. (1980). *A. Rev. Microbiol.* **34**, 497.
Tahmida Khan, Z.N. and Godward, M.B.E. (1977). *J. cell Sci.* **28**, 303.
Van Eykelenburg, C. (1977). *Antonie Van Leewenhoek* **43**, 89.
Van Breemar, C., Fairnas, B.R., Gerba, P., and McNaughton, E.D. (1972). *Circadian Research* **30**, 44.
Volpe, P., Salviati, G., Di Virgilio, F., and Pozzan, T. (1985). *Nature, Lond.* **316**, 347.
Walsby, A.E. (1985). *Proc. R. Soc.* B **226**, 345.
Wilcox, M. (1970). *Nature, Lond.* **228**, 686.
Wilcox, M., Mitchison, G.J., and Smith, R.J. (1973). *J. cell. Science* **13**, 637.
Williams, R.J.P. (1986). In *Calcium and the cell*, (D. Evered and J. Whelan), p. 145. John Wiley, Chichester, Sussex.
Wolf, M., Binder, A., and Bachofen, R. (1981). *Eur. J. Biochem.* **118**, 423.
Wolk, C.P. (1965). *Developmental Biology* **12**, 15.
Wolk, C.P. (1978). *J. Bact.* **96**, 2138.
Wolk, C.P. (1983). *Biology of cyanobacteria* (ed. N.G. Carr and B.A. Whitton) p. 359. Blackwell Scientific Publications, Oxford.
Wolk, C.P., Thomas, J., and Shaffer, P.W. (1976). *J. biol. Chem.* **251**, 5027.

11 Thioredoxin and enzyme regulation

P. ROWELL

AFRC Research Group on Cyanobacteria and Department of Biological Sciences, University of Dundee, Dundee DD1 4HN, UK

A.J. DARLING

MRC Virology Unit, Institute of Virology, Church Street, Glasgow G11 5JR, UK

D.V. AMLA

National Botanical Research Institute, Lucknow 226001, India

W.D.P. STEWART

AFRC Research Group on Cyanobacteria and Department of Biological Sciences, University of Dundee, Dundee DD1 4HN, UK

Introduction

Thioredoxins are ubiquitous, small (M_r approx. 12 000) redox proteins. They have active centres containing two cysteine residues in close proximity, in the sequence -Cys–Gly–Pro–Cys–, which can be reversibly oxidized to form a disulphide bridge. This reversible oxidation–reduction is essential for most of the functions of thioredoxins described to date (for recent reviews, see Holmgren 1985; Holmgren *et al.* 1986). Thioredoxin was first shown to be involved in the reduction of sulphoxide and sulphate in yeast (Black *et al.* 1960; Wilson *et al.* 1961) and to function as a hydrogen donor in ribonucleotide reduction in *Escherichia coli* (Laurent *et al.* 1964). It has since been isolated and characterized from many sources and has been shown to function in various other processes, including protein disulphide reduction (Holmgren 1979*a*), the replication and assembly of certain bacteriophages (Mark and Richardson 1976; Russel and Model 1985), activation of steroid hormone receptors (Grippo *et al.* 1983), and reductive activation or deactivation of several enzymes (Wolosiuk and Buchanan 1977; Buchanan 1980, 1986). In addition, the enzyme protein disulphide isomerase of rat liver and pancreas, which catalyses thiol-disulphide interchange, has two regions of sequence homology with *E. coli* thioredoxin (Edman *et al.* 1985).

Thioredoxin-deficient mutants of *E. coli* are able to grow in laboratory culture but they do not support the replication of certain bacteriophages: DNA polymerase of bacteriophage T7 requires host cell thioredoxin for activity (Mark and Richardson 1976; Huber *et al.* 1986) and there is also a thioredoxin requirement for the assembly of filamentous phages f1 and M13 (Russel and Model 1985, 1986; Lim *et al.* 1985). Analysis of such mutants led to the discovery of glutaredoxin (Holmgren 1976), which is reduced via glutathione reductase and glutathione. Glutaredoxin, like thioredoxin, is a small (M_r approx. 10 000) dithiol-containing protein and is a more effective hydrogen donor for ribonucleotide reductase than thioredoxin (Holmgren 1979*b*). Bacteriophage T4 infection of *E. coli* results in the synthesis of a phage-specific thioredoxin (Berglund 1969) which is more closely related, structurally and functionally, to *E. coli* glutaredoxin than thioredoxin, although glutaredoxin and thioredoxin have similar three-dimensional structures and may have overlapping functions (Eklund *et al.* 1984; Holmgren 1985).

Light functions in photosynthetic organisms not only in the generation of ATP and NADPH by photosynthetic electron transfer, but also in the regulation of enzyme activities. Such regulation involves a number of factors, including changes in H^+ and divalent cation concentrations, substrate and co-factor concentrations, and light-dependent activation and deactivation of enzymes (for reviews, see Anderson 1979; Buchanan 1980, 1986; Preiss 1982). Effective enzyme modulation in response to light is necessary in order to adjust metabolic activities to cellular requirements and to avoid the futile cycling of metabolites. In many examples of light-dependent enzyme modulation, photosynthetically generated reductant is apparently linked to the reduction of disulphides of the target enzymes, either via ferredoxin–thioredoxin reductase and thioredoxin (Buchanan 1980, 1986) or via the light-effect-mediator (LEM) system (Anderson *et al.* 1986).

The ferredoxin–thioredoxin system has been extensively studied and, in this paper, we will review recent findings, for cyanobacteria in particular, and present some recent results obtained in our laboratory.

The occurrence of thioredoxins in photosynthetic organisms

Higher plant chloroplasts

In higher plants, thioredoxins occur in both green and non-green tissues (Bestermann *et al.* 1983), with chloroplasts containing thioredoxins *f* and *m* and a ferredoxin–thioredoxin reductase (Buchanan 1980; Schurmann 1981; Schurmann *et al.* 1981; Soulie *et al.* 1981; Buc *et al.* 1984; Crawford *et al.* 1986; Droux *et al.* 1987). Two chloroplastic enzymes which are routinely used for the assay of thioredoxins are fructose-1,6-bisphosphatase (FBPase), which is activated by thioredoxin *f*, and NADP-dependent malate dehydrogenase (NADP-MDH), which is activated by both thioredoxins *f* and *m*. Other chloroplast enzymes which are activated by reduced thioredoxin include sedoheptulose-1,7-bisphosphatase (SBPase), NADP-dependent glyceraldehyde-3-phosphate dehydrogenase (NADP-G3PDH), phosphoribulokinase (PRuK) and coupling factor ATPase (see Buchanan 1980, 1986). Additionally, glucose-6-phosphate dehydrogenase (G6PDH) is deactivated by reduced thioredoxin. A single thioredoxin *f* (M_r approx. 11 200) and three *m*-type thioredoxins [*m*b, *m*c, and *m*d, which are N-terminal redundant isomers of M_r 11 553, 11 425, and 11 354, respectively (Maeda *et al.* 1986)] have been isolated from spinach chloroplasts (Schurmann 1981; Schurmann *et al.* 1981). Buc *et al.* (1984) have isolated two *f*-type thioredoxins from this source.

Chloroplast ferredoxin–thioredoxin reductase (De la Torre *et al.* 1979) is an iron-containing enzyme (Schurmann 1981; Droux *et al.* 1984, 1987). Non-photosynthetic plant tissues (Cao *et al.* 1984; Buchanan 1986) have an NADP-dependent thioredoxin reductase, as do aerobic heterotrophic organisms (see Holmgren 1985; Buchanan 1986).

Algae

Three thioredoxins of M_r approx. 12 000 (Wagner and Follmann 1977; Wagner *et al.* 1978) and one of M_r approx. 28 000 (Langlotz *et al.* 1986) have been isolated from the green alga *Scenedesmus obliquus*. This larger thioredoxin, designated *f*, stimulated the activities of algal and spinach chloroplast FBPases and algal NADP-MDH, was inactive as a hydrogen donor for algal ribonucleotide reductase, but showed some activity with *E. coli* ribonucleotide reductase. Studies on a mutant strain which did not develop a chloroplast in the dark indicated that it is the main chloroplast thioredoxin of this alga and may be the equivalent of spinach chloroplast thioredoxins *f* and *m* (Langlotz *et al.* 1986). In contrast, M_r 12 000 thioredoxin activated NADP-MDH and, to a lesser extent, FBPase and was also highly active with algal and *E. coli* ribonucleotide reductases.

Two proteins (M_r approx. 14 000), identified as a thioredoxin and a glutaredoxin, and an NADP-dependent thioredoxin reductase have been isolated from *Chlorella pyrenoidosa* (Tsang 1981).

Photosynthetic bacteria

The photosynthetic purple sulphur bacterium *Chromatium vinosum* has a single thioredoxin of M_r approx. 13 000 and a NADP-dependent thioredoxin reductase (Johnson *et al.* 1984). Reduced thioredoxin, however, does not activate FBPase, SBPase, or PRuK of this photosynthetic bacterium (Crawford *et al.* 1984), indicating that thioredoxins may not be involved in the regulation of the reductive pentose phosphate pathway in anoxygenic photosynthetic organisms (Buchanan 1986).

Cyanobacteria

The occurrence of thioredoxin in a wide range of cyanobacteria was demonstrated by Schmidt and Christen (1979) using the thioredoxin-dependent adenosine-3'-phosphate-5'-phosphosulphate (PAPS) sulphotransferase of *Synechococcus* PCC 6301 as an assay system (Wagner *et al.* 1978). Thioredoxins of M_r approx. 12 000, referred to below as thioredoxin *m*, have subsequently been purified and characterized from several cyanobacteria including *Synechococcus* 6301 (Schmidt 1980), *Anabaena* PCC 7119 (*Nostoc muscorum*) (Yee *et al.* 1981; Gleason and Holmgren 1981; Gleason *et al.* 1985), and *Anabaena cylindrica* CCAP 1403/2a (Ip *et al.* 1984). Schmidt (1980) reported the isolation of two thioredoxins (A, M_r >11 800 and B, M_r approx. 11 800) from *Synechococcus* 6301. Both were active in PAPS sulphotransferase and FBPase assays, but only B was active in an adenosine-5'-phosphosulphate (APS) sulphotransferase assay.

Yee *et al.* (1981) isolated, from *N. muscorum*, thioredoxins *f* (M_r approx. 16 000) and *m* (M_r approx. 9 000), based on the activation of chloroplast FBPase and NADP-MDH, and a ferredoxin-thioredoxin reductase, similar to the chloroplast enzyme (De la Torre *et al.* 1979; see also Droux *et al.* 1987). The thioredoxin *f* was shown to selectively activate *N. muscorum* FBPase and SBPase, whereas the thioredoxin *m* selectively activated *N. muscorum* PRuK (Crawford *et al.* 1984; Sutton *et al.* 1984) and FBPase. SBPase and PRuK were shown to be light-activated, *in vitro*, by the ferredoxin–thioredoxin system.

A thioredoxin *m* (Gleason *et al.* 1985), purified from *Anabaena* 7119 using an assay based on the reduction of insulin disulphides (Gleason and Holmgren 1981), was active as a hydrogen donor to *Anabaena* and *E. coli* ribonucleotide reductases and activated spinach FBPase and NADP-MDH (Whittaker and Gleason 1984). A larger (M_r approx. 25 500) thioredoxin, designated thioredoxin *f* and possibly identical to that isolated by Yee *et al.* (1981), was purified from *Anabaena* 7119 (Whittaker and Gleason 1984) using the activation of spinach FBPase as an assay. This thioredoxin consists of a single polypeptide with a single active centre dithiol (Whittaker and Gleason 1984) and thus resembles *S. obliquus* thioredoxin *f* (Langlotz *et al.* 1986), but is larger than higher plant thioredoxin *f* (Schurmann *et al.* 1981; Buc *et al.* 1984). Although the *Anabaena* 7119 thioredoxin *f* activated spinach FBPase, it did not activate FBPase or ribonucleotide reductase of *Anabaena*, was not active in insulin reduction, and activated spinach NADP-MDH only slightly. It is questionable (Whittaker and Gleason 1984) whether activation of spinach FBPase alone is sufficient evidence for the identity of a thioredoxin, since certain unrelated proteins are known to have this ability (Wada and Buchanan 1981).

We have isolated a single thioredoxin *m* (M_r approx. 11 600) from each of several cyanobacteria [*A. cylindrica, Anabaena variabilis* ATCC 29413, *Nostoc* CAN (a free-living isolate of the cyanobiont of the lichen *Peltigera canina*), *Synechocystis* PCC 6714, *Synechococcus* PCC 6803, *Nodularia harveyana* CCAP 1452/1 and *Oscillatoria* sp. CCAP 245B] using conventional protein purification techniques (Ip *et al.* 1984) and immuno-affinity chromatography, adapted from the method of Sjoberg and Holmgren (1973), for rapid purification of small quantities ($< 100\ \mu g$) of thioredoxin. Thioredoxins were assayed using the deactivation of *A. variabilis* G6PDH, the activation of *A. variabilis* FBPase and the activation of *S. obliquus* NADPH-MDH and glutamine synthetase (GS). We have found evidence of only a single thioredoxin *m* in each case. However, failure to detect a larger thioredoxin, in conventional purification procedures, may be due to our choice of enzymes, which did not include spinach FBPase, for the assays. These thioredoxins, purified to homogeneity as judged by electrophoresis and isoelectric focusing (pI values in the range 4.3 to 5.2), were similar to each other, both structurally and in their abilities to activate/deactivate the above enzymes. *A. cylindrica* and *A. variabilis* thioredoxins (the only ones tested in this context) also activated PRuK (see below).

Thioredoxin *m*, from *Anabaena* 7119 (Gleason and Holmgren 1981; Gleason *et al.* 1985), *A. cylindrica* (Ip *et al.* 1984), and *A. variabilis*, has a single polypeptide chain of M_r 11 550, approx. 11 680, and approx. 11 540, respectively. Fig. 11.1 shows a comparison of the partial amino-acid sequences of *A. cylindrica* and *A. variabilis* thioredoxins with the amino acid sequences of thioredoxins from several other sources. Each cyanobacterial thioredoxin has two cysteine residues, at positions 31 and 34, in the sequence -Trp-Cys-Gly-Pro-Cys-Arg-. The available *A. variabilis* sequence (first 88 residues) is identical with the *Anabaena* 7119 sequence. There are 12 differences, mainly in the N-terminal region, between the partial *A. cylindrica* sequence and the other cyanobacterial sequences. The cyanobacterial sequences show extensive homology with those of *E. coli* and *Corynebacterium nephridii* thioredoxins and thioredoxin *m* from spinach chloroplasts, particularly in the region of the active site dithiol, and many of the amino acid changes are conservative. The chloroplast thioredoxin f partial sequence shows much less homology, being identical only in the active site sequence -Trp-Cys-Gly-Pro-Cys-.

Structural differences between the *A. cylindrica* and *A. variabilis* thioredoxins were apparent during the purification and characterization of the two proteins. The isoelectric points differed by 0.8 pH units (4.38 and 5.13 for the *A. cylindrica* and *A. variabilis* thioredoxins, respectively), and the amino acid compositions were different, particularly with respect to the content of lysine residues which was lower in the case of *A. cylindrica* thioredoxin, resulting in fewer tryptic peptides on peptide mapping. There were no marked differences between the thioredoxins in their ability to activate/deactivate the enzymes tested, although they were antigenically distinct. Polyclonal antibodies, raised against each of the purified proteins, showed no cross-reactivity with the other thioredoxin on immuno-electrophoretic analysis. The antiserum against *A. variabilis* thioredoxin *m* cross-reacted

		1	11	21	31	41
A. cylindrica (a)		SAAASVTDDS	FDQDVLQSDV	PVLVDFWAPW	CGPCRMVAPV	VEEIAAQYEG
A. variabilis (b)		SAAAQVTDST	FKQEVLDSDV	PVLVDFWAPW	CGPCRMVAPV	VDEIAQQYEG
Anabaena 7119 (c)		SAAAQVTDST	FKQEVLDSDV	PVLVDFWAPW	CGPCRMVAPV	VDEIAQQYEG
spinach m (d)	KASEAV	KEVQDVNDSG	WKEFVLQSSE	PSMVDFWAPW	CGPCKLIAPV	IDELAKEYSG
C. nephridii (e)		ATVKVDNSN	FQSDVLQSSE	PVVVDFWAEW	CGPCKMIAPA	LDEIATEMAG
E. coli (f)	S	DKIIHLTDDS	FDTDVLKADG	AILVDFWAEW	CGPCKMIAPI	LDEIADEYQG
spinach f (g)				...LNMFTQW	CGPCKANGDK	EATHLGVQQA

	51	61	71	81	91	101
(a)	QLKVVKVNTD	ENPQVAGRYG	IRSIPTLMIF	KGGQKVDMVV	GAVPK...	
(b)	KIKVVKVNTD	ENPQVASQYG	IRSIPTLMIF	KGGQKVDM.. .		
(c)	KIKVVKVNTD	ENPQVASQYG	IRSIPTLMIF	KGGQKVDMVV	GAVPKTTLSQ	TLEKHL
(d)	KIAVTKLNTD	EAPGIATQYN	IRSIPTVLFF	KNGERKESII	GDVSKYQL	
(e)	QVKIAKVNID	ENPELAAQFG	VRSIPTLLMF	KDGELAANMV	GAAPKSRLAD	WIKASA
(f)	KLTVAKLNID	QNPGTAPKYG	IRGIPTLLLF	KNGEVAATKV	GALSKGQLKE	FLDANLA
(g)	M...					

Fig. 11.1. A comparison of the partial amino-acid sequences of *A. cylindrica* and *A. variabilis* thioredoxins. Amino-acid sequences of *E. coli* (Holmgren 1968; Wallace and Kushner 1984), *C. nephridii* (Meng and Hogenkamp 1981), and *Anabaena* 7119 (Gleason *et al.* 1985) thioredoxins, spinach chloroplast thioredoxin *m* (Maeda *et al.* 1986), and the partial sequence of spinach chloroplast thioredoxin *f* (Tsugita *et al.* 1983) are also shown for comparison. Residue 35 (arginine) of the *A. cylindrica* sequence was previously incorrectly identified as threonine (Ip *et al.* 1984). Amino acids are written by one letter abbreviations: A, alanine; C, cysteine; D, aspartic acid; E, glutamic acid; F, phenylalanine; G, glycine; H, histidine; I, isoleucine; K, lysine; L, leucine; M, methionine; N, asparagine; P, proline; Q, glutamine, R, arginine; S, serine; T, threonine; V, valine; W, tryptophan; Y, tyrosine. . . . indicates incomplete sequencing.

with thioredoxins from a wide range of other cyanobacteria (Rowell *et al.*, unpublished data), in contrast to the antiserum against *A. cylindrica* thioredoxin *m* which failed to cross-react with thioredoxins from most other cyanobacteria tested (Ip *et al.* 1984). On immuno-electrophoretic analysis, the antiserum against *A. variabilis* thioredoxin *m* cross-reacted with a second protein in cell-free extracts of *A. variabilis* and *A. cylindrica*. This protein was not, however, detectable in purified thioredoxin preparations by either immuno-electrophoresis or Western blotting of isoelectric focusing gels. The protein (M_r approx. 50 000, by gel filtration on Sephadex G-75, and with a pI value of 4.6) showed thioredoxin-like activity in the deactivation of G6PDH and appeared to be weakly associated with thylakoid membranes (see below).

The concentration, redox state, and localization of thioredoxins in cyanobacteria

Although there is a great deal of information available on the effects of thioredoxins on the activities of enzymes *in vitro*, there have been few studies on the concentration, localization, and redox state of thioredoxins *in vivo*.

Cellular concentrations

Cellular concentrations of thioredoxins have been estimated as 15 μM for *E. coli* (Holmgren *et al.* 1978) and 100–160 μM for isolated pea chloroplasts (Scheibe 1981). However, thioredoxins may be highly localized, effectively increasing concentrations [for example, association of some chloroplast thioredoxin *m* with thylakoid membranes (Ashton *et al.* 1980) or association of thioredoxin with the plasma membrane of *E. coli* (Lunn and Pigiet 1986)].

Using rocket immuno-electrophoresis to quantify thioredoxin *m* in cell-free extracts (Ip *et al.* 1984), together with estimates of intracellular volume (see Reed *et al.* 1987), we have determined intracellular thioredoxin *m* concentrations for several cyanobacteria. These were 60 μM for the unicellular *Synechocystis* 6714, 207 μM for *N. harveyana*, 130 μM for *A. cylindrica*, and 71–135 μM for *A. variabilis*. Irrespective of the differences, which are at least partly attributable to the methods of intracellular volume estimation, the values are all within the range 60–207 μM, indicating that the mean intracellular concentrations are very similar to those in isolated chloroplasts (Scheibe 1981).

Thioredoxin *m* constituted 0.22 per cent and 0.27 per cent of the total protein in cell-free extracts of *A. cylindrica* (Ip *et al.* 1984) and *A. variabilis*, respectively, whether cells were N_2-fixing, NO_3^--grown, or NH_4^+-grown. The lack of an effect of nitrogen source on the level of thioredoxin *m* contrasts with the findings of Schmidt and Christen (1979) who obtained evidence of a higher level of thioredoxin activity in N_2-fixing than in NO_3^--grown *A. variabilis*. Cell-free extracts of isolated heterocysts of *A. cylindrica* had about 30 per cent of the thioredoxin *m* content (as a percentage of total extractable protein) of extracts of whole filaments (Ip *et al.* 1984). It is difficult, however, to take account of losses occurring during preparation and the degree of contamination by vegetative cells. Immuno-cytochemical labelling studies (see below) also indicated that heterocysts have a decreased level of thioredoxin *m*. There is no information available for thioredoxin *f*.

Redox state

In cell-free extracts of *E. coli* (Holmgren and Fagerstedt 1982) and *A. cylindrica* (Darling *et al.* 1986), thioredoxin is largely oxidized. When cells are permeabilized and treated with iodoacetate to alkylate the reduced form

Table 11.1 Effects of light, darkness and carbon and nitrogen sources on the redox state of thioredoxin *m* in *Anabaena cylindrica* and *Anabaena variabilis*.

Cyanobacterium	Carbon or nitrogen sources added	Light/dark	Reduced thioredoxin (% of total)
A. cylindrica	none	light	72
	NO_3^-	light	70
	NH_4^+	light	69
	none	dark (b)	10
	NO_3^-	dark (b)	12
	NH_4^+	dark (b)	10
A. variabilis	none	light	63
	fructose	light	61
	fructose + NO_3^-	light	61
	none	dark (b)	26
	fructose	dark (a)	37
	fructose + NO_3^-	dark (a)	39

Cultures were grown at 25 °C in BG-11_0 medium, aerated at 1 ℓ per min, with fructose (5 mM), nitrate (5 mM) or ammonium (5 mM) added as indicated, in the light at 55 μmol photons m^{-2} s^{-1} incident at the surface of the vessel. Where indicated, cultures were grown in (a) darkness or (b) light followed by incubation for 3 hours in darkness before assay. Oxidized and reduced thioredoxin were estimated as described by Darling *et al.* (1986). Cetyltrimethylammonium bromide, to 0.02 per cent (w/v), and iodoacetate, to 15 mM, at pH 7.5, were added to aliquots of culture treated as described above. After a further 15 min incubation, cells were disrupted by sonication and cell debris removed by centrifugation at 35 000 × *g* for 15 min. The oxidized and reduced (carboxymethylated) forms of thioredoxin were separated and estimated by two-dimensional immuno-electrophoresis using an antiserum raised against the purified thioredoxin (Ip *et al.* 1984).

of thioredoxin, and thus prevent oxidation, a greater proportion of the thioredoxin is present in the reduced form: about 75 per cent in the case of *E. coli* (Holmgren and Fagerstedt 1982) and a variable amount, depending on whether cells had been incubated in the light or in darkness, in the case of *A. cylindrica* (Darling *et al.* 1986) and *A. variabilis*. Table 11.1 summarizes the results obtained and Fig. 11.2 shows time courses of changes in the redox state of *A. cylindrica* and *A. variabilis* thioredoxins which occurred on light–dark transitions. The proportion of reduced thioredoxin *m* was greater in filaments incubated in continuous light than in those incubated in darkness. On transfer from light to dark, the proportion of reduced thioredoxin decreased rather slowly ($t_{½}$ approx. 30 min), whereas, on re-illumination, it increased rapidly to over 80 per cent in less than 15 min and remained high in *A. cylindrica* though it declined to a lower value in *A. variabilis* and did not achieve a steady state within the experimental period.

Thus, *A. cylindrica* and *A. variabilis* show a rapid reduction of thioredoxin *m* on transfer from dark to light, which is consistent with reduction via a ferredoxin–thioredoxin reductase (Yee *et al.* 1981; Droux *et al.* 1987),

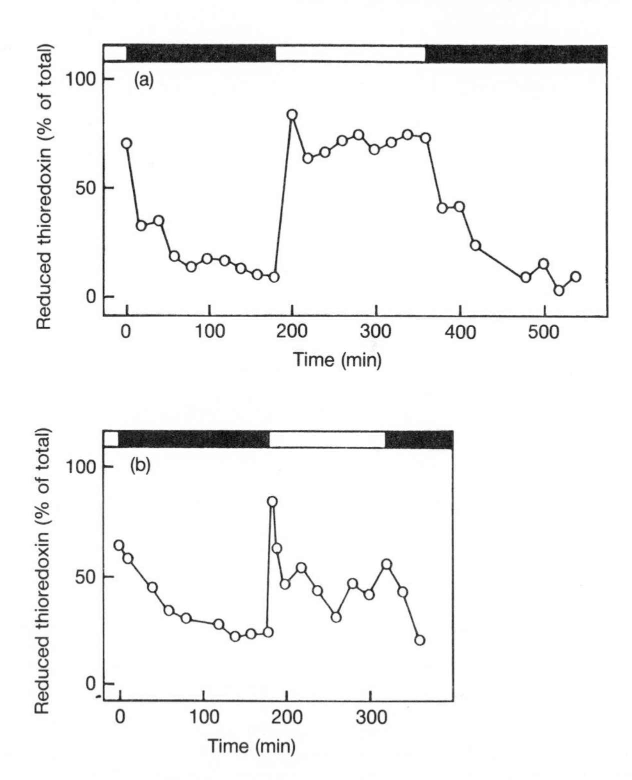

Fig. 11.2. The effects of light–dark transitions on the redox state of thioredoxin *m* in (a) *A. cylindrica* (from Darling *et al.* 1986) and (b) *A. variabilis*. Cultures, grown in continuous light (see Table 11.1) in BG-11_0 medium (2.5 and 1.9 μg chl *a* ml^{-1} for (a) and (b), respectively), were incubated in the light or dark as indicated by the light and dark bars, respectively. Aliquots of culture taken at timed intervals were treated as described in the footnote to Table 11.1.

dependent on photosynthetically reduced ferredoxin. The results are also consistent with the effects of light on the redox state of thioredoxin *m* in chloroplasts of *Pisum sativum* (Scheibe 1981) and *Zea mays* (Rebeille and Hatch 1986). The changes in redox state were less marked in *A. variabilis* than in *A. cylindrica*. The precise mechanisms of thioredoxin reduction have not yet been established and it is possible that, in some cases, mechanisms other than, or in addition to, ferredoxin–thioredoxin reductase are employed. In this context, it should be noted that Gleason (1986) has recently provided evidence of a NADPH-dependent thioredoxin reductase in *Anabaena* 7119.

Anabaena variabilis is capable of heterotrophic growth on fructose (see Smith 1982) and, although this sugar had no effect on the redox state of thioredoxin *m* in *A. variabilis* grown in the light (Table 11.1), the proportion of reduced thioredoxin was greater in filaments grown in darkness in the presence of fructose than in filaments grown in its absence in the light then transferred to darkness for up to 3 hours. However, the proportion of reduced thioredoxin was much smaller in filaments grown heterotrophically in the dark than in those grown in the light, which may possibly be related to the much slower growth rate under heterotrophic conditions.

Localization

E. coli thioredoxin is located in an osmotically sensitive compartment, possibly being associated with the plasma membrane at membrane adhesion sites (Lunn and Pigiet 1986). Earlier immunocytochemical techniques had indicated that the thioredoxin was located in the periplasmic space and the nucleoid region (Nygren *et al.* 1981).

Although extracts of isolated heterocysts contain thioredoxin (Ip *et al.* 1984; Udvardy *et al.* 1984), we have found, on analysis of cell-free extracts and in immuno-cytochemical labelling experiments using antisera against purified thioredoxins from *A. cylindrica* and *A. variabilis*, that the level of thioredoxin *m* is lower than in vegetative cells of the same cyanobacterium (Ip *et al.* 1984; Cossar *et al.* 1985*a*). Whether these cyanobacteria have a second thioredoxin, possibly thioredoxin *f*, located in heterocysts and/or vegetative cells remains to be determined. In vegetative cells of *A. cylindrica*, thioredoxin *m* is located mainly in the nucleoplasm and is clearly not associated with the chromatoplasm, in contrast with our findings for the Calvin cycle enzyme ribulose-1,5-bisphosphate carboxylase/oxygenase (Cossar *et al.* 1985*b*). However, antiserum against the *A. variabilis* thioredoxin, which did not cross-react with *A. cylindrica* thioredoxin but did cross-react with a second thioredoxin-like protein in cell-free extracts of *A. cylindrica* (see above), labelled the chromatoplasm of *A. cylindrica* vegetative cells, indicating that this thioredoxin-like protein is associated with thylakoid membranes of vegetative cells. Further studies to characterize this protein are in progress.

The localization of *A. cylindrica* thioredoxin *m* and the properties of cyanobacterial thioredoxins are consistent with a function, such as the hydrogen donor for ribonucleotide reductase, in the synthesis of precursors for DNA, rather than in the regulation of the activities of Calvin cycle enzymes. The decreased level of thioredoxin *m* in heterocysts may be related to the inability of mature heterocysts to dedifferentiate (Wilcox *et al.* 1973). Synchronous cultures of *Anacystis nidulans* (*Synechococcus* 6301) can be obtained by incubation in the dark followed by re-illumination (see Marino and Asato 1986), and ribonucleotide reductase activity is maximal during periods of DNA synthesis in synchronous cultures of *Synechococcus cedrorum* (Gleason 1979). It will be worthwhile to determine whether there is any involvement of thioredoxin in the effects of light on the growth of such cyanobacteria.

The role of thioredoxins in the light-dependent regulation of enzyme activities in cyanobacteria

Photosynthetic CO_2 fixation in cyanobacteria is via the Calvin cycle and heterotrophic metabolism involves the oxidative pentose phosphate pathway (Pelroy and Bassham 1972; Pelroy *et al.* 1972). The immediate cessation of CO_2 fixation that occurred on transfer of *Aphanocapsa* (*Synechocystis*) 6714 from light to dark (Pelroy and Bassham 1973; Pelroy *et al.* 1976) was attributed to inhibition of PRuK, SBPase, and FBPase activities. Duggan and Anderson (1975) demonstrated the *in vivo* light-dependent activation of PRuK and deactivation of G6PDH in *A. nidulans*, whereas FBPase, SBPase, and NADP-G3PDH were not activated under these conditions. In cell-free extracts, PRuK, SBPase, and NADP-G3PDH activities were stimulated by dithiothreitol, whereas G6PDH activity was decreased. The correlations between the effects of light and those of thiols on enzyme activities suggest the possible involvement of thioredoxins (Buchanan 1980) or the LEM system (Anderson 1979; Anderson *et al.* 1986) in the light-dependent modulation of enzyme activities.

We have so far failed to detect a thioredoxin *f* in extracts of *A. variabilis* or *A. cylindrica* using *Anabaena* FBPase as the target enzyme, in agreement with the findings of Whittaker and Gleason (1984) but in contrast with those of (Crawford et al. 1984); we did detect thioredoxin *m* by this means (Ip *et al.* 1984). However, the activation of *Anabaena* FBPase was rather weak, required relatively high concentrations of thioredoxin *m*, and did not occur during pre-incubation of the enzyme with reduced thioredoxin: it proceeded slowly after substrate addition to initiate the reaction. This is consistent with the mechanism of activation of *Synechococcus leopoliensis* FBPase (Gerbling *et al.* 1985) which involves interconversion of the major form of the enzyme between oxidized and reduced dimers (both inactive) and the Mg^{2+}-and fructose-1,6-bisphosphate (FBP)-promoted conversion of the reduced dimer to an active tetramer. Thioredoxin-dependent activation of FBPase could thus function to maintain a reduced form of the enzyme which is then regulated by changes in Mg^{2+} and FBP concentrations.

We have obtained evidence that PRuK purified from *A. cylindrica* (J.L. Serra *et al.*, unpublished data) and *Anabaena* 7119 is activated by reduced thioredoxin. This agrees with previous findings for *N. muscorum* PRuK (Crawford *et al.* 1984), but contrasts with findings for *Chlorogloeopsis fritschii* PRuK (Marsden and Codd 1984). The purified *Anabaena* 7119 enzyme is a dimer of M_r approx. 87 000 (two M_r approx. 41 600 subunits), as determined by gel filtration and SDS gel electrophoresis, respectively, although we obtained evidence of the presence of a minor higher M_r form of the enzyme in the early stages of purification. Although activation of *Anabaena* 7119 PRuK by *A. variabilis* thioredoxin *m* was observed, equivalent concentrations of spinach thioredoxin *f* were markedly more effective. The reasons for this difference in the extent of activation require further investigation. In relation to the higher M_r form of PRuK, it will be of interest to determine whether cyanobacterial PRuK and NADP-G3PDH can occur in the form of a latent, thioredoxin-activated oligomeric protein as they do in the alga *S. obliquus* (Nicholson *et al.* 1986). However, it has been reported that NADP-G3PDH of *Anabaena* 7119, *A. cylindrica*, and *A. variabilis* is not activated by reduced thioredoxin (Papen *et al.* 1986).

We have been unable to use *Anabaena* NADP-MDH as a target enzyme for detection of thioredoxins, since the activity in extracts of whole filaments is very low. Papen *et al.* (1983), have however demonstrated higher levels of this enzyme in extracts of isolated heterocysts and shown that it and NADP-isocitrate dehydrogenase, from heterocysts and vegetative cells of *A. cylindrica*, are activated by reduced thioredoxin. NAD(P)-dependent glucose dehydrogenase of *Nostoc* MAC is deactivated by reduced thioredoxin (Juhasz *et al.* 1986).

G6PDH is a key regulatory enzyme of the oxidative pentose phosphate pathway and is considered to be important in electron transfer to nitrogenase in cyanobacteria (see Apte *et al.* 1978). G6PDH of *A. variabilis* (Cossar *et al.* 1984) and *Anabaena* 7120 (Udvardy *et al.* 1984) is rapidly deactivated by low concentrations of reduced thioredoxin *m*. This may be an important factor in the inhibition of the oxidative pentose phosphate pathway in vegetative cells in the light (Pelroy *et al.* 1972; Duggan and Anderson 1975; Apte *et al.* 1978). In addition, the effect of thioredoxin is dependent on other factors, such as glucose-6-phosphate and glutamine which protect against deactivation (Cossar *et al.* 1984) and which may be important in regulation of the enzyme (Schaeffer and Stanier 1978). The activity of G6PDH is several-fold higher in heterocysts than in vegetative cells, allowing the provision of reductant to nitrogenase in the light and in the dark (see Apte *et al.* 1978). This may be related to the decreased level of thioredoxin *m* in heterocysts (Ip *et al.* 1984; Cossar *et al.* 1985*a*).

The activity of G6PDH is enhanced in cyanophage AS-1 infected *A. nidulans*, probably due to an effect on the redox modulation of the enzyme (Cseke *et al.* 1981). An increase in the proportion of reduced thioredoxin *m* in cyanophage N-1 infected *N. muscorum* (Amla *et al.* 1987) may indicate a function in cyanophage N-1 development, although it did not correlate with changes in G6PDH activity.

In *Anabaena cylindrica* and *Anacystis nidulans*, GS is reversibly deactivated on transfer from the light to darkness (Rowell *et al.* 1979). Reactivation, *in vivo*, was observed on re-illumination and *in vitro* on adding

2-mercaptoethanol. A similar light-dependent stimulation of GS activity, and of nitrite reductase and NADP-isocitrate dehydrogenase activities, *in vivo*, has been reported for *S. leopoliensis* and, in each case, enzyme activity was also stimulated by reduced thioredoxin *in vitro* (Tischner and Schmidt 1984). Although cyanobacterial GS has been reported to be activated by reduced thioredoxin (Papen and Bothe 1984; Tischner and Schmidt 1984), we have, so far, been unable to confirm this finding using either cell-free extracts or purified GS from *A. cylindrica* (Ip *et al.* 1984) or *A. variabilis*. We have recently obtained further evidence, from studies on the metabolism via GS of the ammonium analogue, methylammonium, that GS activity *in vivo* is inhibited on transfer of *A. variabilis* from light to dark (Kerby *et al.* 1986). The mechanism of inhibition/deactivation of GS requires further investigation and, in addition to thioredoxin, factors such as cytoplasmic pH changes, ATP limitation, reduced substrate availability and altered availability of certain divalent cations (Ip *et al.* 1983) need to be considered.

Cyanobacterial thioredoxin *m* is structurally similar to *E coli* thioredoxin and chloroplast thioredoxin *m*, and a large number of potential functions have been demonstrated *in vitro*. However, the thioredoxin system of cyanobacteria has not yet been fully elucidated, particularly in relation to thioredoxin *f*, and the physiological functions of cyanobacterial thioredoxins have not yet been firmly established. Further investigation of the thioredoxin system of cyanobacteria, particularly in relation to the effects of light on growth and development, should prove rewarding.

Acknowledgements

We thank the Agricultural and Food Research Council, the Natural Environment Research Council, and the Indian Ministry of Science and Technology for financial support. We are also indebted to Dr D. Pappin, Department of Biochemistry, University of Leeds, UK, for amino acid sequencing of thioredoxins and Dr P. Schurmann, Laboratoire de Biochimie, Universite de Neuchatel, Switzerland, for the generous gift of spinach chloroplast thioredoxin *f*. We are grateful to Dr N.W. Kerby for critical comments on the manuscript.

References

Amla, D.V., Rowell, P., and Stewart, W.D.P. (1987). *Arch. Microbiol.* **148,** 321.

Anderson, L.E. (1979). In *Photosynthesis II. Photosynthetic carbon metabolism and related processes* (ed. M. Gibbs and E. Latzko) p. 271. Springer-Verlag, Berlin.

Anderson, L.E., Jablonski, P., and Mohamed, H. (1986). In *Thioredoxin and glutaredoxin systems: structure and function* (ed. A. Holmgren, C.I. Branden, H. Jornvall, and B-M. Sjoberg) p. 267. Raven Press, New York.

Apte, S.K., Rowell, P., and Stewart, W.D.P. (1978). *Proc. R. Soc. Lond.* B **200**, 1.
Ashton, A.R., Brennan, T., and Anderson, L.E. (1980). *Plant Physiol.* **66**, 605.
Berglund, O. (1969). *J. biol. Chem.* **244**, 6306.
Bestermann, A., Vogt, K., and Follmann, H. (1983). *Eur. J. Biochem.* **131**, 339.
Black, S., Horte, E.M., Hudson, B., and Wartzsky, L. (1960). *J. biol. Chem.* **235**, 2910.
Buc, J., Riviere, M., Gontero, B., Sauve, P., Meunier, J-C., and Ricard, J. (1984). *Eur. J. Biochem.* **140**, 199.
Buchanan, B.B. (1980). *A. Rev. plant Physiol.* **31**, 341.
Buchanan, B.B. (1986). In *Thioredoxin and glutaredoxin systems: structure and function* (ed. A. Holmgren, C.I. Branden, H. Jornvall, and B-M. Sjoberg) p. 233. Raven Press, New York.
Cao, Q., Johnson, T.C., Kung, J.E., and Buchanan, B.B. (1984). In *Advances in photosynthesis research* IV (ed. B. Sybesma) p. 877. Martinus Nijhoff, The Hague.
Cossar, J.D., Rowell, P., and Stewart, W.D.P. (1984). *J. gen. Microbiol.* **130**, 991.
Cossar, J.D., Darling, A.J., Ip, S.M., Rowell, P., and Stewart, W.D.P. (1985*a*). *J. gen. Microbiol.* **131**, 3029.
Cossar, J.D., Rowell, P., Darling, A.J., Murray, S., Codd, G.A., and Stewart, W.D.P. (1985*b*). *FEMS Microbiol. Lett.* **28**, 65.
Crawford, N.A., Yee, B.C., Hutcheson, S.W., Wolosiuk, R.A., and Buchanan, B.B. (1986). *Arch. Biochem. Biophys.* **244**, 1.
Crawford, N.A., Sutton, C.W., Yee, B.C., Johnson, T.C., Carlson, D.E., and Buchanan, B.B. (1984). *Arch. Microbiol.* **139**, 124.
Cseke, C., Balogh, A., and Farkas, G.L. (1981). *FEBS Lett.* **126**, 85.
Darling, A.J., Rowell, P., and Stewart, W.D.P. (1986). *Biochem. biophys. Acta* **850**, 116.
De la Torre, A., Lara, C., Wolosiuk, R.A., and Buchanan, B.B. (1979). *FEBS Lett.* **107**, 141.
Droux, M. Jacquot, J., Suzuki, A., and Gadal, P. (1984). In *Advances in photosynthesis research* (ed. C. Sybesma) p. 533. Martinus Nijhoff, The Hague.
Droux, M., *et al.* (1987). *Arch. Biochem. Biophys.* **252**, 426.
Duggan, J.X. and Anderson, L.E. (1975) *Planta* **122**, 293.
Edman, J.C., Ellis, L., Blacher, R.W., Roth, R.A., and Rutter, W.J. (1985). *Nature, Lond.* **317**, 267.
Eklund, H., *et al.* (1984). *EMBO J.* **3**, 1443.
Gerbling, K-P., Steup, M., and Latzko, E. (1985). *Eur. J. Biochem.* **147**, 207.
Gleason, F.K. (1979). *Arch. Microbiol.* **123**, 15.
Gleason, F.K. (1986). In *Thioredoxin and glutaredoxin systems: structure and function* (ed. A. Holmgren, C.I. Branden, H. Jornvall, and B-M. Sjoberg) p. 21. Raven Press, New York.
Gleason, F.K. and Holmgren, A. (1981). *J. biol. Chem.* **256**, 8306.
Gleason, F.K., Whittaker, M.M., Holmgren, A., and Jornvall, H. (1985). *J. biol. Chem.* **260**, 9567.
Grippo, J.F., Tienrungroj, W., Dahmer, M.K., Housley, P.R., and Pratt, W.B. (1983). *J. biol. Chem.* **258**, 13658.

Holmgren, A. (1968). *Eur. J. Biochem.* **6**, 475.
Holmgren, A. (1976). *Proc. natn. Acad. Sci. USA* **73**, 2275.
Holmgren, A. (1979*a*). *J. biol. Chem.* **254**, 3672.
Holmgren, A. (1979*b*). *J. biol. Chem.* **254**, 9627.
Holmgren, A. (1985). *A. Rev. Biochem.* **54**, 237.
Holmgren, A. and Fagerstedt, M. (1982). *J. biol. Chem.* **257**, 6926.
Holmgren, A., Ohlsson, I., and Grankvist, M-L. (1978). *J. biol. Chem.* **253**, 430.
Holmgren, A., Branden, C-I., Jornvall, H., and Sjoberg, B-M. (ed.) (1986). *Thioredoxin and glutaredoxin systems: structure and function*. Raven Press, New York.
Huber, H.E., Russel, M., Model, P., and Richardson, C.C. (1986). *J. biol. Chem.* **261**, 15006.
Ip, S.M., Rowell, P., and Stewart, W.D.P. (1983). *Biochem. biophys. Res. Commun.* **114**, 206.
Ip, S.M., Rowell, P., Aitken, A., and Stewart, W.D.P. (1984). *Eur. J. Biochem.* **141**, 497.
Johnson, T.C., Crawford, N.A., and Buchanan, B.B. (1984). *J. Bact.* **158**, 1061.
Juhasz, A., Csizmadia, V., Borbely, G., Udvardy, J., and Farkas, G.L. (1986). *FEBS Lett.* **194**, 121.
Kerby, N.W., Rowell, P., and Stewart, W.D.P. (1986). *Arch. Microbiol.* **143**, 353.
Langlotz, P., Wagner, W., and Follmann, H. (1986). *Z. Naturf.* **41**, 275.
Laurent, T.C., Moore, E.C., and Reichard, P. (1964). *J. biol. Chem.* **239**, 3436.
Lim, C-J., Haller, B., and Fuchs, J. (1985). *J. Bact.* **161**, 799.
Lunn, C.A. and Pigiet, V. (1986). In *Thioredoxin and glutaredoxin systems: structure and function* (ed. A. Holmgren, C.I. Branden, H. Jornvall, and B-M. Sjoberg) p. 165. Raven Press, New York.
Maeda, K., Tsugita, A., Dalzoppo, D., Vilbois, F., and Schurmann, P. (1986). *Eur. J. Biochem.* **154**, 197.
Marino, G.T. and Asato, Y. (1986). *J. gen. Microbiol.* **132**, 2123.
Mark, D.F. and Richardson, C.C. (1976). *Proc. natn. Acad. Sci. USA* **73**, 780.
Marsden, W.J.N. and Codd, G.A. (1984). *J. gen. Microbiol.* **130**, 999.
Meng, M. and Hogenkamp, H.P.C. (1981). *J. biol. Chem.* **256**, 9174.
Nicholson, S., Easterby, J.S. and Powls, R. (1986). *FEBS Lett.* **202**, 19.
Nygren, H., Rozell, B., Holmgren, A., and Hansson, H-A. (1981) *FEBS Lett.* **133**, 145.
Papen, H. and Bothe, H. (1984). *FEMS Microbiol. Lett.* **23**, 41.
Papen, H., Neuer, G., Refraian, M., and Bothe, H. (1983) *Arch. Microbiol.* **134**, 73.
Papen, H., Neuer, G., Sauer, A., and Bothe, H. (1986). *FEMS Microbiol. Lett.* **36**, 201.
Pelroy, R.A. and Bassham, J.A. (1972). *Arch. Mikrobiol.* **86**, 25.
Pelroy, R.A. and Bassham, J.A. (1973). *J. Bact.* **115**, 943.
Pelroy, R.A., Levine, G.A., and Bassham, J.A. (1976). *J. Bact.* **127**, 633.
Pelroy, R.A., Rippka, R., and Stanier, R.Y. (1972). *Arch. Mikrobiol.* **87**, 303.
Preiss, J. (1982). *A. Rev. Plant Physiol.* **33**, 431.
Rebeille, F. and Hatch, M.D. (1986). *Arch. Biochem. Biophys.* **249**, 171.
Reed, R.H., Kerby, N.W., and Stewart, W.D.P. (1987). *Phycologia* **26**, 391.

Rowell, P., Sampaio, M.J.A.M., Ladha, J.K., and Stewart, W.D.P. (1979). *Arch. Microbiol.* **120**, 195.
Russel, M. and Model, P. (1985), *Proc. natn. Acad. Sci. USA* **82**, 29.
Russel, M. and Model, P. (1986). *J. biol. Chem.* **261**, 14997.
Scheibe, R. (1981). *FEBS Lett.* **133**, 301.
Schaeffer, F. and Stanier, R.Y. (1978). *Arch. Microbiol.* **116**, 9.
Schmidt, A. (1980). *Arch. Microbiol.* **127**, 259.
Schmidt, A. and Christen, U. (1979). *Z. Natur.* **34**, 1272.
Schurmann, P. (1981) In *Photosynthesis IV. Regulation of carbon metabolism* (ed. G. Akoyunoglou) p. 273. Balaban, Philadelphia, Penn.
Schurmann, P., Maeda, K., and Tsugita, A. (1981). *Eur. J. Biochem.* **116**, 37.
Sjoberg, B-M. and Holmgren, A. (1973). *Biochem. biophys. Acta.* **315**, 176.
Smith, A.J. (1982). In *The biology of cyanobacteria* (ed. N.G. Carr and B.A. Whitton) p. 47. Blackwell Scientific Publications, Oxford.
Soulie, J-M., Buc, J., Meunier, J.C., Pradel, J., and Ricard, J. (1981). *Eur. J. Biochem.* **119**, 497.
Sutton, C.W., Crawford, N.A., Yee, B.C., Carlson, D.C., and Buchanan, B.B. (1984). In *Advances in photosynthesis research III* (ed. B. Sybesma) p. 633. Martinus Nijhoff, The Hague.
Tischner, R. and Schmidt, A. (1984). *Arch. Microbiol.* **137**, 151.
Tsang, M.L-S. (1981). *Plant Physiol.* **68**, 1098.
Tsugita, A., Maeda, K., and Schurmann, P. (1983). *Biochem. biophys. Res. Commun.* **115**, 1.
Udvardy, J., Borbely, G., Juhasz, A., and Farkas, G.L. (1984). *J. Bact.* **157**, 681.
Wada, K. and Buchanan, B.B. (1981). *FEBS Lett.* **124**, 237.
Wagner, W. and Follmann, H. (1977). *Biochem. biophys. Res. Commun.* **77**, 1044.
Wagner, W., Follmann, H., and Schmidt, A. (1978). *Z. Naturf.* **33**, 517.
Wallace, B.J. and Kushner, S.R. (1984). *Gene* **32**, 399.
Whittaker, M.M. and Gleason, F.K. (1984). *J. biol. Chem.* **259**, 14088.
Wilcox, M., Mitchison, G.J., and Smith, R.J. (1973). *J. cell. Sci.* **13**, 637.
Wilson, I.G., Asaki, T., and Bandurski, R.S. (1961). *J. biol. Chem.* **236**, 1822.
Wolosiuk, R.A. and Buchanan, B.B. (1977). *Nature, Lond.* **266**, 565.
Yee, B.C., *et al.* (1981). *Arch. Microbiol.* **130**, 14.

12 The responses of cyanobacteria to salt stress

R.H. REED

Department of Biological Sciences, University of Dundee, Dundee DD1 4HN, Scotland, UK

W.D.P. STEWART

AFRC Research Group on Cyanobacteria and Department of Biological Sciences, University of Dundee, Dundee DD1 4HN, UK

Introduction

Cyanobacteria are prokaryotic phototrophs of world-wide distribution (Fogg *et al.* 1973). Free-living forms are often abundant in habitats where there are rapid and substantial fluctuations in a range of environmental conditions, including salinity and water status (e.g. the intertidal and supralittoral regions of marine rocky shores, marine stromatolites, estuaries, and salt marshes); they are also major components of the microbial flora in oceanic waters of constant salinity, and in many freshwater habitats (Stewart 1983). Furthermore, cyanobacteria in lichen symbioses are frequently subjected to severe desiccation while maritime and marine forms may also encounter salinity stress.

Until recently, the responses of cyanobacteria to salinity stress were poorly documented (Waterbury and Stanier 1981), in contrast to heterotrophic bacteria (Measures 1975; Imhoff 1986) and phototrophic eukaryotic algae (Ben-Amotz and Avron 1983). However, in recent years, substantial progress has been made towards a better understanding of the physiological mechanisms responsible for salinity tolerance and osmotic adjustment in cyanobacteria, largely as a result of the initial studies of Borowitzka and co-workers (Borowitzka *et al.* 1980; Mackay *et al.* 1983, 1984) and, more recently, of other research groups (e.g. Blumwald and Tel-Or 1982; Erdmann 1983; Richardson *et al.* 1983*a*; Mohammad *et al.* 1983; Reed and Stewart 1983, 1985; Blumwald *et al.* 1983*a*, 1983*b*; Reed *et al.* 1984*a*, 1984*b*, 1986*a*; Warr *et al.* 1984*a*, 1985*a*, 1987; Moore *et al.* 1985; Tel-Or *et al.* 1986). This paper provides an overview of recent studies on osmotic adjustment in salt-stressed cyanobacteria, with particular emphasis on the work of our group at Dundee.

The discovery of low molecular weight organic osmotica in cyanobacteria

Borowitzka *et al.* (1980) used natural abundance ^{13}C nuclear magnetic resonance (NMR) spectroscopy to provide the first unequivocal evidence of organic solute-mediated osmotic adjustment in the marine unicellular cyanobacterium *Synechococcus* RRIMP/N/100, showing that the heteroside glucosyl-glycerol [O-α-D-glucopyranosyl-(1→2)-glycerol; lilioside] was accumulated in cells grown in seawater. This technique clearly showed, first that glucosyl-glycerol was the only organic solute present in osmotically significant amounts within the cells of this isolate, second that the intracellular glucosyl-glycerol concentration varied as a direct function of the external salinity over the salinity range for optimal growth, and third that the intracellular accumulation of glucosyl-glycerol was able to counterbalance over 20 per cent of the external osmolality in hypersaline media. Later studies confirmed that the heteroside was freely mobile in intact cells, although the rotational motion was slower by a factor of approx. 2.4, due to an increase in intracellular viscosity (Norton *et al.* 1982). These observations served to demonstrate the usefulness of ^{13}C NMR techniques in studies of organic solute-mediated osmotic adjustment and provided a stimulus for subsequent research into the role of organic osmotica in salt-stressed cyanobacteria.

The diversity of organic osmotica in cyanobacteria

In recent years, numerous cyanobacteria from several distinct habitats have been screened for organic solute-mediated osmotic adjustment, using ^{13}C NMR, ^{1}H NMR, and chemical assay procedures. These surveys have demonstrated that the accumulation of a single organic osmolyte as a principle intracellular solute is a feature that is shared by isolates from a wide variety of environments, including freshwaters, estuaries, seawaters, and hypersaline habitats. However, the nature of the principal organic solute was found to vary between isolates: three classes of solute have been identified: (i) the disaccharides, sucrose and trehalose, (ii) the heteroside, glucosyl-glycerol, and (iii) the quaternary ammonium compounds, glycine betaine and glutamate betaine. The preliminary observations of Mackay *et al.* (1983) suggested that glucosyl-glycerol accumulation was unique to marine cyanobacteria and that the presence of this osmolyte could be used to distinguish marine isolates from non-marine strains. Subsequent studies (Mackay *et al.* 1984; Reed *et al.* 1984*a*, 1984*b*; Reed and Stewart 1985) have shown that the accumulation of specific organic solutes is not linked solely to the habitat specificity of the isolate. Taken together, the results available to date clearly show that the type of organic solute is more closely related to the upper salinity limit for growth, dividing the cyanobacteria into three distinct groups. Thus the least halotolerant strains (group 1) accumulate disaccharides, while cyanobacteria of intermediate halotolerance (group 2) synthesize glucosyl-glycerol, and the most halotolerant forms (group 3) accumulate either glycine betaine or glutamate betaine in response to salinity stress.

The accumulation of disaccharides is most common in salt-sensitive (stenohaline) freshwater and brackish isolates, although some strains of marine origin contain sucrose (e.g. *Nodularia harveyana*, Warr *et al.* 1984*a*, and *Anabaena* CA, Reed *et al.* 1986*a*) or trehalose (e.g. *Rivularia atra*, Reed and Stewart 1983) as their principal organic solute. Glucosyl-glycerol is the major low molecular weight organic osmolyte in several marine strains and in certain euryhaline freshwater isolates which are capable of growth in full-strength seawater and in hypersaline media (Richardson *et al.* 1983*a*; Reed and Stewart 1985). The presence of glucosyl-glycerol-accumulating cyanobacteria in freshwater environments and disaccharide-accumulating forms in marine waters suggests that organic solute profiles are not the sole determinants of survival and growth in these habitats (Reed *et al.* 1986*a*). However, the accumulation of quaternary nitrogen compounds in several halotolerant/halophilic isolates from hypersaline environments (e.g. salt lakes, supralittoral stromatolites, and algal mats) indicates that there may be a more sharply defined relationship between organic solute (betaine) accumulation and habitat under conditions of severe salinity stress (Reed *et al.* 1984 *b*).

No clear relationships between the accumulation of specific organic solutes and taxonomic groupings have been observed, since different strains of the genera *Calothrix, Dermocarpa, Lyngbya, Myxosarcina, Nodularia, Oscillatoria, Phormidium, Plectonema, Pseudanabaena, Spirulina, Synechococcus*, and *Synechocystis* show dissimilar organic solute accumulation profiles. However, all isolates belonging to the genera *Anabaena* and *Nostoc* which have been screened to date have accumulated sucrose as their sole organic osmolyte (Reed *et al.* 1986*a*; L.J. Stal and R.H. Reed, unpublished observations).

The compatibility of organic osmotica in cyanobacteria

The division of cyanobacteria into (1) least, (2) intermediate, and (3) most halotolerant forms is in accord with observations of the relative compatibility of (1) disaccharides, (2) polyol-derivatives, and (3) betaines with metabolic activity. The studies of Warr *et al.* (1984*b*) and Warr (unpublished thesis, 1985) have shown that the activity of the enzyme glutamine synthetase is sensitive to the osmolyte concentration *in vitro*, with sucrose causing the greatest decrease in activity while glycine betaine was non-inhibitory. These results suggest that there is a rank order of compatibility (non-inhibition) that correlates with the concentrations at which these organic solutes occur in cyanobacteria, i.e. glycine betaine > polyol-derivatives (glucosyl-glycerol) > disaccharides. This, in turn, may influence the upper salinity limit for growth, since the inhibitory effects of the least-compatible osmolyte (sucrose) may limit the intracellular accumulation of this metabolite, thus restricting the growth of disaccharide-accumulating strains in hypersaline media, in contrast to betaine-synthesizing forms (Warr *et al.* 1984*b*). This proposal is also consistent with the observation that intracellular levels of disaccharides fall within a lower range than glucosyl-glycerol, while glycine betaine is accumulated to the highest intracellular concentrations (Warr *et al.* 1984*a*; Reed *et al.* 1984*b*; Reed and Stewart 1985) and may explain, in part at least, some of the trends in organic solute accumulation, habitat and halotolerance noted by Mackay *et al.* (1984) and Reed *et al.* (1984*a*, 1986*a*).

The synthesis of secondary organic osmotica

While the screening programmes of Mackay *et al.* (1984) and Reed *et al.* (1984*a*; 1984*b*) confirmed that a single, strain-specific solute tends to dominate the low molecular weight organic solute profile of salt-stressed cyanobacteria, the existence of smaller quantities of other (carbohydrate) osmotica has been reported. Warr *et al.* (1985*a*) provided additional information for *Spirulina platensis*, which accumulates glucosyl-glycerol as a primary organic osmolyte at concentrations in excess of 240 mol m^{-3} when grown in a hypersaline medium (150 per cent seawater). Natural abundance ^{13}C NMR spectroscopy of cells grown in 100 per cent seawater confirmed that glucosyl-glycerol was the only organic solute present in osmotically significant quantities. However, smaller amounts of trehalose were also detected in cells subjected to salinity shock (48 hours incubation). The highest levels of this disaccharide were observed in a hyposaline medium (50 per cent seawater), in contrast to glucosyl-glycerol (150 per cent seawater). Furthermore, the intracellular concentration of trehalose in *S. platensis* was temperature-sensitive, increasing as the growth temperature was raised (Warr *et al.* 1985*a*).

The effects of temperature and salinity on the low molecular weight carbohydrate accumulation profiles of a range of unicellular cyanobacteria have been studied by Warr *et al.* (1985*b*). Four glucosyl-glycerol-accumulating isolates of *Synechocystis* also synthesized sucrose as a secondary osmoticum when grown at high temperature (35°C). In contrast, three disaccharide-(sucrose-) accumulating strains of *Synechococcus* produced no additional organic osmotica when grown over a range of temperatures and salinities. Detailed studies of the interaction between temperature and salinity on carbohydrate accumulation in *Synechocystis* PCC6714 have shown that the synthesis of sucrose as a secondary osmoticum is favoured by a combination of high temperature and low salinity. Furthermore, a temporal effect was noted: sucrose was accumulated rapidly on upshock (salinity increase), reaching a maximum value at 12–24 hours and declining thereafter at a faster rate at 20 °C than at 37 °C (Warr *et al.* 1985*b*). These observations suggest that in certain glucosyl-glycerol-accumulating cyanobacteria, osmotic adjustment is not limited to a single carbohydrate and that a disaccharide may be produced as a secondary organic osmoticum in response to changes in salinity and/or temperature.

The synthesis of organic osmotica in darkness

Several reports have noted the lack of accumulation of organic osmotica in exponentially growing cyanobacteria subjected to salinity increase in darkness (e.g. *Microcystis firma*, Erdmann 1983; *Synechocystis* spp., Warr *et al.* 1985*b*; *Agmenellum quadruplicatum*, Tel-Or *et al.* 1986; *Synechococcus* RRIMP/N/100, Mackay and Norton 1987), in contrast to the responses of several eukaryotic microalgae (e.g. *Dunaliella tertiolecta*, Muller and Wegmann 1978; *Chlorella emersonii*, Setter and Greenway 1983). This phenomenon has been examined in detail in *Synechocystis* PCC6714 by Warr (unpublished thesis, 1985). Glycogen enrichment (incubation for 6 days in a medium free of combined nitrogen, BG-11_0) raised the intracellular glycogen level from 17.5 to 178.0 kg m^{-3}. When glycogen-rich (NO_3^--starved) cells were subsequently upshocked from freshwater BG-11_0 medium to a seawater-based medium in the light, the glycogen level fell rapidly within the first hour after upshock to a level similar to that of the control (NO_3^--grown) cells and the cells accumulated significantly greater amounts of low molecular weight carbohydrates (i.e. glucosyl-glycerol plus sucrose) suggesting that glycogen may contribute to organic osmolyte accumulation in light-incubated cells.

Glycogen-enriched cells upshocked in the dark also increased their intracellular carbohydrate levels, in contrast to exponentially growing (glycogen-deficient) cells. These results suggest that the synthesis of organic osmolytes in dark-incubated cyanobacteria occurs only in cells with substantial intracellular reserves of glycogen (Warr, unpublished thesis, 1985). Similar observations have been made for sucrose and proline synthesis in starch-enriched, dark-incubated *Enteromorpha intestinalis* in response to hypersaline treatment (Edwards *et al.* 1987).

Rates of synthesis of organic osmotica

The dynamics of the accumulation of low molecular weight carbohydrates in several isolates of cyanobacteria subjected to hypersaline stress at 25 °C have been studied by Warr *et al.* (1987). In all case, the synthesis of the disaccharides sucrose or trehalose was more rapid than heteroside (glucosyl-glycerol) biosynthesis at this temperature. Thus, 6 hours after upshock, disaccharides were always present at more than 60 per cent (and, in most cases, at more than 80 per cent) of their maximum value, not only in the six strains which accumulated a disaccharide as the sole organic osmoticum but also in the four heteroside accumulators which produced a disaccharide as a secondary osmolyte. In contrast, glucosyl-glycerol concentrations did not exceed 35 per cent of the maximum value over the same time period. A similar trend was observed in the time required to achieve an osmolyte concentration equivalent to 90 per cent of the maximum value: disaccharides were synthesized to more than 90 per cent of their maximum value within 8 hours, in contrast to the 24–48 hours required for synthesis of a comparable level of heteroside. Based on these observations, Warr *et al.* (1987) have suggested that certain disaccharide-accumulating cyanobacteria may have a selective advantage over heteroside-accumulating forms in intertidal, estuarine, and other habitats where rapid changes in salinity may occur, due to their capacity for rapid synthesis of organic solutes in response to upshock. Conversely, strains accumulating glucosyl-glycerol, with slower rates of organic solute biosynthesis after upshock, are more likely to be favoured in non-brackish marine habitats. These observations may serve to explain, in part at least, the success of certain disaccharide-accumulating cyanobacteria in brackish habitats *sensu* den Hartog (1967), e.g. *Rivularia atra* (Reed and Stewart 1983) and *Nodularia harveyana* (Warr *et al.* 1984*a*), and the occurrence of glucosyl-glycerol in picoplanktonic *Synechococcus* spp. from open ocean waters (Reed and Stewart 1985; R.H. Reed, unpublished observations). However, the capacity of certain glucosyl-glycerol accumulating strains to synthesize a variable amount of sucrose as a secondary osmolyte (Warr *et al.* 1985*a*, 1985*b*; Reed *et al.* 1985) may weaken this hypothesis.

Moore (unpublished thesis, 1987) has shown that the synthesis of glycine betaine in *Aphanothece halophytica* (*Synechococcus* PCC7418) subjected to hypersaline treatment is less rapid than disaccharide (sucrose) accumulation in *Anacystis nidulans* (*Synechococcus* PCC6301), since the former required over 48 hours to achieve a new steady state, while the latter required only 6 hours. The short-term accumulation of carbohydrates (glucosyl-glycerol and trehalose) was noted for salt-stressed cells of *A. halophytica*, reaching maximum values at 12–24 hours and declining to osmotically insignificant levels after 48 hours of hypersaline treatment. Accumulation of carbohydrates was also observed in dark-incubated, salt-stressed *A. halophytica*, while glycine betaine levels showed no increase; intracellular levels of glycogen were not monitored during these experiments (Moore, unpublished thesis, 1987). The metabolism of glycine betaine in higher plants is also known to be light-sensitive, since betaine synthesis is linked to photosynthetic electron transport (Hanson *et al.* 1985).

The release of organic osmotica in response to salinity shock

Reed and Stewart (1983) demonstrated that 10–12 per cent of the internal pool of the osmolyte trehalose was recovered from the medium when colonies of *Rivularia atra* were transferred from full-strength seawater to a freshwater-based medium (i.e. downshock), while the major fraction was metabolized (presumably by conversion to glycogen). Moore (unpublished thesis, 1987) has also shown that intracellular glycine betaine is liberated when *A. halophytica* is subjected to hypo-osmotic shock. A similar loss of organic solutes in response to downshock has been reported for certain eukaryotic algae (e.g. *Phaeodactylum tricornutum*, Schobert 1980). In contrast, studies with other cyanobacteria, including *N. harveyana* (Warr *et al.* 1984*a*) and *Synechocystis* PCC6803 (Richardson, unpublished thesis, 1986), have found no evidence for organic solute release in response to hypo-osmotic stock.

The liberation of intracellular metabolites in response to osmotic downshock has been studied in detail by Reed *et al.* (1986*c*) for the unicellular cyanobacteria *Synechocystis* PCC6714 and *Synechococcus* PCC6311. Release of low molecular weight carbohydrates and amino acids was initiated by rapid transfer of free-living or immobilized cells from a saline medium (containing 490 mol m^{-3} NaCl) to a freshwater-based medium; up to 50 per cent of the low molecular weight carbohydrates and 73 per cent of the amino acids were released within 2 min, with no significant loss of viability. Such release of organic osmolytes may be involved in the development of heterotrophic communities associated with cyanobacteria in habitats of fluctuating salinity and/or lichen symbioses (Reed and Stewart 1983). A similar proposal has been made by Jones and Stewart (1969*a*) with regard to the enhanced liberation of fixed nitrogen from *Calothrix scopulorum* in environments where conditions are in constant flux, where the nitrogen released may become available to associated non-N_2-fixing organisms (Jones and Stewart 1969*b*).

The biosynthesis and turnover of organic solutes

Tel-Or *et al.* (1986) have studied low molecular weight carbohydrate turnover in response to abrupt changes in salinity in the glucosyl-glycerol-accumulating marine unicell *Agmenellum quadruplicatum* (*Synechococcus* PCC7002) using ^{13}C NMR techniques. For intact cells, enrichment of sodium bicarbonate with ^{13}C at 30 per cent (w/w) gave NMR spectra that were suitable for quantitative work, enabling intracellular concentrations of glucosyl-glycerol to be measured without cell disruption. Pulse-chase experiments showed that (^{13}C-enriched) glycogen, which had been accumulated by the cells under nitrogen-limited growth at low salinities, could be used for the light-dependent synthesis of glucosyl-glycerol in response to hypersaline shock treatment. It was also shown that glucosyl-glycerol increase in response to upshock was less rapid than glucosyl-glycerol depletion on down-shock—the former occurred on a time scale similar to that of cell doubling (approx. 12–24 hours) while the latter required 2–8 hours. Similar observations have been reported for other cyanobacteria, including *N. harveyana*, which accumulated sucrose over a 48 hour period on transfer from fresh-water to 100 per cent seawater, while transfer from 100 per cent seawater to freshwater decreased the intracellular sucrose concentration (by conversion to glycogen) within 2 hours (Warr *et al.* 1984*a*).

Mackay and Norton (1987) have used ^{13}C-enrichment procedures to investigate the synthesis and turnover of glucosyl-glycerol during osmotic shock in cells of *Synechococcus* RRIMP/N/100 grown at high temperature. A rapid increase in the intracellular concentration of glucosyl-glycerol occurred in response to hyperosmotic shock, with approximately 90 per cent of the solute being synthesized from newly fixed carbon (photosynthesis) and up to 10 per cent from intracellular reserves (glycogen). The time period for glucosyl-glycerol increase (approx. 100 min at 38–40 °C) was broadly equivalent to the doubling time for the organism of 160–200 min at this high temperature, and corresponded to the lag period for cells in hypersaline medium (Mackay and Norton 1987). Somewhat similar findings were reported by Richardson *et al.* (1983*a*) for the euryhaline freshwater cyanobacterium *Synechocystis* PCC6803 grown at a lower temperature. In this organism, glucosyl-glycerol accumulation required approximately 24 hours to reach a new steady state value (at 28 °C) when cells were transferred from a freshwater medium to seawater, with an extended lag period on transfer to marine medium of 24 hours. It is possible that the increased lag period resulting from hyperosmotic stress is due, in part at least, to a requirement for optimum intracellular concentrations of organic osmotica (Richardson, unpublished thesis, 1986).

The pathways for the biosynthesis of organic solutes in cyanobacteria have not been elucidated. However, glucosyl-glycerol is structurally (Kollmann *et al.* 1979) and functionally (Reed *et al.* 1980), comparable to the heteroside floridoside [O-α-D-galactopyranosyl-(1→2)-glycerol; galactosyl-glycerol], which occurs as an intracellular osmolyte in certain red algae (Reed 1985) together with isofloridoside [O-α-D-galactopyranosyl-(1→1)-glycerol]: the latter metabolite is also found in osmotically stressed cells of the chrysophyte *Poterioochromonas malhamensis* (Kauss 1977). The pathway for biosynthesis of galactosyl-glycerols has been studied in red algae (Kremer and Kirst 1981) and in *P. malhamensis* (Kauss 1979); the condensation of UDP-galactose and glycero-3-phosphate gives galactosyl-glycerol-phosphate with subsequent dephosphorylation to yield galactosyl-glycerol. For glucosyl-glycerol synthesis to proceed in a similar manner would require the combination of UDP-glucose and glycero-3-phosphate. It is noteworthy that the osmotic regulation of isofloridoside synthesis in hyperosmotically stressed *P. malhamensis* has been studied in detail (Kauss 1979) and appears to involve the activation of pre-existing enzymes of heteroside synthesis via protease action, mediated by changes in calcium/calmodulin (Kauss and Thomson 1982) and/or polar lipids (Kauss 1982). However, the mechanism whereby a change in external osmotic pressure is sensed and transformed into a regulatory signal remains unclear.

The biosynthetic and catabolic pathways of disaccharide formation and degradation have been well characterized in higher plants (Ap Rees 1984) and in certain heterotrophic micro-organisms (Theuvelin 1984). However, the pathways and their metabolic regulation in cyanobacteria are not known.

The pathway for glycine betaine formation in higher plants involves the sequential methylation of phosphoryl–ethanolamine (formed from serine) to give phosphoryl-choline which is hydrolysed to free choline prior to oxidation via betaine aldehyde to glycine betaine: the oxidation steps are physiologically irreversible (Hanson and Grumet 1985). Betaine formed by this pathway appears to be quite metabolically inert and, on a time scale of days, it is not catabolized. The route of glycine betaine biosynthesis in halophilic/halotolerant cyanobacteria remains to be determined. However, Sibley and Yopp (1986) have suggested that S-adenosylmethionine-mediated transmethylations may be involved in glycine betaine formation in *A. halophytica*.

A betaine uptake system in betaine-synthesizing cyanobacteria

Glycine betaine is synthesized as an intracellular osmotic effector in halophilic and halotolerant cyanobacteria from hypersaline habitats (Mohammad *et al.* 1983; Mackay *et al.* 1984; Reed *et al.* 1984*b*), in salt-tolerant higher plants (Wyn Jones and Gorham 1986) and in certain halophilic photosynthetic bacteria (Galinski and Trüper 1982; Galinski *et al.* 1985). However, exogenous glycine betaine is known to stimulate the growth of salt-sensitive enteric bacteria (e.g. *Escherichia coli, Salmonella typhimurium*) in minimal media of inhibitory osmotic strength (Le Rudulier and Valentine 1982) due to the existence of an osmotically induced betaine transport system (Perroud and Le Rudulier 1985; Cairney *et al.* 1985*a*, 1985*b*). These organisms are unable to synthesize glycine betaine in response to osmotic stress, and their endogenous organic osmolytes include glutamic acid (Measures 1975) and trehalose (Imhoff 1986; Larsen *et al.* 1987). However, the intracellular accumulation of osmotically significant amounts of glycine betaine by the uptake of exogenous betaine or its direct precursors (i.e. choline or glycine betaine aldehyde: Landfald and Strφm 1986) raises the upper salinity limit for growth, presumably because the role of betaine as a highly compatible, haloprotective organic solute (Pollard and Wyn Jones 1979; Warr *et al.* 1984*b*). Furthermore, Bouillard and Le Rudulier (1983) have shown that nitrogenase activity of *Klebsiella pneumoniae* is enhanced by exogenous betaine, at high osmolality. Similar experiments have failed to increase the upper salinity limit for growth or N_2 fixation of salt-sensitive disaccharide- and heteroside- accumulating cyanobacteria (D.L. Richardson *et al.*, unpublished observations), and recent studies on glycine betaine uptake in a range of these cyanobacteria have shown that they are unable to accumulate exogenous betaine to osmotically significant internal levels (Moore *et al.* 1987). In contrast, the salt-tolerant, glycine-betaine-synthesizing unicellular strains *A. halophytica, Dactylococcopsis salina* and *Synechocystis* DUN52 possess a betaine transport system which is responsible for the rapid, net uptake of exogenous ^{14}C-labelled glycine betaine, equilibrating at osmotically significant intracellular concentrations within 30 min (e.g. 120 mol m^{-3} ^{14}C-betaine in seawater-grown *A. halophytica*). Higher ^{14}C-glycine betaine levels were observed for cells grown in media of elevated osmotic strength, in agreement with studies on enteric bacteria. However, it is envisaged that the active betaine uptake system in betaine-synthesizing cyanobacteria from hypersaline habitats may be involved in the re-acquisition of biosynthetically produced quaternary ammonium compounds, lost from the cell interior due to 'leakage' or as a consequence of changes in membrane permeability due to salinity variation (Reed 1984; Reed *et al.* 1986*b*). A somewhat similar role has been proposed for the periplasmic amino-acid transport systems of bacteria (Ames 1986).

The role of inorganic ions in short-term osmotic adjustment

Early studies on the significance of Na^+ as an intracellular osmolyte suggested that cellular levels of this cation were low in cyanobacteria as a consequence of active Na^+ extrusion (Batterton and Van Baalen 1971; Dewar and Barber 1974). Subsequent research has confirmed this view for several isolates grown in freshwater- and seawater-based media; such studies, using cells that were pre-equilibrated for at least 24 hours in hypersaline and hyposaline media, have given no indication of any significant involvement of intracellular Na^+ in the long-term osmotic adjustment processes of *Synechocystis* PCC6803 (Richardson *et al.* 1983*b*), *N. harveyana* (Warr *et al.* 1984*a*), and several halophilic and halotolerant strains (Reed *et al.* 1984*b*). However, recent studies on the effects of hyperosmotic shock treatment on the cell volume and solute content of unicellular cyanobacteria have demonstrated that short-term changes in plasma membrane permeability, due to osmotic imbalance between the cell and its surrounding fluid, may lead to a decrease in cell volume (measured using a particle size analyser) that is smaller than predicted from the Boyle–van't Hoff relationship for non-turgid cells (Reed *et al.* 1986*b*). Thus the initial decrease in cell volume caused by osmotically driven extrusion of water from the cell interior is followed by a rapid entry of extracellular Na(C1), reaching a maximum value at approximately 2 min in *Synechocystis* PCC6714. Subsequent incubation of cells in hypersaline media containing up to 500 mol m^{-3} NaC1 led to a marked reduction in cell Na^+ within 20 min, indicating an efficient, active Na^+ extrusion system. Spin-labelling methods have been used by Candau *et al.* (1983) and Blumwald *et al.* (1983*b*) to show a similar shrinkage and subsequent swelling in *Anacystis nidulans* (*Synechococcus* L1402-1), indicating permeability to extracellular NaC1 on upshock. The rapid entry and subsequent extrusion of Na^+ has also been observed in *A. nidulans* (*Synechococcus* PCC6311, Blumwald *et al.* 1983*a*; *Synechococcus* PCC6301, Moore *et al.* 1985).

The extrusion of Na^+ from hyperosmotically stressed cells of *Synechocystis* PCC6714 was found to be markedly sensitive to the external K^+ concentration, with maximal net Na^+ extrusion in media containing K^+ at 1–10 mol m^{-3}. The possibility that such K^+ dependence was due to active K^+ uptake (Reed and Stewart 1984) during Na^+ extrusion has been investigated in detail by Reed *et al.* (1985) and the following sequence of events has been established for *Synechocystis* PCC6714:

1. Initially (i.e. within 5–15 sec of upshock), cell shrinkage occurs as a result of the osmotically driven extrusion of water from the cell.

2. This initial shrinkage is followed by a period (up to 2 min) of rapid recovery of cell volume towards the original value, caused by the entry of exogenous Na^+ and $C1^-$ as a consequence of a substantial, transient increase in membrane permeability. During this period, photosynthetic activity was undetectable.

3. After 2 min, cell Na^+ decreased rapidly while the intracellular K^+ concentration increased by an osmotically equivalent amount; both of these processes appeared to operate at similar rates, with maximum cell K^+ and minimum Na^+ concentration achieved after 20–30 min. Over this time, the intracellular concentration of $C1^-$ remained high and stable, presumably acting as the counter-ion for K^+, during Na^+ extrusion. Maximum recovery of photosynthetic O_2 evolution was observed during this period, reaching over 70 per cent of its pre-upstock value at 30 min.

4. After approximately 1 hour, the intracellular levels of K^+ and $C1^-$ declined as the organic osmolytes sucrose and glucosyl-glycerol were synthesized, reaching new steady state values similar to the original ion concentrations at 24 hours.

5. Sucrose accumulation was faster initially than that for glucosyl-glycerol, reaching a lower maximum value at 24 hours and declining there-

after and accounting for less than 10 per cent of the low molecular weight carbohydrate fraction after 15 days in 500 mol m^{-3} NaCl (Warr *et al.* 1985*b*), in agreement with the proposal that glucosyl-glycerol is the major osmolyte in this isolate (Reed and Stewart 1985).

It appears that the temporal changes in intracellular solute levels described above are due primarily to isotonic substitution (iso-osmotic regulation: Wyn Jones and Gorham 1983), whereby a solute is replaced by another metabolite with no net increase in cell osmolality. However, this proposal requires further detailed study, using pressure nephelometry and with a gas-vacuolate isolate, to determine the role of isotonic substitution in the osmotic adjustment processess of cyanobacteria. Despite this, it is clear that significant changes in the cellular complement of inorganic ions may precede organic solute accumulation, and may be involved in short-term osmotic adjustment in *Synechocystis* PCC6714, prior to organic osmolyte synthesis as in the eukaryotic alga *Tetraselmis subcordiformis* (Kirst 1977).

Turgor regulation in cyanobacteria

The effects of NaCl treatment on cell turgor have been studied by Reed and Walsby (1985) in the freshwater unicell *Microcystis* sp. BC 84/1 using the gas vesicles of this isolate as built-in pressure probes. Technical aspects of the use of pressure nephelometry to measure cell turgor pressure and related parameters are described in detail by Walsby (1980). Transfer of *Microcystis* sp. BC 84/1 to culture medium containing up to 270 mol m^{-3} NaCl caused an initial decrease in turgor which was followed by a partial recovery (up to 61 per cent of the original value to 60 min), providing the first direct evidence of turgor regulation due to osmotic adjustment in a prokaryotic micro-organism; a limited regain in turgor was observed at higher NaCl concentrations. The partial recovery of turgor in low-NaCl media was linked to short-term increases in cell K^+ content, and estimation of internal concentrations of K^+ in cells transferred to 135 mol m^{-3} Na Cl showed that the accumulation of K^+, together with a counter-ion, could account for the observed turgor recovery. However, longer term experiments showed that this isolate was extremely salt-sensitive, in contrast to the response of many euryhaline freshwater cyanobacteria (Reed *et al.* 1984*a*). Similar stenohaline responses have been obtained for a range of gas-vacuolate and non-gas-vacuolate freshwater isolates of *Microcystis* spp. (R.H. Reed and G.A. Codd, unpublished observations) and further studies are envisaged to identify a suitable euryhaline gas-vacuolate cyanobacterium which may be used to establish, first the contribution of individual organic and inorganic osmotica to turgor regulation, and second the significance of turgor regulation to the long-term growth of cyanobacteria in saline media.

Acknowledgements

Research support for the authors' studies of the effects of salinity stress on cyanobacteria has been provided by the Royal Society, Agriculture and Food Research Council, Natural Environment Research Council, and Science and Engineering Research Council. RHR is a Royal Society Research Fellow.

References

Ames, G.F.L. (1986). *A. Rev. Biochem.* **55**, 397.
Ap Rees, T. (1984). In *Storage carbohydrates in vascular plants* (ed. D.H. Lewis) p. 53. Cambridge University Press.
Batterton, J.C. and Van Baalen, C. (1971). *Arch. Mikrobiol.* **76**, 151.
Ben-Amotz, A. and Avron, M. (1983). *A. Rev. Microbiol.* **37**, 95.
Blumwald, E. and Tel-Or, E. (1982). *Arch. Microbiol.* **132**, 168.
Blumwald, E., Mehlhorn, R.J., and Packer, L. (1983*a*). *Plant Physiol.* **73**, 377.
Blumwald, E., Mehlhorn, R.J., and Packer, L. (1983*b*). *Proc. natn. Acad. Sci. USA* **80**, 2599.
Borowitzka, L.J., Demmerle, S., Mackay, M.A., and Norton, R.S. (1980). *Science* **210**, 650.
Bouillard, L. and Le Rudulier, D. (1983). *Physiol. Veg.* **21**, 447.
Cairney, J.C., Booth, I.R., and Higgins, C.F. (1985*a*). *J. Bact.* **164**, 1218.
Cairney, J.C., Booth, I.R., and Higgins, C.F. (1985*b*). *J. Bact.* **164**, 1224.
Candau, P., Mehlhorn, R.J., and Packer, L. (1983). In *Photosynthetic prokaryotes: cell differentiation and function* (ed. G.C. Papageorgiou and L. Packer) p. 91. Elsevier, Amsterdam.
den Hartog, C. (1967). *Blumea* **15**, 31.
Dewar, M.A. and Barber, J. (1974). *Planta* **117**, 163.
Edwards, D.M., Reed, R.H., and Stewart, W.D.P. (1987). *Marine Biol.* **95**, 583.
Erdmann, N. (1983). *Z. Pflanzenphysiol.* **110**, 147.
Fogg, G.E., Stewart, W.D.P., Fay, P., and Walsby, A.E. (1973). *The blue green algae*. Academic Press, London.
Galinski, E.A. and Trüper, H.G. (1982). *FEMS Microbiol. Lett.* **13**, 357.
Galinski, E.A., Pfeiffer, H.P., and Trüper, H.G. (1985). *Eur. J. Biochem.* **149**, 135.
Hanson, A.D. and Grumet, R. (1985). In *Cellular and molecular biology of plant stress* (ed. J.L. Key and T. Kosuge) p. 71. Alan R. Liss, New York.
Hanson, A.D., May, A.M., Grumet, R., Bode, J., Jamieson, G.C., and Rhodes, D. (1985). *Proc. natn. Acad. Sci. USA* **82**, 3678.
Imhoff, J. (1986). *FEMS Microbiol. Rev.* **39**, 57.
Jones, K. and Stewart, W.D.P. (1969*a*). *J. mar. Biol. Ass. UK* **49**, 475.
Jones, K. and Stewart, W.D.P. (1969*b*). *J. mar. Biol. Ass. UK* **49**, 701.
Kauss, H. (1977). In *International review of biochemistry, plant biochemistry II* Vol. 13, (ed. D.H. Northcote), p. 119. University Park Press, Baltimore, MD.
Kauss, H. (1979). *Prog. Phytochem.* **5**, 1.
Kauss, H. (1982). *Plant Physiol.* **71**, 169.
Kauss, H. and Thomson, K.S. (1982). In *Plasmalemma and tonoplast: their*

functions in the plant cell (ed. D. Marme, E. Marre, and R. Hertel) p. 255. Elsevier, Amsterdam.
Kirst, G.O. (1977). *Planta* **135**, 69.
Kollmann, V.H., Hanners, J.L., London, R.E., Adame, E.G., and Walker T.E. (1979). *Carbohydr. Res.* **73**, 193.
Kremer, B.P. and Kirst G.O. (1981). *Plant Sci. Lett.* **23**, 349.
Landfald, B. and Strφm, A.R. (1986). *J. Bact.* **165**, 849.
Larsen, P.I., Sydnes, L.K., Landfald, B., and Strφm, A.R. (1987). *Arch. Microbiol.* **147**, 1.
Le Rudulier, D. and Valentine, R.C. (1982). *Trends biochem. Sci.* **7**, 431.
Mackay, M.A. and Norton, R.S. (1987). *J. gen. Microbiol.* **133**, 1535.
Mackay, M.A., Borowitzka, L.J., and Norton, R.S. (1983). *Mar. Biol.* **73**, 301.
Mackay, M.A., Borowitzka, L.J., and Norton, R.S. (1984). *J. gen. Microbiol.* **30**, 2177.
Measures, J.C. (1975). *Nature, Lond.* **257**, 398.
Mohammad, F.A.A., Reed, R.H., and Stewart, W.D.P. (1983). *FEMS Microbiol. Lett.* **16**, 287.
Moore, D.J. (1987). The osmotic adjustment of the halophilic cyanobacterium *Aphanothece halophytica*. Unpublished Ph.D. thesis. University of Dundee.
Moore, D.J., Reed, R.H., and Stewart, W.D.P. (1985). *J. gen. Microbiol.* **131**, 1267.
Moore, D.J., Reed, R.H., and Stewart, W.D.P. (1987). *Arch. Microbiol.* **147**, 399.
Muller, W. and Wegmann, K. (1978). *Planta* **141**, 155.
Norton, R.S., Mackay, M.A., and Borowitzka, L.J. (1982). *Biochem. J.* **202**, 699.
Perroud, B. and Le Rudulier, D. (1985). *J. Bact.* **161**, 393.
Pollard, A. and Wyn Jones, R.G. (1979). *Planta* **144**, 291.
Reed, R.H. (1984). *J. Membrane Biol.* **82**, 83.
Reed, R.H. (1985). *Br. Phycol. J.* **20**, 211.
Reed, R.H. and Stewart, W.D.P. (1983). *New Phytol.* **95**, 595.
Reed, R.H. and Stewart, W.D.P. (1984). *Biochim. Biophys. Acta* **812**, 155.
Reed, R.H. and Stewart, W.D.P. (1985). *Mar. Biol.* **88**, 1.
Reed, R.H. and Walsby, A.E. (1985). *Arch. Microbiol.* **143**, 290.
Reed, R.H., Collins, J.C., and Russell, G. (1980). *J. exp. Bot.* **31**, 1539.
Reed, R.H., Richardson, D.L., and Stewart, W.D.P. (1986*b*). *Plant Cell Environ.* **9**, 25.
Reed, R.H., Chudek, J.A., Foster, R., and Stewart, W.D.P. (1984*b*). *Arch. Microbiol.* **138**, 333.
Reed, R.H., Richardson, D.L., Warr, S.R.C., and Stewart, W.D.P. (1984*a*). *J. gen. Microbiol.* **130**, 1.
Reed, R.H., Warr, S.R.C., Kerby, N.W., and Stewart, W.D.P. (1986*c*). *Enz. microbiol. Technol.* **8**, 101.
Reed, R.H., Warr, S.R.C., Richardson, D.L., Moore, D.J., and Stewart, W.D.P. (1985). *FEMS Microbiol. Lett.* **28**, 225.
Reed, R.H., *et al.* (1986*a*). *FEMS Microbiol. Rev.* **39**, 51.
Richardson, D.L. (1986). *Osmotic responses of the cyanobacterium Synechocystis PCC6803*. Unpublished Ph.D. thesis, University of Dundee.
Richardson, D.L., Reed, R.H., and Stewart, W.D.P. (1983*a*). *FEMS Microbiol. Lett.* **18**, 99.

Richardson, D.L., Reed, R.H., and Stewart, W.D.P. (1983*b*). *Br. phycol. J.* **18**, 209.

Schobert, B. (1980). *Physiol. Plant.* **50**, 37.

Setter, T. and Greenway, H. (1983). *Plant Cell Environ.* **6**, 227.

Sibley, M.H. and Yopp, J.H. (1986). *Plant Physiol.* (Suppl.) **80**, 124.

Stewart, W.D.P. (1983). *Symp. Soc. gen. Microbiol.* **15**, 1.

Theuvelin, J.M. (1984). *Microbiol. Rev.* **48**, 42.

Tel-Or, E., Spath, S., Packer, L., and Mehlhorn, R.J. (1986). *Plant Physiol.* **82**, 646.

Walsby, A.E. (1980). *Proc. R. Soc. Lond.* B **208**, 73.

Warr, S.R.C. (1985). Low molecular weight organic solutes: their role in the osmotic adjustment of cyanobacteria. Unpublished Ph.D. thesis, University of Dundee.

Warr, S.R.C., Reed, R.H., and Stewart, W.D.P. (1984*a*). *Mar. Biol.* **79**, 21.

Warr, S.R.C., Reed, R.H., and Stewart, W.D.P. (1984*b*). *J. gen. Microbiol.* **130**, 2169.

Warr, S.R.C., Reed, R.H., and Stewart, W.D.P. (1985*a*). *Planta* **163**, 424.

Warr, S.R.C., Reed, R.H., and Stewart, W.D.P. (1985*b*). *New Phytol.* **100**, 285.

Warr, S.R.C., Reed, R.H., and Stewart, W.D.P. (1987). *Br. phycol. J.* **22**, 175.

Waterbury, J.B. and Stanier, R.Y. (1981). In *The prokaryotes* (ed. M.P. Starr, H. Stolp, H.G. Trüper, A. Balows, and H.G. Schlegel) p. 221. Springer-Verlag, Berlin.

Wyn Jones, R.G. and Gorham, J. (1983). In *Encyclopedia of plant physiology 12C: physiological plant ecology III: responses to the chemical and biological environment* (ed. O.L. Lange, P.S. Nobel, C.B. Osmond, and H. Ziegler) p. 35. Springer, Berlin.

Wyn Jones, R.G. and Gorham, J. (1986). *Outlook Agric.* **15**, 33.

Poster abstracts

The effect on photosynthesis in *Anacystis nidulans* of varying the calcium concentration of growth medium

L. Roberts and E.H. Evans
School of Applied Biology, Lancashire Polytechnic, Preston, Lancashire, UK

The cyanobacterium *Anacystis nidulans* was grown in medium which contained differing concentrations of calcium. Maximal oxygen evolution and phycocyanin:chlorophyll *a* ratio was observed in cells grown at 20 μM Ca^{2+}, and decreased at higher concentrations. Thylakoids isolated from cells grown at differing [Ca^{2+}] showed a different response to Ca^{2+}. Thylakoids from cells grown at low [Ca^{2+}] showed increased light-induced oxygen evolution on addition of Ca^{2+} to concentrations comparable to the growth medium. Thylakoids from cells grown at higher [Ca^{2+}] were stimulated similarly but by higher levels of Ca^{2+}. This is comparable with data we have obtained with grasses collected from calcicole and calcifuge sites. Subsequent addition of Ca^{2+} in excess of that stimulating oxygen evolution in *Anacystis* thylakoids inhibited the activity. This inhibition was abolished in the presence of Mg^{2+}. This confirms our proposals for at least two sites of action of Ca^{2+} (England and Evans 1983); one is associated with Mg^{2+} binding, and one solely associated with photosystem II which binds Ca^{2+} in relation to the previous Ca^{2+} growth conditions of the cells.

England, R.R. and Evans, E.H. (1983). *Biochem. J.* **210**, 473.

Cellular Ca^{2+} content and incident irradiance in *Nostoc* 6720

R.J. Smith and A. Wilkins
University of Lancaster, Department of Biological Sciences, Bailrigg, Lancester LA1 4YQ, UK

The proportion of heterocysts in *Anabaena cylindrica* cultures varies according to the incident irradiance such that the proportion falls as irradiance is decreased. However, at low irradiance the trend reverses and the proportion increases again (Fogg 1949). The proportion of heterocysts induced by nitrate removal in cultures of *Nostoc* 6720 increases with treatments which raise the cellular Ca^{2+} content (Smith *et al.* 1987).

As the incident irradiance is reduced below 30 $Em^{-2}s^{-1}$ the Ca^{2+} content of *Nostoc* 6720 increased. This increase appears to result from decreased Ca^{2+} efflux as the rate of loss of $^{45}Ca^{2+}$ from the EGTA-resistant fraction declines when illumination of the culture is decreased. A similar effect is observed in *Anacystis nidulans* (Becker and Brand 1985)

The inverse relationship between Ca^{2+} content and incident irradiance allows the hypothesis that Ca^{2+}-mediated regulation may occur in cyanobacteria. The increased proportion of heterocysts in *A. cylindrica* may be an example of such regulation.

Becker, D. and Brand, J.J. (1985). *Plant Physiol.* **79**, 552.
Fogg, G.E. (1949). *Ann. Bot.* **13**, 241.
Smith, R.J., Hobson, S., and Ellis, E. (1987). *New Phytol.* **105**, 531.

Induction of a Ca^{2+} binding protein in *Nostoc* 6720

G. Jackson, S. Temple, and R.J. Smith
University of Lancaster, Department of Biological Sciences, Bailrigg, Lancester, LA1 4YQ, UK

In the presence of lanthanum (0.1 mM) the cell content and external binding of Ca^{2+} in *Nostoc* 6720 cultures increases two- to threefold and ten- to fifteenfold, respectively (Smith *et al.* 1987). When chloramphenicol is also present the increase in cell content still occurs, but the increase in the external binding is inhibited. A similar, though less pronounced effect, is observed with Ca^{2+}-depleted cultures.

The increase in cellular Ca^{2+} content during lanthanum inhibition may be attributed to increased Ca^{2+} influx. The rate of Ca^{2+} incorporation is substantially increased relative to untreated controls. In contrast no significant effect on efflux has been detected.

The increase in external binding in both lanthanum-inhibited and Ca^{2+}-depleted cultures requires protein synthesis. A protein of M_r approx. 16 Kd has been detected in the cell membrane/wall fraction and is enhanced by both treatments. The protein may provide an external binding site. Calcium depletion or inhibition of binding appears to promote gene expression which is required to increase external sites for Ca^{2+} binding.

Smith, R.J., Hobson, S., and Ellis, I. (1987). *New Phytol.* **105**, 531.

A phytohormone effect on heterocyst differentiation in *Nostoc* 6720

Smith, R.J., Hobson, S. and Ellis, I.
University of Lancaster, Department of Biological Sciences, Bailrigg, Lancaster, LA1 4YQ, UK

The action of the phytohormone abscisic acid (ABA) in inhibiting the opening of stomata has been shown to require Ca^{2+} and to be inhibited by Ca^{2+} channel antagonists (De Silva *et al.* 1985*a*, 1985*b*). ABA also acts as a Ca^{2+}

agonist in rat muscle preparation and *Nostoc* 6720 (Huddart *et al.* 1986), raising the question of whether the phytohormone has a universal action.

The presence of ABA, in cultures of *Nostoc* 6720 induced to form heterocysts by removal of a nitrate supplement, accentuates heterocyst differentiation in a dose-dependent manner. The reduced effect at concentrations of ABA above the optimum of 1 μM may reflect some disruption of membrane function (Huddart *et al.* 1986).

The effect of ABA relative to the Ca^{2+} concentration present in the growth medium demonstrates a Ca^{2+} dependency, the interaction between ABA and Ca^{2+} being significant ($p > 0.001$, analysis of variance). Comparable effects are observed on treatment of cultures with the Ca^{2+} ionopore A23187.

De Silva, D.L.R., Cox, C.R., Hetherington, A.M., and Mansfield, T.A. (1985*a*). *New Phytol.* **101**, 555.

De Silva, D.L.R., Hetherington, A.M., and Mansfield, T.A. (1985*b*). *New Phytol.* **100**, 473.

Huddart, H., Smith, R.J., Langton, P.D., Hetherington, A.M. and Mansfield, T.A. (1986). *New Phytol.* **104**, 161.

Light co-activates NADPH-dependent glyceraldehyde 3-phosphate dehydrogenase and phosphoribulokinase *in vivo* by promoting the dissociation of a multimeric protein complex

R. Powls, S. Nicholson, and J.S. Easterby

Department of Biochemistry, University of Liverpool, PO Box 147, Liverpool L69 3BX, UK

A chloroplast multimeric protein with latent activities of NADPH-dependent glyceraldehyde 3-phosphate dehydrogenase (NADPH-G3PDH) and phosphoribulokinase (PRK) has been purified to homogeneity from the green unicellular alga, *Scenedesmus obliquus* (Nicholson *et al.* 1987). Co-activation of NADPH-G3PDH and PRK occurred on incubation with dithiothreitol (10 mM) and NADPH (1 mM). The multimeric protein (M_r 560 000) was composed of two polypeptides (42 000 and 39 000) and had a subunit composition 8G6R, where G is a subunit (39 000) conferring G3PDH activity and R that with PRK activity (42 000). Co-activation was accompanied by depolymerization to give the active forms of NADPH-G3PDH (4G) and PRK (2R).

The co-activation promoted by incubation of the latent multimeric protein with dithiothreitol (1 mM) and NADPH (0.5 mM) was stimulated by algal thioredoxin. In extracts from heterotrophically grown alga the large multimeric enzyme predominated, whereas in those from photoheterotrophic alga the proportion of the multimeric enzyme was decreased and the low molecular weight active forms of the enzymes were predominant. These findings are consistent with the *in vivo* activation by light being due to the dissociation of the large multimeric enzyme promoted by photoreduced thioredoxin.

Nicholson, S., Easterby, J.S., and Powls R. (1987). *Eur. J. Biochem.* **162**, 423.

Organic osmoregulatory solutes of marine eukaryotic unicellular algae

D.M.J. Dickson
Department of Botany and Microbiology, University College of Swansea, Swansea, SA2 8PP UK

G.O. Kirst
Algal Physiology Unit, University of Bremen, Bremen, FRG

In a survey of the major organic osmotica of marine unicellular algae the following compounds paralleled the increases of external salinity: the quaternary ammonium compound (QAC), glycine betaine, and the amino acid, proline, in the diatoms *Cyclotella cryptica, Cyclotella meneghiniana*, and *Phaeodactylum tricornutum*; the QAC, homarine (*N*-methyl picolinic acid betaine), in *C. cryptica* and *C. meneghiniana*; and the tertiary sulphonium compound 3-dimethylsulphoniopropionate (DMSP) and glycerol in *P. tricornutum*. The rhodophyte *Porphyridium aerugineum* synthesized proline, glycine betaine, floridoside, glycerol, and galactose. The prasinophyte algae *Platymonas subcordiformis, Tetraselmis chui*, and *Prasinocladus* synthesized mannitol, glycine betaine, homarine, and DMSP. In these prasinophytes proline was an insignificant osmoticum. The prymnesiophyte algae *Prymnesium parvum* and *Ruttnera spectabilis* synthesized DMSP and an unidentified polyol, which by reference to the prymnesiophyte *Pavlova* (*Monochrysis*) *lutheri* may be the cyclitol 1,4/2,5 cyclohexanetetrol (see Craigie 1969). QACs were not detected in these two prymnesiophytes, and proline was an insignificant osmoticum. Homarine exhibits maximum absorption at 270 nm and may influence the absorption of light in photosynthesis by inhibiting UV-B-irradiation (290–320 nm). Our findings are fully discussed in Dickson and Kirst (1987).

Craigie, J.S. (1969). *J. Fish. Res. Bd. Can.* **26**, 2959.
Dickson, D.M.J. and Kirst, G.O. (1987). *New Phytol.* **106**, 657.

Energization of Na^+ extrusion in cyanobacteria: respiratory or photosynthetic?

R. Jeanjean and F. Joset
Universite Aix-Marseille, II Luminy Case 901, 13288 Marseille Cedex 9, France

In two faculative phototrophic, salt-tolerant *Synechocystis* spp. (6714 and 6803), adaptation to high salt needs 5–6 hours for completion. During this time photosynthetic O_2 evolution remained constant, while dark O_2 uptake increased by a factor of 1.5–2.0. In the presence of high salt (0.3 M NaCl), growth slows down, more so under energy-limiting conditions (photoheterotrophy or chemoheterotrophy) than in phototrophic conditions in which energy is not limited. These results show the involvement of an O_2-consuming process in Na^+ extrusion, in agreement with the possible involvement of (part of) the respiratory chain for the provision of energy; they do not, however, exclude photosynthetic sources.

Mutants unable to extrude Na^+ were obtained with a frequency of about 10^{-7} to 10^{-8} by penicillin enrichment as a result of their inability to grow in the presence of high salt concentrations.

Five clones sensitive to Na^+ (Nas) were tested for their growth responses and O_2 exchange capacities. All showed no growth in the presence of 0.3 M NaCl, but their growth rates in the absence of Na^+ were similar to those of the wild type, except for two mutants (6803-Nas 41 and 43), which were unable to grow in the dark. However, all of the mutants analysed had a lower O_2-uptake capacity.

In mutants 6803-Nas 41 and 43, the mutation may affect a step common to respiration and Na^+ extrusion with the residual O_2-uptake capacity being too small to sustain both processes. In the other mutants (6714-Nas 21 and 22; 6803-Nas 21), because heterotrophic growth was not modified, the mutations may affect either a step similar to the above, but to a lesser extent, or a step specific to Na^+ extrusion.

Part 5: Interactions

13 Nutrient interactions in the marine environment

ANTHONY G. DAVIES

Marine Biological Association, Citadel Hill, Plymouth PL1 2PB, UK

Introduction

There is a substantial body of evidence that the specific growth rate of a monospecific phytoplankton population is a hyperbolic function of the cell quota (cellular concentration) of the rate-limiting nutrient: e.g. nitrate (Caperon 1968), vitamin B_{12} (Droop 1968), phosphate (Fuhs 1969; Rhee 1973), silicate (Paasche 1973), iron (Davies 1970). Where the effects of pairs of nutrients have been investigated, it has been concluded that, in accord with Liebig's law of the minimum, only one of the nutrients is ever rate-limiting and that no interaction takes place between them. The specific growth rate is then only related to the cell quota of a single nutrient rather than being an additive or multiplicative function of two (or more) cell quotas: e.g. vitamin B_{12} and phosphate (Droop 1974), nitrate and phosphate (Rhee 1978; Terry 1980).

This conclusion, based on experiments with continuous cultures, is, however, seemingly at odds with the results of studies of the responses of natural populations of microalgae to nutrient enrichment. Here it has been found that the joint addition of phosphate with nitrate or ammonium often has a synergistically stimulatory effect upon the rate of photosynthesis or the production of chlorophyll (chl) *a* in phytoplankton assemblages, both in lakes (Fuhs *et al.* 1972; Gerhart 1975; Lean and Pick 1981) and in the sea (Granéli 1978, 1981; Vince and Valiela 1973).

A further, and as yet incompletely explained, complexity is that, in the first few hours after a nutrient spike has been added, photosynthesis has frequently been observed to be inhibited by the enrichment (Healey 1979; Turpin 1983; Falkowski and Stone 1975; Lean and Pick 1981). In most of these studies, nutrient uptake by the phytoplankton and therefore changes in the cell quotas were occurring concurrently with the measurement of the rate of photosynthesis. In view of this, and also since there is evidence that nitrate and phosphate uptake are interactive (Rhee 1974; Terry 1982), the observations are particularly difficult to interpret. Healey (1979) found that the depression of photosynthesis usually ceased once the nutrient uptake that followed the enrichment was complete. It was thus clear that a better understanding of the effect of nutrient additions upon the growth of phytoplankton would be obtained by allowing the uptake processes to attain equilibrium before measuring the photosynthetic response of the population. We had successfully used this approach to study how toxic metals inhibited carbon fixation in coastal phytoplankton populations (Davies and Sleep 1979, 1980) and have now applied the same technique to determine how increments in nutrient concentrations influence photosynthesis in marine microalgae. This paper describes some of our findings and their relationship to other research.

Experiments with English Channel populations

Briefly, for studies with natural phytoplankton assemblages, a sample of seawater containing its indigenous population was collected in the evening, passed through a 170 μm nylon mesh to remove the larger zooplankton, then divided into 2 l aliquots. These were spiked with a nitrogenous nutrient without and with phosphate, all at environmentally realistic concentrations to simulate the increases in nutrient levels which occur due to the temporary breakdown of the thermocline, frontal mixing, or tidal excursions. The aliquots acting as controls received additions of water only to correspond to that in which the nutrients were dissolved. Following the additions, the bottles containing the water were kept overnight at 15 °C on a slowly rotating wheel in an incubator. To simulate natural conditions, they were maintained in darkness between 2000 hours and 0600 hours and subsequently received 50 $\mu E\ m^{-2}\ sec^{-1}$ irradiance from Northlight fluorescent tubes for about 3 hours. Carbon fixation rates in the aliquots were then measured over a period of 5 to 7 hours in nominally 125 ml glass bottles in the same incubator at a constant irradiance of 150 $\mu E\ m^{-2}\ sec^{-1}$. Full details of the practical aspects of the work will be published elsewhere.

A typical set of results obtained at Station F in mid-Channel south of the Lizard peninsula is shown in Fig. 13.1 where the data have been normalized to the mean of the carbon fixation rates in the controls. Although the phosphate addition alone had no effect, the slight depression being within the experimental error of ± 15 per cent indicated by the dashed lines, it had a marked synergistic interaction with both ammonium and nitrate. Surprisingly, urea was not utilized by this particular assemblage. A more quantitative illustration of the synergism is shown in Fig. 13.2 for Station M, just to the west of Guernsey. Here the addition of 0.5 μM ammonium caused about a 40 per cent increase in the carbon fixation rate, the effects of higher concentrations being slightly smaller. The additional presence of 0.5 μM phosphate, however, caused a very large stimulation of photosynthesis.

Explanation of these results clearly required information on the changes in cell quotas which had resulted from the enrichments. Due to the impossibility of separating detrital matter from phytoplankton assemblages, reliable

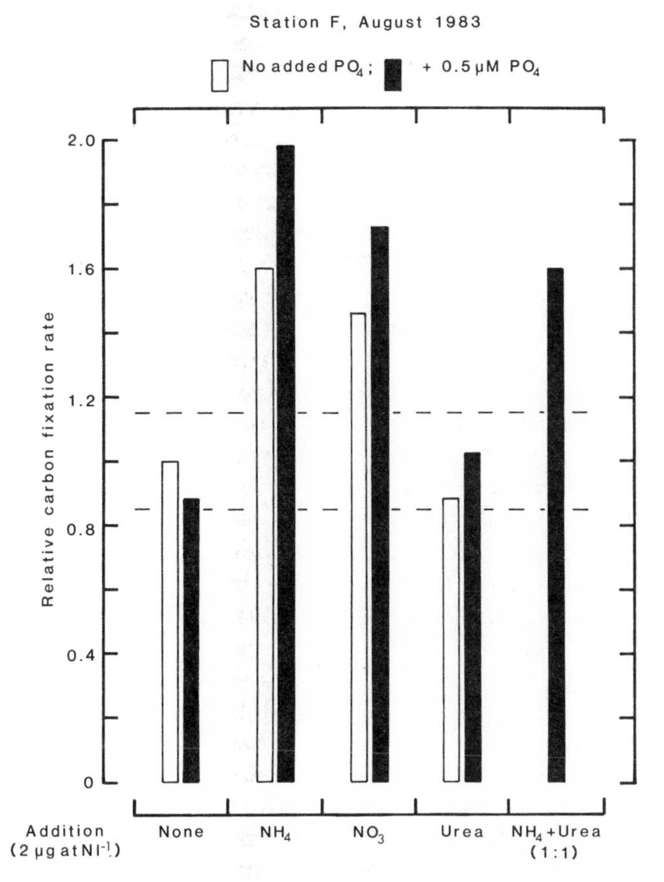

Fig. 13.1. The effect of nitrogen enrichments, in the absence and presence of 0.5 μM phosphate, upon the rate of carbon fixation in the phytoplankton assemblage present at Station F (49°15′ N, 05°02′ W) in August 1983. The population consisted mainly of *Chrysochromulina* sp. and contained 1.9 μg chl a l^{-1}. Nutrient levels before enrichment were <0.2 μM nitrate and 0.06 μM reactive phosphate. The mean of the carbon fixation rates in the controls was 3.9 μg C $l^{-1}h^{-1}$; other values have been normalized to this. The dotted lines indicate the range of the experimental error. Phosphate alone, and urea with and without phosphate, had no effect. Ammonium and nitrate both stimulated photosynthesis and phosphate interacted synergistically with each of these nutrients.

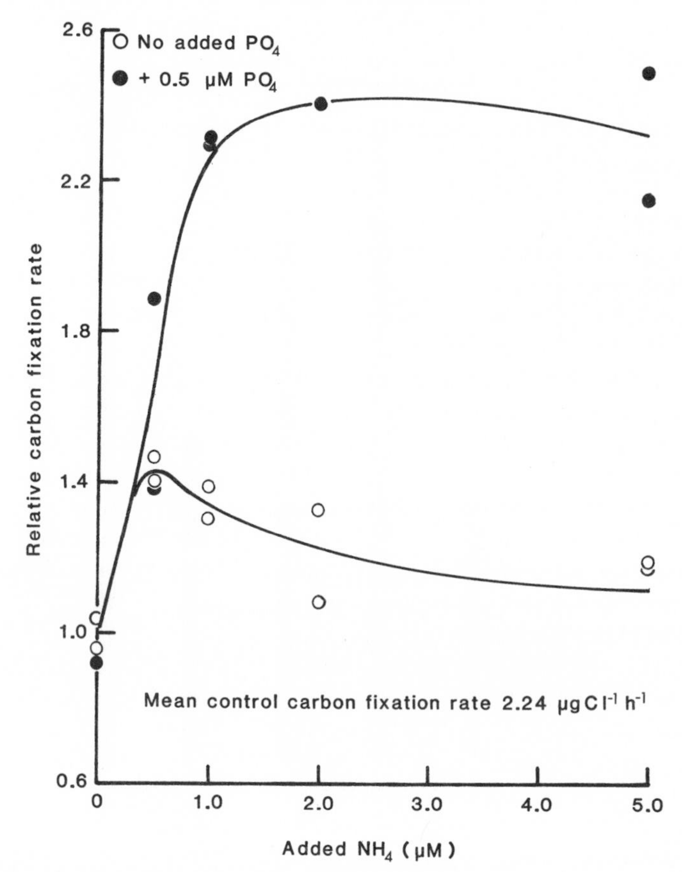

Fig. 13.2. The effect of ammonium enrichments, in the absence and presence of 0.5 μM phosphate, upon the rate of carbon fixation in the phytoplankton assemblage present at Station M (49°23′ N, 03°20′ W) in July 1984. The population consisted mainly of diatoms, *Rhizosolenia delicatula* and *Leptocylindrus danicus* being predominant, and contained 1.0 μg chl *a* l^{-1}. Nutrient levels before enrichment were 0.2 μM nitrate and 0.05 μM reactive phosphate. The carbon fixation rates have been normalized to the mean of those in the unspiked controls. In the absence of added phosphate, 0.5 μM ammonium caused a 40 per cent stimulation of photosynthesis; higher concentrations had a smaller effect. The presence of 0.5 μM phosphate alone had little influence but caused very large increases in carbon fixation when added with the ammonium enrichments.

analyses for the carbon, nitrogen, and phosphorus contents of natural populations cannot be obtained. We therefore resorted to using low density monospecific laboratory cultures to pursue the problem.

Experiments with low density nutrient-limited cultures of *Skeletonema costatum*

This has proved to be an extremely advantageous way of bridging the gulf between the dense cultures normally employed in physiological work in the laboratory and field studies of sparse natural populations; measurement of the effect of environmental levels of nutrients upon photosynthesis in cell suspensions of a density similar to natural populations can be made using the normal radiocarbon fixation technique while the cellular nutrient concentrations can be determined without the risk of contamination. *Skeletonema costatum* was chosen as the test species because it is a much studied, reasonably representative, and easily cultured diatom. The nutrient-deficient suspensions were obtained by inoculating unenriched, autoclaved, filtered seawater with cells obtained from an axenic stock culture and allowing growth to proceed for a period extending several days beyond the point where nitrate and phosphate had disappeared from the water. The culture was then filtered to remove the cells, the nutrient-free filtrate being re-autoclaved, and some of the cells which had already grown in the water returned to it to give chl *a* levels approximating to those present in the English Channel in the summer-time. This suspension was then treated in exactly the same way as a natural population except that, following the overnight incubation with the nutrient enrichments, samples were removed on to filters for elemental analysis.

The outcome of two experiments involving ammonium and phosphate additions is reported here. In experiment 1, the enrichments used were the same as those employed for the work at Station M (Fig. 13.2). The results for *S. costatum* [Fig. 13.3(a)] were slightly different from those for the natural assemblage in that the ammonium additions without phosphate repressed the carbon fixation rate. Here, 0.5 μM phosphate alone caused a small but significant stimulation of photosynthesis, but, as with the population at Station M, the joint addition of ammonium and phosphate caused very large increases in the rate of carbon fixation. The changes in the cellular levels of phosphorus and nitrogen (calculated as P:C and N:C atomic ratios) resulting from the nutrient additions are illustrated in Fig. 13.3(b) and (c). Whereas the increment in the P:C ratio was constant and independent of the ammonium enrichment [Fig. 13.3(b)], the N:C ratio increased linearly with the ammonium additions and much more steeply in the presence of phosphate [Fig. 13.3(c)]. In Fig. 13.4, the carbon specific growth rate (calculated from the measured cellular carbon content) is plotted as a function of the N:C ratio in

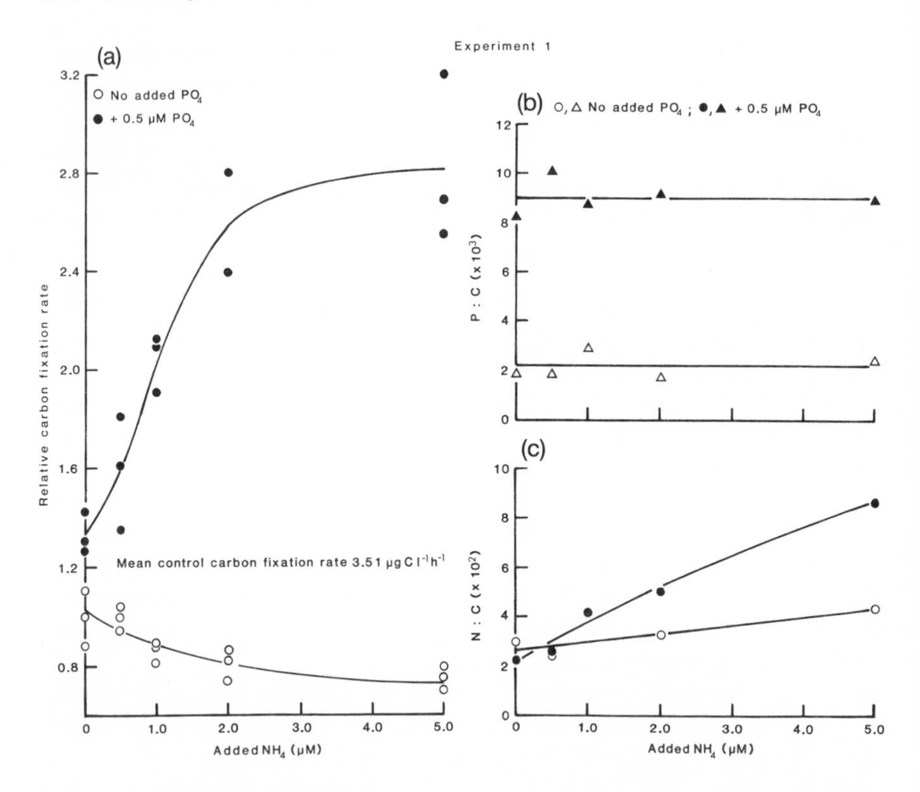

Fig. 13.3. (a) The same experiment as that carried out at Station M (Fig. 13.2) but using a low density laboratory population of nutrient deficient *S. costatum* containing 1.9 μg chl *a* l^{-1}. Nutrient levels before enrichment were <0.2 μM nitrate and <0.02 μM phosphate. Here, ammonium enrichments, in the absence of added phosphate, inhibited photosynthesis. The presence of 0.5 μM phosphate alone had a small positive effect but, as with the natural population, caused a very large stimulation of carbon fixation when added jointly with the ammonium enrichments. (b) The P:C ratios in *S. costatum* in the absence and presence of 0.5 μM phosphate. The increase in the P:C value caused by the addition was independent of the ammonium enrichment. (c) The N:C ratios in *S. costatum* as a function of the ammonium enrichments without and with 0.5 μM phosphate. The ratios increased considerably more steeply when phosphate was added indicating the coupling between ammonium uptake and phosphate availability.

the absence and presence of the phosphate. This shows clearly how, at the lower P:C ratio, the small increases in the N:C ratio repressed the growth, while the increases in N:C associated with the higher P:C ratio resulted in a typically hyperbolic response in the carbon specific growth rate. The curve shown in Fig. 13.4 is the 'best fit' of the equation

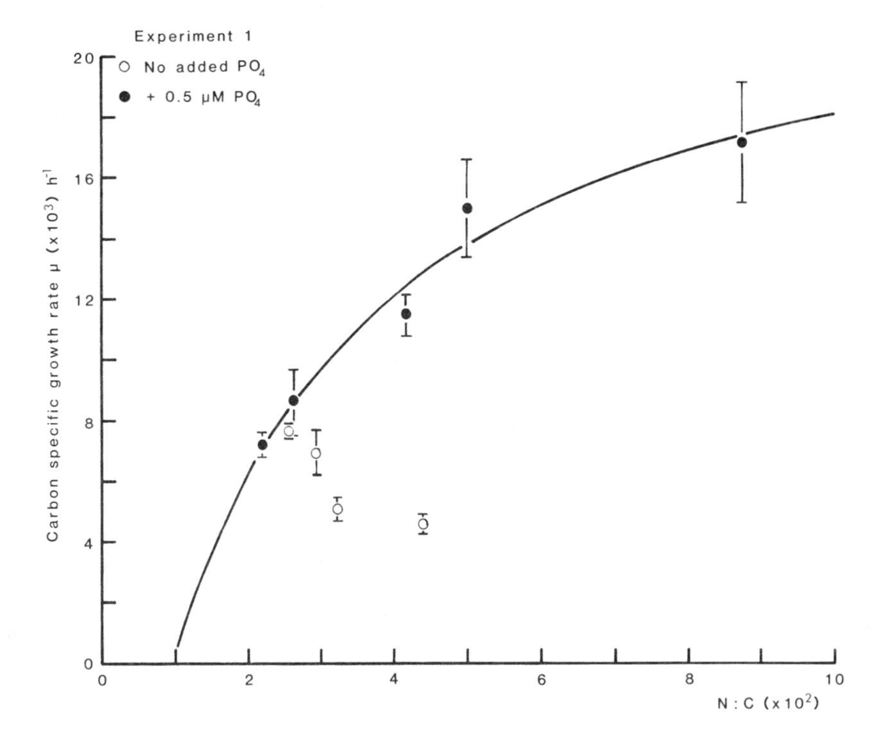

Fig. 13.4. The carbon specific growth rate as a function of the N:C ratio in *S. costatum* in experiment 1. The vertical bars represent ±1 Standard Error. The increasing N:C ratios, in the absence of phosphate, repressed the growth rate, but when 0.5 μM phosphate was present the data conformed to a typical hyperbola. The equation of the 'best fit' curve shown is given in the text.

$$\mu = \frac{\mu_{max}[(N:C) - (N:C)_0]}{K_{N:C} + [(N:C) - (N:C)_0]}$$

where the maximal value of the specific growth rate, $\mu_{max} = 0.024\ h^{-1}$, the half-saturation constant, $K_{N:C} = 0.029$, and the subsistence value of N:C, $(N:C)_0 = 0.0097$.

In experiment 2, the phosphate additions were varied in the absence and presence of 5 μM ammonium; the results are shown in Fig. 13.5. All of the phosphate enrichments significantly stimulated carbon fixation, though the increase in rate was about the same when the enrichment was 1.0 μM as when it was only 0.125 μM, suggesting that nitrogen had then become limiting [Fig. 13.5(a)]. This was confirmed by the much higher carbon fixation rates in the samples to which 5 μM ammonium was also added, though, as before, ammonium in the absence of phosphate caused a marked repression of

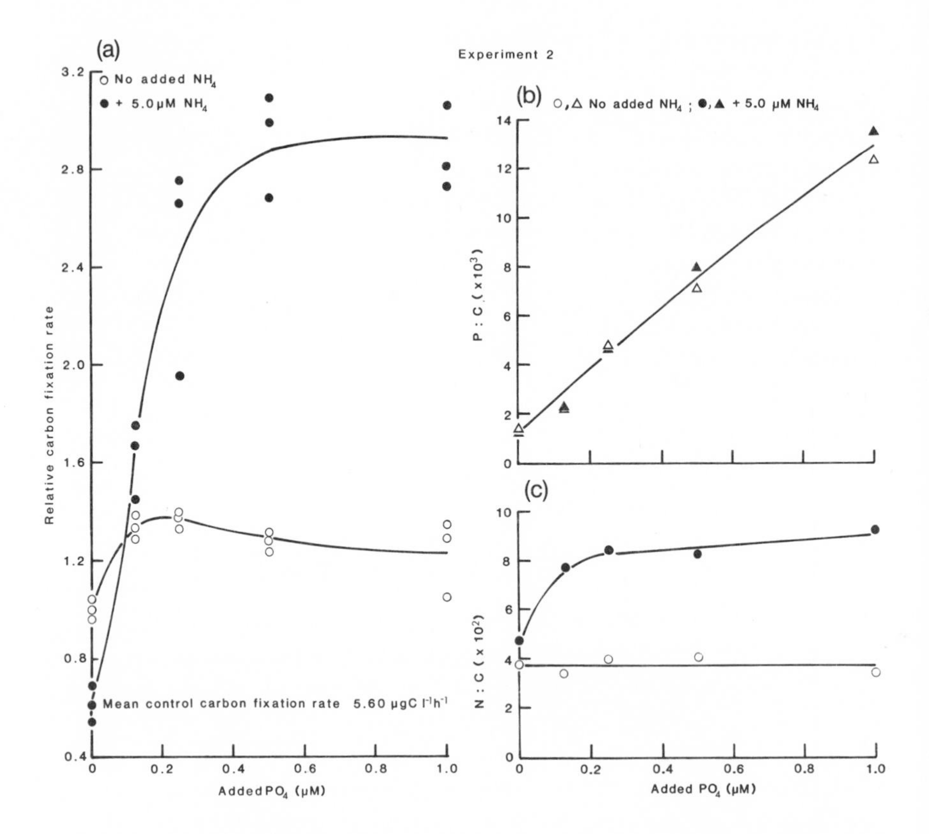

Fig. 13.5. (a) The effect of phosphate enrichments in the absence and presence of 5 μM ammonium upon the rate of carbon fixation in a low density population of nutrient deficient *S. costatum* containing 2.8 μg chl *a* 1^{-1}. Nutrient levels before enrichment were <0.2 μM nitrate and <0.02 μM reactive phosphate. The carbon fixation rates have been normalized to those in the unspiked controls. In the absence of added ammonium, the phosphate enrichments caused small increases in carbon fixation. Ammonium alone inhibited photosynthesis but caused a very large stimulatory effect when added jointly with the phosphate enrichments. (b) The P:C ratios in *S. costatum* as a function of the phosphate enrichment in the absence and presence of 5 μM ammonium. The linear increase in the P:C ratio caused by the phosphate additions was independent of the ammonium concentration. (c) The N:C ratios in *S. costatum* as a function of the phosphate enrichments in the absence and presence of 5 μm ammonium. The ratio initially increased steeply with phosphate addition but levelled off at 0.25 μM phosphate, even though only about 50% of the available ammonium had been incorporated. This suggested that the nitrogen cell quota was then approaching its saturation value.

photosynthesis [Fig. 13.5(a)]. Increases in the P:C ratio were linearly correlated with the concentration of phosphate added and independent of the presence or absence of ammonium [Fig. 13.5(b)]. Cellular nitrogen uptake from the 5 μM ammonium enrichment also increased with the phosphate concentration, though the N:C ratio approached a constant value at phosphate levels above 0.25 μM even though only about 50 per cent of the added ammonium had been incorporated by the cells [Fig. 13.5(c)]. This suggests that the nitrogen cell quota was then close to its saturation value.

Conversion of the carbon fixation rates into carbon specific growth rates showed the dramatic effect of the elevated nitrogen cell quotas caused by the ammonium enrichment (Fig. 13.6). The curves in the diagram are based on

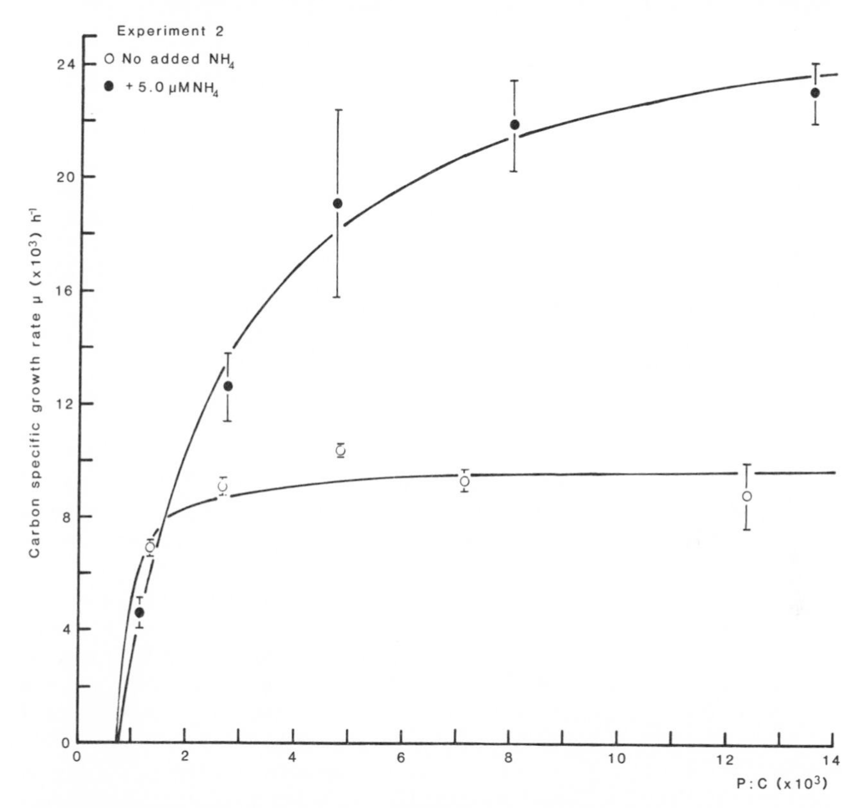

Fig. 13.6. The carbon specific growth rate as a function of the P:C ratio in *S. costatum* in experiment 2. The vertical bars represent ± 1 standard error. In the absence of added ammonium, the increasing P:C ratio made little difference to the specific growth rate suggesting that nitrogen was limiting. This was confirmed by the large increases in growth rate which occurred when 5 μM ammonium was also present. The constants for the equations to the 'best fit' curves shown are given in the text.

the same equation as that used in Fig. 13.4 but with P:C replacing N:C. The corresponding values of the constants were:

	No added NH_4	5 μM NH_4
μ_{max} (h^{-1})	0.0098	0.027
$K_{P:C}$	0.0002	0.0021
$(P:C)_0$	0.00075*	0.00075

*Assumed value.

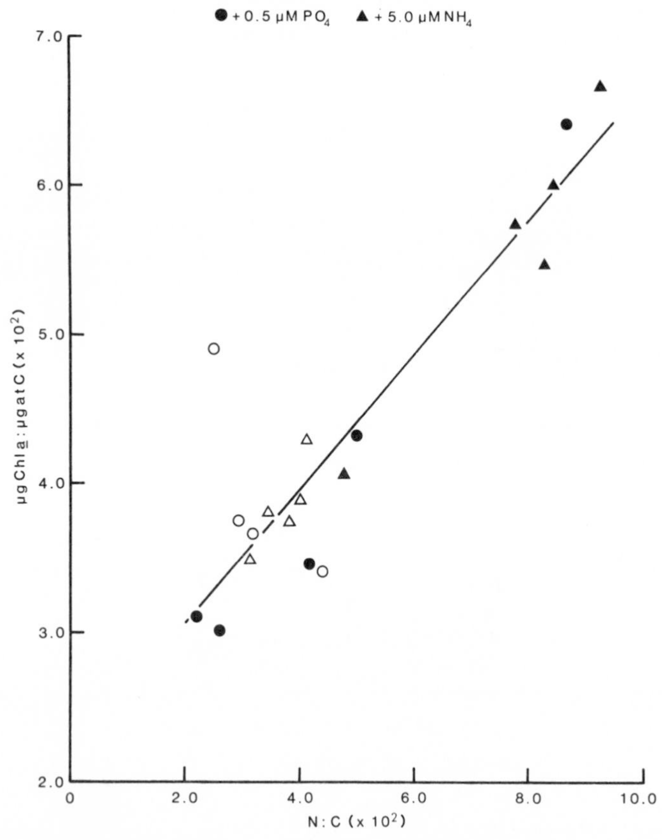

Fig. 13.7. The regression of the chl *a*:C ratio upon the N:C ratio in *S. costatum*. The parameters were highly correlated (correlation coefficient 0.90) indicating the dependence of chlorophyll production upon the nitrogen cell quota. The equation for the 'best fit' line shown is given in the text.

A significant feature of both experiments was that where, following the overnight incubation, there were increases in the N:C ratio, these were associated with enhanced chlorophyll concentrations. Riper *et al.* (1979) found that chlorophyll biosynthesis in *S. costatum* is extremely rapid and can occur in the dark. The chl *a*:C ratios in our work were highly correlated with the N:C ratios; Fig. 13.7 shows the combined data from experiments 1 and 2, with the regression line (correlation coefficient 0.90) being described by the equation:

$$\mu\text{g chl } a{:}\mu\text{g atom C} = 0.022 + 0.45\text{N:C}$$

Although this relationship adequately describes the data over the range of N:C ratios observed in the experiments, there must presumably be an increase in slope at lower N:C values if the relationship is to pass through the origin or intersect the abscissa.

A further important observation was that the carbon specific growth rate was a linear function of the chl *a*:C ratio (correlation coefficient 0.76) (Fig. 13.8). It thus appeared that part, at least, of the observed increases in the rates of photosynthesis was due to the presence of higher chlorophyll levels.

The chlorophyll specific carbon fixation rate (assimilation number) is equivalent to the carbon specific growth rate divided by the chl *a*:C ratio and thus normalizes growth rates for differences in cellular chlorophyll concentrations. The values calculated for experiments 1 and 2 are plotted together against the P:C and the N:C ratios in Fig. 13.9(a) and (b), respectively. The data in Fig. 13.9(a) were obviously consistent with a hyperbolic relationship between the chlorophyll specific carbon fixation rate and the P:C ratio. In view of the potential sources of error in the parameters involved, the agreement of the data with the 'best fit' hyperbola was extremely good. The equation of the curve shown is:

$$\underset{(\mu\text{gC }\mu\text{gchl } a^{-1}\text{ h}^{-1})}{\text{Chlorophyll specific carbon fixation rate}} = \frac{3.61[(\text{P:C}) - 0.001]}{[(\text{P:C}) - 0.00057]}.$$

The figure of 0.00057 in the denominator is the difference between the half saturation constant (0.00043) and the subsistence value of P:C (0.001).

The dashed curve in Fig. 13.9(b) has no quantitative significance but illustrates the trend of the data. As will be discussed later, it is probably relevant that the negatively correlated portion of this curve, which corresponds to the repressive effect caused by cellular nitrogen increments, is associated with P:C ratios which are close to the subsistence value. Elrifi and Turpin (1985) found a similar C-shaped relationship between the specific growth rate of phosphorus-limited *Selenastrum minutum* and its cellular nitrogen content.

The above expression relating the chlorophyll specific carbon fixation rate to the P:C ratio may be used to demonstrate how this rate varied with the P:N

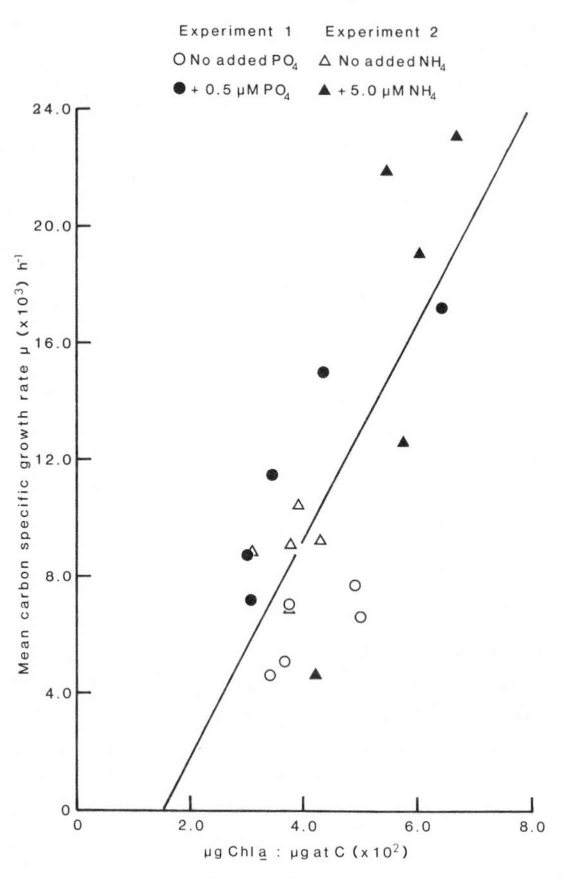

Fig. 13.8 The regression of the carbon specific growth rate upon the chl *a*:C ratio in *S. costatum*. The good correlation between the parameters (correlation coefficient 0.76) suggested that part, at least, of the observed increases in the carbon specific growth rate caused by the nutrient enrichments were due to elevated chlorophyll levels in the phytoplankton.

ratios in *S. costatum* using the measured P:C and N:C ratios. The calculated values are plotted together with the experimental data in Fig. 13.10. The curve (drawn by eye through the calculated points) shows that the carbon fixation rate normalized to unit chlorophyll concentration is independent of the P:N ratio above a value of about 0.1. This indicates that, when the cellular nitrogen quota limits the rate of carbon fixation, it does so by governing the cellular chlorophyll concentration. The chlorophyll specific carbon fixation rate attains 90 per cent of its maximal value when the P:C ratio is 0.0049 [calculated from the curve in Fig. 13.9(a)]. Below this value, the cellular

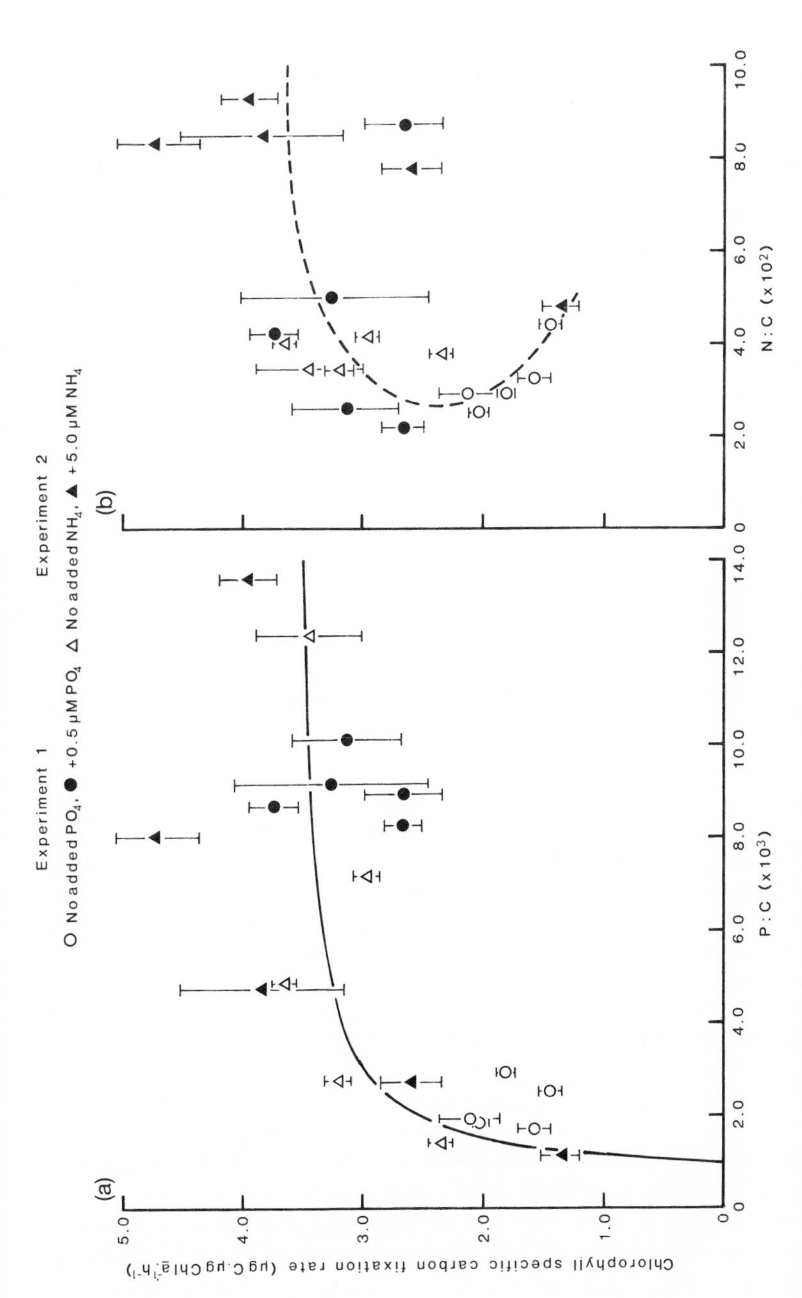

Fig. 13.9. The chlorophyll specific carbon fixation rates (assimilation numbers) plotted against the P:C ratios (a), and the N:C ratios (b) in *S. costatum*. The vertical bars represent ±1 standard error. The data in (a) conformed well to a hyperbolic relationship; the equation for the curve shown is given in the text. The dashed curve in (b) has no quantitative significance but illustrates the trend of the data. The negatively correlated portion of this curve which corresponds to the inhibition of photosynthesis caused by ammonium enrichments in the absence of added phosphate is associated with P:C ratios near the subsistence value. It is postulated that the repression is due to the sequestration of phosphate in the cytoplasm.

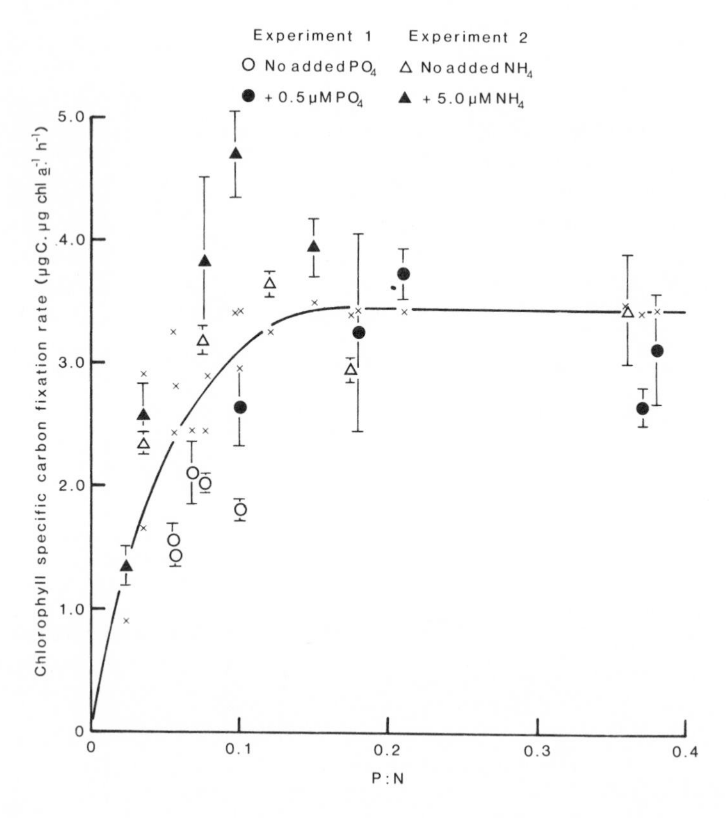

Fig. 13.10. The chlorophyll specific carbon fixation rate plotted as a function of the P:N ratio in *S. costatum*. The crosses are points calculated from the hyperbola shown in Fig. 13.9(a) and the measured values of P:C and N:C. The line has been drawn by eye through the calculated points. Above a P:N value of about 0.1, the line became horizontal, indicating that when nitrogen is limiting it regulates the rate of carbon fixation by governing the cell chlorophyll quota. When phosphorus is limiting, carbon fixation is a function not only of the P:C ratio but, because of the implication of nitrogen in chlorophyll production and the inhibition of photosynthesis, also of the N:C ratio. The relationship of the chlorophyll specific growth rate to the P:N ratio is then much more complex.

phosphorus may be regarded as limiting, but since the chl *a*:C ratio is correlated with the N:C ratio over the whole range of observed values (Fig. 13.7), the carbon specific growth rate (the product of the chlorophyll specific carbon fixation rate and the chl *a*:C ratio) is then a multiple of functions of the P:C ratio and the N:C ratio.

Discussion

Our experiments have shown that ammonium and phosphate interactions occur in the nutrition of microalgae, (1) because of the implication of nitrogen in chlorophyll biosynthesis so that the specific rate of carbon fixation is a multiple of functions of the phosphorus and the nitrogen cell quotas when phosphorus is limiting, and (2) because ammonium incorporation by the phytoplankton is regulated by the phosphorus cell quota.

Sivak and Walker (1986) have pointed out that, although the traditional way of representing the stoichiometry of photosynthesis is

$$CO_2 + 2H_2O \xrightarrow{h\nu} CH_2O + H_2O + O_2$$

a better chemical description of the processes taking place in the chloroplast is given by

$$3CO_2 + 6H_2O + PO_4^{3-} \xrightarrow{h\nu} \text{Triose phosphate} + 3H_2O + 3O_2$$

If, as in our experiments, the carbon dioxide partial pressure and the light intensity remain constant during the period of measurement, the rate of photosynthesis when phosphorus is limiting would be expected to be dependent upon both the intracellular phosphorus pool and the light-capturing ability, i.e. the chlorophyll content, of the cells. This we have, in fact, observed. The export of triose phosphate from the chloroplast is coupled, on a one-to-one basis, with the import of phosphate, while the addition of compounds such as mannose, which sequester phosphate in the cytoplasm, reduces the rate of phosphate entry and hence the rate of departure of triose phosphate. This causes a build-up of starch in the chloroplast and leads to a depression in the rate at which photosynthesis takes place (Herold and Lewis 1977; Chen-She *et al.* 1975; Herold *et al.* 1976). The inhibitory effect upon carbon fixation of the increase in cellular nitrogen at limiting cell phosphate quotas may be due to a similar sequestration.

One piece of evidence used to support the non-interactive interpretation of the dual nutrient effects in phytoplankton is that there is a discontinuity in numerous parameters (cell volume, protein, lipid, DNA) when plotted against the cell N:P ratio; Rhee (1978) found that in chemostat cultures of *Scenedesmus* sp., this occurred at a N:P ratio of about 30 which was interpreted as being the point at which growth switched from being nitrogen- to phosphorus-limited. Significantly, though, there was also a marked increase in the chlorophyll content of the cells at the same ratio, apparently related to rises in the nitrogen cell quotas as the N:P ratio in the inflowing medium increased. Carbon fixation rates in *Scenedesmus*, when normalized to unit chlorophyll, were independent of the N:P ratio over the whole of the range studied.

Falkowski and Stone (1975) have pointed out that higher cellular chlorophyll levels should enhance the light-capturing ability of the phytoplankton and thereby increase ATP synthesis. This would presumably have repercussions on the production of most other cellular constituents, as observed by Rhee (1978). It is possible that a better understanding of how nutrients influence the growth of phytoplankton would be obtained if specific growth rates, instead of being based solely on increases in cell numbers or biomass, were normalized in terms of chlorophyll concentrations.

Acknowledgements

I wish to express my thanks to Prof. P.J. Syrett for his helpful comments about this work and particularly for advice on the literature relating to photosynthesis. I am particularly indebted to Jill Sleep who assisted with the experimental work and helped to prepare this paper.

References

Caperon, J. (1968). *Ecology* **49**, 866.
Chen-She, S.H., Lewis, D.H., and Walker, D.A. (1975). *New Phytol.* **74**, 383.
Davies, A.G. (1970). *J. mar. biol. Ass. UK* **50**, 65.
Davies, A.G. and Sleep, J.A. (1979). *J. mar. biol. Ass. UK* **59**, 937.
Davies, A.G. and Sleep, J.A. (1980). *J. mar. biol. Ass. UK* **60**, 841.
Droop, M.R. (1968). *J. mar. biol. Ass. UK* **48**, 689.
Droop, M.R. (1974). *J. mar. biol. Ass. UK* **54**, 825.
Elrifi, J.R. and Turpin, D.H. (1985). *J. Phycol.* **21**, 592.
Falkowski, P.G. and Stone, D.P. (1975). *Mar. Biol.* **32**, 77.
Fuhs, G.W. (1969). *J. Phycol.* **5**, 312.
Fuhs, G.W., Demmerle, S.D., Canelli, E., and Min Chen (1972). *Am. Soc. Limnol. Oceanogr. Spec. Symp.* **1**, 113.
Gerhart, D.Z. (1975). *Verh. int. verein. Limnol.* **19**, 1013.
Granéli, E. (1978). *Vatten* **34**, 117.
Granéli, E. (1981). *Kieler Meeresforsche, Sonderh.* **5**, 82.
Healey, F.P. (1979). *J. Phycol.* **15**, 289.
Herold, A. and Lewis, D.H. (1977). *New Phytol.* **79**, 1.
Herold, A., Lewis, D.H., and Walker, D.A. (1976). *New Phytol.* **76**, 397.
Lean, D.R.S. and Pick, F.R. (1981). *Limnol. Oceanogr.* **26**, 1001.
Paasche, E. (1973). *Mar. Biol.* **19**, 117.
Rhee, G-Y. (1973). *J. Phycol.* **9**, 495.
Rhee, G-Y. (1974). *J. Phycol.* **10**, 470.
Rhee, G-Y. (1978). *Limnol. Oceanogr.* **23**, 10.
Riper, D.M., Owens, T.G., and Falkowski, P.G. (1979). *Plant Physiol.* **64**, 49.
Sivak, M.N. and Walker, D.A. (1986). *New Phytol.* **102**, 499.
Terry, K.L. (1980). *Botanica Marina* **23**, 757.
Terry, K.L. (1982). *J. Phycol.* **18**, 79.
Turpin, D.H. (1983). *J. Phycol.* **19**, 70.
Vince, S. and Valiela, I. (1973). *Mar. Biol.* **19**, 69.

14 Algal extracellular products—antimicrobial substances

A.K. JONES

Department of Botany and Microbiology, University College of Wales, Aberystwyth Dyfed SY23 3DA, UK

Introduction

Except in axenic culture, micro-organisms rarely occur alone. Throughout Nature interactions of many kinds, including chemical interactions, are a fundamental facet of the various communities within an ecosystem. Of those interactions involving algae and bacteria, one of the earliest descriptions is that of Englemann (1884), who utilized the attraction of bacteria to oxygen produced by algae to demonstrate photosynthetic light responses.

In aquatic environments, surfaces of all sorts are rapidly colonized in a sequential manner by a variety of micro-organisms. Seaweeds may be inhabited by characteristic epiphytic bacteria, as described by Sieburth *et al.* (1974), Sieburth (1975, 1979), Mitchell and Cundell (1977), Cundell *et al.* (1977), and Moss (1982). However, some algae are always apparently more heavily colonized than others; additionally, there may be variation in colonization of different parts of an individual seaweed thallus. It is therefore interesting to speculate on the possibility that some algae may possess a mechanism(s) for the control of growth of potential microbial epiphytes.

Living marine diatoms were found by Waksman *et al.* (1937) to exhibit some resistance to bacterial attack. Later, Droop and Elsen (1966) observed that actively growing populations of *Skeletonema costatum* were colonized at only a low level, whereas a week after the maximum growth had occurred the bacterial numbers attached to the *Skeletonema* had increased ten-fold. Further observations by Sieburth (1968, 1979) using phase contrast and scanning electron microscopy also suggested that the bacteria associated with healthy marine phytoplankton were free rather than attached to the algal cells, but that as the algae became less viable bacterial attachment increased.

Using both direct and viable count methods, Jones (1972) established in freshwater algae that the degree of bacterial attachment diminished along the host series ranging from colonial green algae and cyanobacteria to filamentous green algae and filamentous cyanobacteria and diatoms, and finally to dinoflagellates. Jones (1982) concluded that interactions between mixtures of algae and bacteria in the laboratory could variously be stimulatory, antagonistic, and competitive. These relationships would appear to be influenced by algal extracellular products in a zone around the algae termed the 'phycosphere' and originally described by Bell and Mitchell (1972). Healthy algae may release a proportion of the carbon they fix in photosynthesis (Jones and Cannon 1986) and thus contribute positively to the development of a nutrient-enriched zone around themselves which could enhance attached bacterial growth. Franko and Wetzel (1980, 1981) have found that cyclic adenosine 3',5'-monophosphate (cAMP) is excreted by some freshwater algae, and Azam and Ammerman (1984*a*, 1984*b*) have speculated that cAMP may act as a metabolic cue in algal–bacterial interactions, suggesting that if cAMP is exuded with other bacterial nutrients by algae it may signal the favourable nutrient status of the phycosphere to bacteria.

The nature of algal extracellular products and their release has been discussed by Fogg (1962, 1966, 1971, 1983), Hellebust (1974), Aaronson *et al.* (1980), and Jensen (1984). Such substances include carbohydrates, lipids, peptides, organic phosphates, volatile substances, vitamins, toxins and antibiotics (Jones 1986; Jones and Cannon 1986).

Thus the phycosphere may contain substances which may be stimulatory or inhibitory to associated organisms. We therefore have to consider the possibility that the presence or absence of bacteria from algal surfaces may, amongst other factors, be affected by the occurrence of appropriate specific bacterial nutrients and by the production of antibiotic substances by the algae. Antibiotic properties of seawater towards staphylococci have been observed (Saz *et al.* 1963), apparently mediated by a non-dialyzable compound which was heat-labile and possibly produced by phytoplankton. Similarly, Sieburth (1965) attributed the anticoliform activity observed in Naragansett Bay to a substance released by senescing blooms of *S. costatum*. Increased illumination and lysis of the diatom cells enhanced the antibiotic effect, suggesting that this natural antibiosis was caused by light-activated autolytic products of *Skeletonema*.

The present review will consider the general nature of antimicrobial substances produced by algae. There would appear to be two major rationales for studying antibiotic production by algae; first, the search for novel antibiotics by pharmacological groups for therapeutic medicinal and veterinary exploitation, and second, an elucidation of the role of putative 'antibiotics' in ecology in preventing disease in the algae or fouling by bacteria and other micro-organisms (Reichert and Borowitzka 1984). The main aim of this discussion is to evaluate the second of these two points. It is not the intention to review fully the chemistry of marine natural products and pharmaceutical applications as these have relatively recently been covered in more detail than is possible here in the volumes edited by Scheur (1978–81), Faulkner and Fenical (1977), Shield and Rinehart (1978), Hoppe *et al.* (1979), Hoppe and Levring (1982), Berdy *et al.* (1982), and Metting and Pyne (1986).

Antibiotics from microalgae

The first partly identified antimicrobial compounds isolated from algae were obtained from unicellular green algae, particularly *Chlorella*, which contained a substance termed chlorellin (Pratt *et al.* 1944) composed of peroxides of unsaturated fatty acids (Spoehr *et al.* 1949, Scutt, 1964). Chlorellin exhibited inhibitroy activity against both Gram-positive and Gram-negative bacteria, including *Staphylococcus aureus, Streptococcus pyogenes, Bacillus subtilis*, and *Pseudomonas aeruginosa*. Derivatives of fatty acids are also suggested (Glombitza 1979, citing Olfers-Weber 1977) to be the inhibitors present in *Hydrodictyon reticulatum*.

Pesando (1972) found that photoxidation products of an eikosapentaenic acid, present in extracts of the diatom *Asterionella japonica*, enhibited strong antibiotic activity after a short period of illumination. It was claimed by Aubert and Gauthier (1967) that between 20 to 25 per cent of Mediterranean diatoms may produce inhibitors. Gauthier *et al.* (1978) also found a similar lipid antibiotic in *A. japonica* and *Chaetoceros lauderi*, whose activity was also increased by light. When these two diatoms were grown with the dinoflagellate *Prorocentrum micans* they contained more antibiotic and less carotenoid pigments. It was therefore suggested that *Prorocentrum* produced an extracellular substance which inhibited pigment synthesis by the diatoms with a consequent increase in the photo actively induced lipid antibiotics. Sieburth (1964, 1965) reported that senescing blooms of *S. costatum* exhibited anticoliform activity and also found that concentrated acetone extracts of this diatom inhibited marine isolates of *Vibrio* and *Pseudomonas* as well as being active against *Escherichia coli* and *Sarcina lutea*, whereas other Gram-positive and Gram-negative terrestrial bacteria were virtually unaffected. The inhibitory action of *Skeletonema* against the growth of marine *Vibrio* and *Pseudomonas* isolates was confirmed by Kogure *et al.* (1979). They also showed that inhibition was greatest during the exponential phase of diatom growth. Conversely, they found that the growth of some *Flavobacterium* spp. was stimulated by *Skeletonema*. Cooper *et al.* (1983) obtained both aqueous and organic phase extracts from both *Skeletonema* and another diatom *Phaeodactylum tricornutum*. In the case of *Skeletonema* aqueous extracts from stationary phase cells were more inhibitory against marine bacteria than were extracts from lag or early log phase cultures; the reverse was true for *Phaeodactylum*. Also, the aqueous extracts of *Skeletonema* showed a broader antibiotic spectrum, being active against *P. aeruginosa, Mycoplasma orale*, and the Gram-positive *B. subtilis*, as well as against the marine isolates. However, the lipophilic or organic phase extracts from both diatoms were active against only the Gram-positive and marine bacteria, and showed an antimicrobial spectrum resembling that previously reported by Gauthier *et al.* (1978) and Gauthier (1980) for other planktonic algae. Thus the same alga may contain more than one antibiotic substance, soluble in either water or an organic solvent, each with different spectra of antibiotic activity. It was first suggested by Blaauw-Jensen in 1954 (cited by Sieburth 1965) that *Chlorella* might produce more than one antibiotic substance. The second substance(s) implicated was a chlorophyllide(s) which is either a precursor of chlorophyll or a degradation product resulting from the hydrolysis of the phytol group by the enzyme chlorophyllase. Jorgensen (1962) subsequently demonstrated *in vitro* activity of similar substances extracted from paper chromatograms of cultures of *Chlorella vulgaris, Scenedesmus quadricauda*, and *Chlamydomonas reinhardii*. Sieburth (1965), after similar studies with *Phaeocystis pouchettii*, commented (Sieburth 1968) that chlorophyllide-type materials were unlikely to function as inhibitors to potential algal pathogens as they were only likely to be present in significant amounts after senescence and autolysis of the alga; at that stage they may reach inhibitory concentrations and possibly affect a number of both Gram-positive and Gram-negative bacteria. According to Glombitza (1979) it is possible that this group of substances may initially be antibiotically inactive. However, on illumination they become modified photochemically, either being converted into inhibitors or acting as catalysts for the production of inhibitory compounds (Moreth and Yentsch 1970).

The production of 'antibiotics' by phytoplankton was discussed by Aubert *et al.* (1979), who recorded antibiotic activity in extracts from several groups including diatoms, chrysophytes, dinoflagellates, and cyanobacteria; they also noted that some phytoplankton may contain at least two different antibiotic substances. In addition to lipid substances, they detected an acid polysaccharide of high molecular weight from *Chaetoceros lauderi* and noted the presence of a peptide in the diatom *Fragillaria prinata* and a chrysophyte (*Stichochrysis immobilis*).

Older batch cultures of *C. vulgaris* have been found to inhibit Gram-positive bacteria present in earlier stages of growth (Matusiak *et al.* 1971, Chrost 1972), but members of the genera *Pseudomonas* and *Flavobacterium* were still able to survive. Chrost (1975*a*, 1975*b*) also observed selective inhibition of Gram-positive bacteria in natural freshwater plankton populations. Ethanol extracts from blooms of diatoms, dinoflagellates, and cyanobacteria were inhibitory to a *Bacillus* sp. and *Bacterium* sp. isolated from the lake, whereas Gram-negative bacteria were resistant. The extracts inhibited both growth and respiration of the Gram-positive bacteria, with a much greater effect in the light than in the dark. Some slight inhibition of the Gram-negative bacteria was also noted in light but the inhibitor was not identified. Again an unidentified, light-activated bacterial inhibitor was postulated by Chrost and Siuda (1978) to account for a negative relationship between bacterial heterotrophic activity and chlorophyll *a* concentrations in the photic zones of the lakes they studied compared with a positive relationship between them in the metalimnion. Another unidentified antibiotic with activity against the fungus *Cryptococcus neoformans*, *Bacillus* spp., *E. coli*, *S. lutea*, and *S. aureus* was shown to be produced by *Chlorococcum humicolum* (Pande and Gupta 1977), and ethanol extracts of 14 out of 24 species of green algae exhibited significant activity against *S. aureus* and/or *B. subtilis* (Debro and Ward 1979). Again, variation in antibiotic activity was related to the algal growth phase. In some cases they observed antibacterial activity only in lag phase algal cultures, in others only in the stationary phase. A relatively low molecular weight, heat- and acid-labile substance produced by the colonial green alga *Pandorina morum* was found by Patterson and Harris (1983) to be active against a range of other algae and higher plants as well as to the bacteria *Mycobacterium smegmatis*, *E. coli*, and an *Alcaligenes* sp. This 'toxin' was found to inhibit photosynthetic eukaryotes at a site near the reaction centre of photosystem II and it also inhibited the oxidation of succinate, malate, and pyruvate by isolated potato mitochondria (Patterson *et al.* 1979).

Antibiotic activity ascribable to acrylic acid produced by phytoplankton was first demonstrated by Sieburth (1959*a*, 1959*b*) in his studies of Antarctic penguins. Some of these birds were found to have 'bacteriologically sterile' gastrointestinal tracts, which was attributed to the euphausid ('krill') diet of the penguin. The euphausids had accumulated acrylic acid from the haptophyte *P. pouchetti* on which they had been grazing. The active substance was isolated and identified as acrylic acid by Sieburth (1960, 1961). Acrylic acid-related antibacterial activity was also observed in seaweeds (Katayama 1962). This appears to be also related to the production of substances similar to the allyl sulphides in the scale tissues of onions, which have bactericidal and fungicidal properties. Bywood and Challenger (1953) and Challenger *et al.* (1957) confirmed that dimethyl sulphide was produced by the red alga *Polysiphonia lanosa*. The precursor of this substance is dimethyl propiothetin (DMPT) which has also been detected in *Enteromorpha intestinalis* and *Acrosiphonia centralis*. Enzymatic cleavage of the DMPT upon death of the alga or exposure to air yields dimethyl sulphide and acrylic acid (Cantoni and Anderson 1956):

$$(H_3C)_2S^+{-}CH_2{-}CH_2{-}COOH \longrightarrow (H_3C)_2S + H{-}CH{=}CH{-}COOH + H^+$$

Dimethyl propiothetin   Dimethyl sulphide   Acrylic acid

Dimethyl propiothetin was also detected in six of the 14 species of phytoplankton investigated by Ackman *et al.* (1966). Later, Berland *et al.* (1972) examined the antibacterial substance produced in cultures of *S. immobilis* (a chrysophyte) from which they isolated two antibiotic substances. One of these was a peptide consisting of 11 amino acids linked to a larger molecule and thought to be related to humic acids, but no acrylic acid was detected. Glombitza (1979) has tabulated the widespread occurrence of acrylic acid, dimethyl sulphide, and dimethyl propiothetin in members of eight classes of algae.

Antimicrobials from seaweeds (other than derivatives of DMPT)

More is known of the chemical nature of 'antibiotics' in seaweeds than in microalgae and an indication of their variety is given in Table 14.1; early reports date from Pratt *et al.* (1951), Kamimoto (1955), Chesters and Stott (1956), Roos (1957), Allen and Dawson (1960), Burkholder *et al.* (1960), and Starr *et al.* (1962). Sieburth (1964, 1968) reviewed much of this early work and produced a useful tabular summary of the antibiotics known, the algae from which they were extracted, and the antimicrobial activity of the algae or their extracts towards test Gram-negative or Gram-positive bacteria. More recently, Glombitza (1979) and Berdy *et al.* (1982) have produced a comprehensive review of algal antibiotics, and Jones (1986) has also discussed these compounds in relation to the distribution of the bacteria associated with algae.

As one of the main motivating factors behind the investigation of algal antibiotics has been that of pharmaceutical exploitation, many of the antibacterial assays consisted of 'zone of inhibition' tests on agar plates against the faecal indicator *E. coli* or the animal pathogen *S. aureus*. The test alga or extracts of the alga on paper discs were placed on surface lawns of bacteria and any zones of growth inhibition measured. However, even when tested against marine isolates, the fact that inhibition occurs in such tests is not definitive proof that such inhibitory concentrations could occur in Nature. The ecological significance of activity revealed against these particular organisms is even more debatable.

Of the compounds isolated from seaweeds, phenolic substances, possibly because of the well-documented antimicrobial properties and use of free phenols as antiseptics, attracted early attention. Crato (1892) was the first to note phenolic-tannin-like substances in *Chaetopteris plumosa*. In the brown algae these tannins take the form of numerous refractile inclusions or vesicles, called 'physodes' (Sieburth 1968, Ragan 1976) which exhibit reactions suggestive of phenols. These are particularly prevalent in cells in outer tissues (stipe cortex) and meristems of many brown algae (Ragan and Glombitza 1986), but they have also been found in other algal classes, for example the Xanthophyceae (Craigie and McLachlan 1964). Haug and Larsen (1958) reported that polyphenols with tannin-like activity constituted 9 per cent of the dry weight of *Ascophyllum nodosum*, and Craigie and McLachlan (1964) showed that a catechin-type polyphenol was excreted in

Table 14.1 Examples of antibiotic substances extracted from algae.

Antibiotic	References
Algal Phylum: Rhodophyta	
Acrylic acid CH_2-CH-COOH	Chesters and Stott (1956), Pratt *et al.* (1951), Kamimoto (1955), Allen and Dawson (1960), Sieburth and Conover (unpublished observations, cited in Sieburth 1964)
Fatty acids, carbonyl, and terpene fractions	Katayama (1962)
Brominated phenols, e.g. lanosol	Mautner *et al.* (1953), Saito and Sameshima (1955), Fenical (1975), Phillips and Towers (1982*a*, 1982*b*)
Brominated terpenephenols, e.g. laurinterol	Fenical (1975), Ohta (1979), Rinehart *et al.* (1981)
Isolaurinterol allolaurinterol	Fenical (1975), Ohta (1979), Rinehart *et al.* (1981)
Halogenated ketones	Fenical (1975)
1, 1, 3, 3-tetrabromo-2-nonanone	Fenical (1979)

Table 14.1 *Continued*

Antibiotic	References
p-hydroxybenzaldehyde 	Ohta (1979)
dichloroacetamide $Cl_2CH\ CONH_2$	Ohta (1979)
Ethyl dibromoacrylate Bromo- and chloro-beckerelide 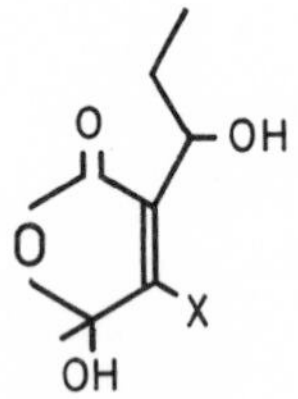 X = Br or X = Cl	Fenical *et al.* (1979), Ohta (1979)
Fimbrolides 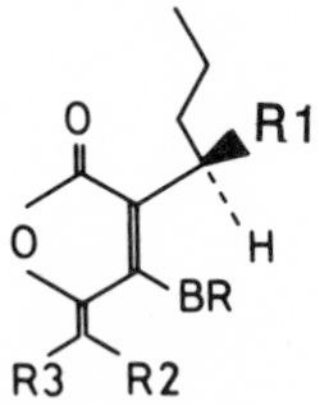	Kazlauskas *et al.* (1977)
R_1 = H R_2, R_3 = H, Cl, Br, I Also Acetoxyfimbrolides, where R_1 = $COOCH_3$ R_2, R_3 = H, Cl, Br, I	Pettus *et al.* (1977)
Elatol 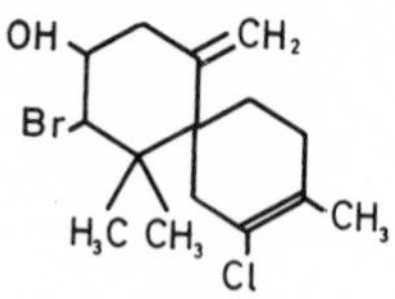	Rinehart *et al.* (1981)

Table 14.1 *Continued*

Antibiotic	References
Laurencienyne 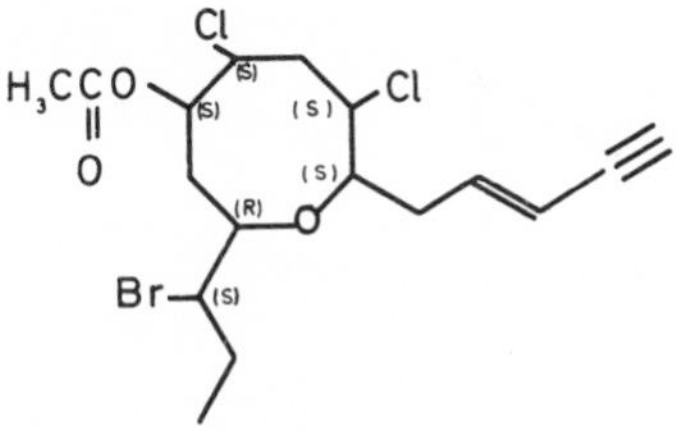	Rinehart *et al*. (1981)
3-5 dinitroguiacol 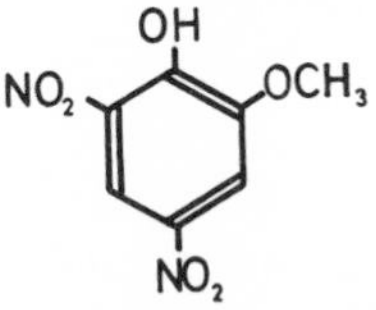	Ohta (1979)
γ-pyroglutamic acid 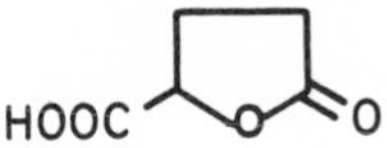	Ohta (1979)
Thiepane 	
Thialane 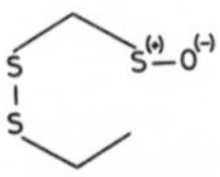	Wratten and Faulkner (1976)
Algal Phylum: Phaeophyta	
Difucol 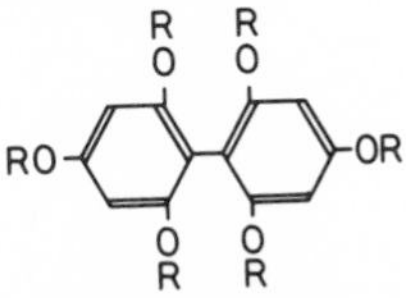	Diphlorethol 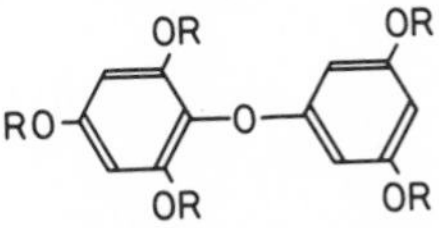
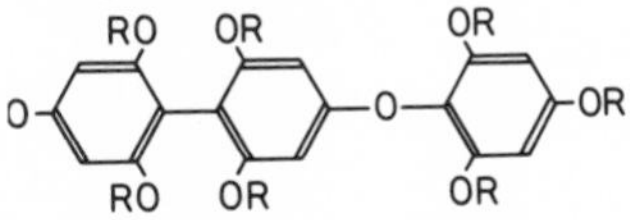 Fucophlorethol A	Glombitza (1979), Glombitza and Klapperich (1985), Higa (1981), Ragan & Glombitza (1986)
R = H_3CCO	

Table 14.1 *Continued*

Antibiotic	References
Bifuhalol	
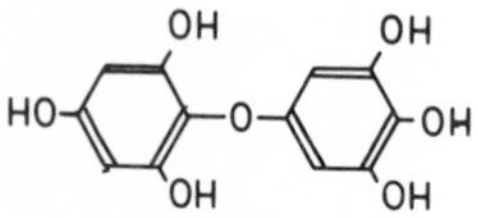	
Sulphuric acid Acrylic acid, fatty acids, carbonyl, and terpene fractions	Roos (1957), Saito and Sameshima (1955), Chesters and Stott (1956), Katayama (1962), Conover and Sieburth (1964), Kamimoto (1955), Challenger *et al.* (1957)
Diterpenoid monoacetates with the spatane skeleton	Da Silva (1982)
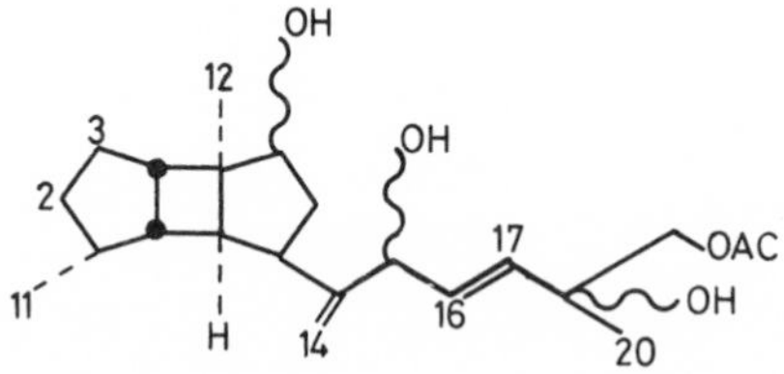	
(-19-acetoxy-5, 15, 18-trihydroxyspata-13, 16-diene)	
Zonarol	Rinehart *et al.* (1981)
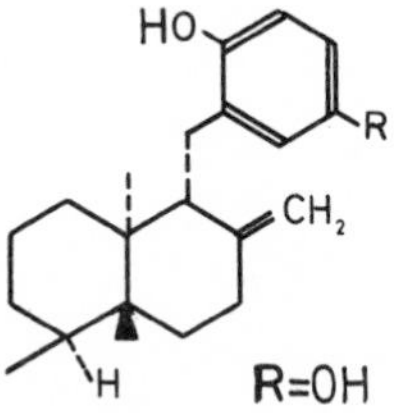	
Dictyodial	Rinehart *et al.* (1981)
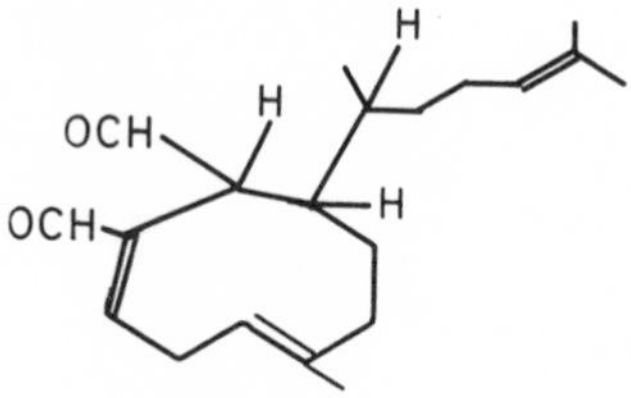	

Table 14.1 *Continued*

Antibiotic	References
Pachydictyol-A-epoxide 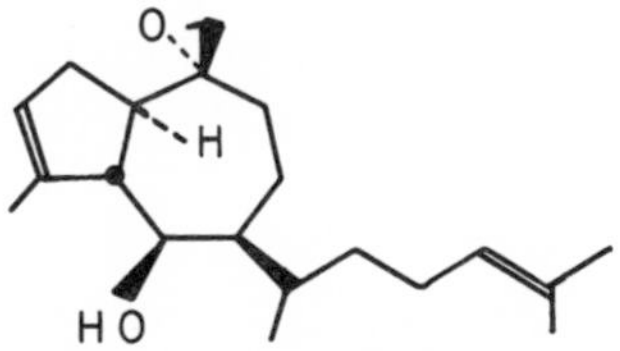	Rinehart *et al.* (1981)
Algal Phylum: Pyrrophyta (in zooxanthellae)	
Terpenes and hydrocarbons	Burkholder and Burkholder (1958), Ciereszko (1962)
Algal Phylum: Bacillariophyta	
Peptides	Berland *et al.* (1974)
Aqueous phase and lipophilic substances	Cooper *et al.* (1983), Aubert *et al.* (1970)
Nucleosides	
Photoactivated lipid antibiotic	Gauthier *et al.* (1978)
Acid polysaccharide	Pesando *et al.* (1979)
Fatty acids	Aubert *et al.* (1968)
Algal Phylum: Chrysophyta	
Acrylic acid	Sieburth (1960)
Algal Phylum: Chlorophyta	
Chlorellin	Pratt *et al.* (1944)
Acrylic, fatty acid, and terpenoid fractions	Starr *et al.* (1962), Allen and Dawson (1960), Roos (1957), Challenger *et al.* (1957), Harder and Opperman (1953), Kamimoto (1955), Katayama (1962)
Chlorophyllides	Jorgensen (1962), Levina (1961), Telitchenko *et al.* (1962)
Ethanol soluble extracts	Debro and Ward (1979)
Udoteal 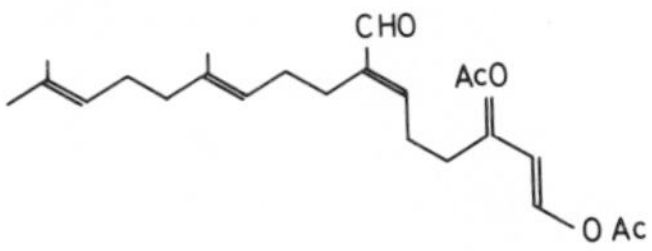	Paul *et al.* (1982), Nakatsu *et al.* (1981)

large amounts by *Fucus vesiculosus*; this substance had a similar absorption spectrum (peak at 280 nm) to yellow-coloured ultraviolet-absorbing materials in seawater ('Gelbstoff'). Of the phenolic substances, free phloroglucinol was first unambiguously detected by Glombitza *et al.* (1973) but it probably only accounts for a small fraction of the total phenolic material, the greater part of which consists of higher molecular weight dehydropolymers of phloroglucinol (Glombitza 1979, Glombitza and Klapperich 1985, Higa 1981). Amongst the lower molecular weight oligomers of phloroglucinol are the oligoaryls (fucols) with one to four phloroglucinol units, the oligoaryl ethers (phlorethols), and the fucophlorethols which have diaryl as well as diaryl ether linkages. Another related group of compounds, the fuhalols, have additional hydroxyl groups (Table 14.1). Other low molecular weight phenolics include the isofuhalols, eckols, and halogenated and sulphated phlorotannins (Ragan and Glombitza 1986). The majority of these phenolic substances are sensitive to oxidation, particularly the fuhalols, and consequently their isolation has usually required prior acetylation or methylation. Glombitza (1979) has discussed the free phenols found in various algal classes and given some idea of their antibiotic activity, which is probably ascribable to the lower molecular weight oligomers rather than to the polymeric phlorotannins. The phenols probably act as non-specific enzyme inhibitors and may be active in sea water. Conover and Sieburth (1964) also noted that the actively growing tips of the brown seaweed *Sargassum muticum* were relatively bacteria free; furthermore, antibacterial activity against a *Vibrio* isolate was exhibited only by the actively growing tips of *Sargassum* and not at all by the older bacterially colonized portions of the weed. Sieburth and Conover (1965) found that tannins were detected only where there was no frond-tip fouling and this correlated with antibacterial activity.

Halogenated derivatives of amino acids, phenols, terpene phenols, monoterpenes, sesquiterpenes, and diterpenes, acetylenes, and miscellaneous organic compounds formed by red algae have been investigated by Fenical (1975). The biological function of many of these compounds is still a matter of speculation (Fenical 1975), but some may well function as chemical messengers in an exocrine system, possibly acting as a defence or discouragement to invertebrate predators and/or parasitic components of the microflora. The bromophenols, including lanosol (Table 14.1), were found to be generally antibacterial by a number of workers (quoted by Fenical 1975, Phillips and Towers 1982*a*, 1982*b*), as were also the terpene phenols of which laurinterol (Table 14.1) was found to show strong activity towards *S. aureus*. Ohta (1979) also showed it to be active against *B. subtilis*. Apparently both laurinterol and allolaurinterol actively inhibit reverse transcriptase (cited in Rinehart *et al.* 1981).

Halogenated ketones were isolated from the red algae *Asparagopsis* and

Bonnemaisonia by Fenical (1975) and Fenical *et al.* (1979). The ketones from *Asparagopsis* were extremely toxic, small polyhalogenated compounds, and are potentially powerful antibacterials. However, as they are related to the tear gas component, monobromoacetone, commercial exploitation seems unlikely. The genera *Bonnemaisonia, Delisea*, and *Ptilonia* produced halogen-containing compounds with C_7 and C_9 components, which, although toxic, may have some potential for development as antiseptic compounds; for example, 1,1,3,3-tetrabromo-2-nonanone had a minimum inhibitory concentration against *S. aureus* which compared very favourably with penicillin and tetracycline. Ohta (1979) also found the following halogen- and non-halogen containing substances in methanol extracts of the corraline red alga *Marginosporum aberrans*: laurinterol, isolaurinterol, *p*-hydroxybenzaldehyde, dichloroacetamide, and 3,5-dinitroguaiacol (Table 4.1); and from the red alga *Beckerella subcostatum*: bromobeckerelide, chlorobeckerelide, and γ-pyroglutamic acid (Table 4.1), all new compounds with commercial potential. Small lactones, other than the beckerelides, exhibiting antimicrobial activity have been isolated from *Delisea fimbriata* by dichloromethane extraction; these include the fimbrolides (Table 14.1), hydroxyfimbrolides, and acetoxyfimbrolides (Kazlauskas *et al.* 1977; Pettus *et al.* 1977).

Wratten and Faulkner (1976) found cyclic polysulphides, namely several thiepanes and thiolanes (Table 14.1), in the red alga *Chondria californica*. Glombitza (1979) quotes reports of similar cyclic polysulphides with antimicrobial activity in species of the brown seaweed *Dictyopteris* and the fungus *Lentinus edulis*. Rinehart *et al.* (1981) also identified a number of antimicrobials from a large variety of marine organisms, including algae, and some of these are listed in Table 14.1. However, they doubted the usefulness of substances such as zonarol, yahazunol, dictyodial, and pachydictyol-A-expoxide, as these often had minimum inhibitory concentrations too large to suggest clinical potential. Reichelt and Borowitzka (1984) reported an initial screening of a large range of brown, green, and red algal extracts for *in vitro* antimicrobial activity against Gram-positive and Gram-negative bacteria, yeasts, fungi, and a protozoan, and *in vivo* activity in mice. The extracts showing highest activity were further investigated. Much of the antibiotic activity in the extracts of red algae was due to halogenated compounds and that in the extracts of brown algae to various phenolics. The compounds thought most likely to be of pharmacological potential were phenols containing an alkyl group—which might render them less toxic to animals. However, Reichelt and Borowitzka (1984) commented that algae seemed a rather disappointing source of therapeutically relevant compounds, in so far as only a very small number showing activity in *in vitro* tests were also active *in vivo* in mice. Toxicity to the test animal was a major problem, particularly with the red algal extracts. Of those compounds which were sufficiently non-toxic many became bound to serum proteins and were either inactivated, not absorbed and circulated to the site of infection or were metabolized to inactive forms or excreted. However, limited potential for some compounds from brown algae in wound- and skin-cleansing preparations was noted.

There are relatively fewer reports of antimicrobials from green seaweeds, but Fenical and Paul (1984) have described antimicrobial and cytotoxic terpenoids from some tropical green algae from the family Udotaceae and cite a number of other recent reports of such compounds, the majority of which are related sesquiterpenoids and diterpenoids, mostly in the linear or uncyclized forms. These all contain the conjugated bis-enol acetate functional groups; an example is udoteal (Table 14.1), originally isolated by Paul *et al.* (1982) and Nakatsu *et al.* (1981).

Seasonal variation in the production of antimicrobials by seaweeds

Chesters and Stott (1956) found that the seaweeds they extracted and tested each exhibited a somewhat different seasonal variation in their antibiotic activities. A more detailed survey of 151 species of British algae was carried out by Hornsey and Hide (1974, 1976*a*) and they showed widespread antibacterial activity throughout the Chlorophyta, Phaeophyta, and Rhodophyta. Sometimes, however, one species was active although a closely related one was inactive; e.g. *Laminaria digitata* was inhibitory to bacteria whilst *Laminaria hyperborea* was not; similarly, in the Rhodophyta, *Chondrus crispus* was active although the related *Mastocarpus stellatus* (*Gigartina stellata*) showed no antimicrobial properties. They are also unable to find any activity in the Fucaceae although inhibitory polyphenols in the actively growing parts of fucoid thalli have been recorded, as already discussed. Some seasonal variation in antimicrobial activity was also noted (Hornsey and Hide 1974); this was discussed in more detail (Hornsey and Hide 1976*a*) with reference to 11 seaweeds, and four main types of seasonal variation were proposed:

(a) the *Polysiphonia* type—uniformly active throughout the year;
(b) the *Laminaria* type—with maximum activity during winter months (also found in *C. crispus, Laurencia pinnatifida*, and *Ulva lactuca*);
(c) the *Dictyota* type—having a summer peak of antibiotic production (also noted in *Dilsea carnosa* and *A. nodosum*);
(d) the *Codium* type—where there is a spring peak of activity (also in *Halidrys siliquosa*).

No convincing data were obtained to support a suggestion that some algae had an autumn production peak. Seasonal variations in the antibiotic activity of seaweeds were also reported by Pratt *et al.* (1951), Vacca and Walsh

(1954), Roos (1957), Burkholder *et al.* (1960), and Almodovar (1964). It is interesting that Hornsey and Hide (1974) detected no activity in many of the Fucales as they are rich in phenolic substances. However, *A. nodosum* (Hornsey and Hide 1976*a*) showed activity in June, July, and August only, which exemplifies the need for seasonal sampling if the possibility of missing antibiotic production is to be minimized. Ragan and Jensen (1978), however, reported that the highest levels of polyphenols in *A. nodosum* and *F. vesiculosus* were in the autumn and winter months, a pattern similar to that found by Phillips and Towers (1982*a*) in the red alga *Rhodomela larix* where the winter and summer levels of lanosol were 3.8 per cent and 1.2 per cent, respectively, of the dry weight. It is possible that the lower summer concentrations were the result of environmental conditions causing higher exudation rates than in winter.

Other explanations for the failure of Hornsey and Hide (1974, 1976*a*) to detect antibiotic activity in species of *Fucus* may be deduced from personal observations (Jones and Tidswell, unpublished observations) that polyphenol-containing methanol extracts of the *Fucus* species exhibit variable antimicrobial activity in inhibiting marine bacterial isolates, but not *E. coli*. We also observed that apical portions of the *Fucus* thalli or acetone extracts of thalli showed no antimicrobial activity. Thus failure to detect activity could result from the use of inappropriate bacteria to cover the spectrum of activity exhibited by these polyphenols, or in the case of thallus apices the possibility exists that inhibitory concentrations of polyphenols were not reached on the test plates.

Distribution of the bacterial flora and antimicrobial activity in algae

Even though the growth of many marine bacteria may be supported by organic matter liberated by phytoplanktonic algae (Jones 1982, Cole 1982, Jones and Cannon 1986), actively growing diatoms are often relatively free of attached bacteria, as noted earlier. Sieburth (1968) reported that when attachment did occur it was usually brief and the bacteria were perpendicular to the algal cell: he attributed this to an adverse acid microzone around the diatom cells caused by the excretion of organic acids, the perpendicular bacterial attachment being an ameliorative response. However, some algal products likely to be implicated, for example glycollic acid, acrylic acid, and dimethyl sulphide, act as attractants for *Vibrio alginolyticus* at concentrations of 100 μM, 10 μM, and 1mM, respectively (Sjobald and Mitchell 1979). Because acrylate reduced bacterial viability only at 0.1 M, one may conclude that, instead of repelling bacteria at concentrations likely to be found in the sea, these substances may contribute positively to the formation of a phycosphere. This nicely illustrates the difficulty of ascribing an antibiotic role to algal substances in Nature as opposed to a laboratory situation.

The same putative 'antibiotic' could be inhibitory to some bacteria at one concentration and stimulatory or neutral at another; furthermore, at different antibiotic concentrations the same range of bacteria may be affected differently.

The genus *Vibrio* often seems to be associated with marine algae. Sieburth (1968) found no definite differences between the bacterial flora of the red alga *Polysiphonia* and the brown alga *Sargassum*; *Vibrio* was dominant and little difference was observed between active and inactive segments of the seaweeds. Sieburth attributed this partly to artefacts arising from the culture procedure used to estimate bacterial numbers since polyphenols and other antibiotics extracted when the algae were ground up to obtain the bacterial suspensions for plating could have caused inhibition of some bacterial species. Direct observations were needed and the advent of the scanning electron microscope (SEM) facilitated this, although reliable quantitative estimates of bacterial colonization remain difficult and direct identification of the bacteria is impossible. However, using the SEM, Sieburth (1975, 1979) confirmed that in *U. lactuca* the bacteria were usually attached perpendicular to the thallus; in *Monostroma* only one morphological bacterial type was observed, growing preponderantly on the thallus edge. Mitchell and Cundell (1977) and Cundell *et al.* (1977), also using the SEM, reported that only microcolonies of yeasts were present on the tips of *A. nodosum*; the main stipe above the holdfast was covered by a lawn of end-attached bacterial rods, and the bacterial population density was greater in the internodal region which represented the fourth year of growth, where a dense lawn of end-attached bacteria was overlaid by filamentous bacteria. Sieburth *et al.* (1974) found that although the surfaces of *Ascophyllum* were relatively free from bacteria when it was actively growing in spring, delimited microzones with surface slime containing adhering microbes could be observed. Also attached were short filaments of *Leucothrix mucor* which, with pennate diatoms and yeasts, formed dense microzonal areas in winter when the *Ascophyllum* was relatively dormant. Sieburth (1968, 1976, 1979) and Sieburth *et al.* (1974) concluded that polyphenols inhibited colonization seasonally in *Ascophyllum* and the bacteria-free apical tips resulted from the secretion of tannin-(polyphenol-) like compounds, desiccation during exposure to air, and active meristematic growth (Mitchell and Cundell 1977; Cundell *et al.* 1977). Tannic acid may reach concentrations of 1 to 2 mg l^{-1} in seawater around *Ascophyllum* and such concentrations elicit negative chemotactic responses from motile marine bacteria (Mitchell and Cundell 1977); these authors proposed, therefore, that negative chemotaxis may deter the colonization of surfaces by marine bacteria. However, Hornsey and Hide (1985) were unable to detect antibiotic activity in species of *Fucus*.

In a survey of six seaweeds, Hornsey and Hide (1976*b*) distinguished three main patterns of antibiotic distribution within algal thalli. These were:

(a) maximum activity in meristematic or young tissues (shown by *C. crispus, D. carnosa*, and *Codium fragile*);

(b) uniform activity throughout the thallus (*U. lactuca*);

(c) maximum activity in the older parts of the thallus (*Laminaria saccharina*).

Unlike the earlier conclusion of Conover and Sieburth (1964), Hornsey and Hide (1976*a*, 1976*b*) found that, whereas antibiotic activity was at a maximum in the young tissues of *C. crispus*, monthly assays of antibiotic activity indicated a winter maximum when the metabolism was probably minimal; however, in *D. carnosa* the pattern was more like that suggested by Conover and Sieburth (1964), with maximum activity in plants extracted in the spring and in the youngest parts of the plants, suggesting a correlation between antibiotic production and high metabolic rate. Phillips and Towers (1982*a*) also found that the highest concentrations of the bromophenol lanosol in *Rhodomela larix* were in the youngest and most rapidly growing parts of the algal thallus; i.e. the apical tips and the regions producing dense lateral branches. It is noteworthy that this is consistent with the findings of Ballantine (1979), wherein the older regions of the *Rhodomela* thallus were most heavily colonized with epiphytes.

As lanosol had been shown by Silva and Bittner (1979) to be an effective antibacterial and antifungal agent, it was suggested that its ecological significance might be in the control of host pathogens: a possible fungal pathogen of *Audouinella purpurea* was reported by Hooper and Titley (1985). Phillips and Towers (1982*a*) were of the opinion that these compounds are unlikely to be the waste products of metabolism as, according to Swain (1977), biochemical evolutionary trends would be unlikely to lead to the formation of functionless by-products more complex and potentially more toxic than their starting materials. Fenical (1975) emphasized the speculative nature of these ideas and noted that, although lanosol was toxic towards cultures of microalgae as well as bacteria, the naturally occurring sulphate salts of lanosol show no antialgal activity, and may even be mildly stimulatory to microalgal growth. He also quoted the case of bromophenol-rich *Polysiphonia lanosa* upon which Chan and McManus (1967) showed a considerable microflora, and the SEM studies of Sieburth *et al.* (1974) which revealed that, even when apparently less well colonized in summer, this alga had localized coatings of bacterial cells, particularly between bifurcations of the thallus and between the outer pericentral cells; presumably, despite a high internal concentration the bromophenols do not reach all parts of the algal surface in a sufficient concentration to inhibit the microflora completely.

Extracts from the fronds of marine algae were more antibacterial than the stipes, which usually gave negative results, according to Pratt *et al.* (1951). Caccamese *et al.* (1980), using toluene-methanol extracts of marine algae, also found more antibacterial substances in the frond tissues; furthermore, they realized that the phase in the life history also affected the antibacterial activity detected in *Laurencia obtusa*, because the lipid extract of the haploid female gametophyte (including the diploid carposporophyte) was more active than that from diploid tetrasporic plants. Similarly, Phillips and Towers (1982*a*) reported higher lanosol concentrations in vegetative compared with tetrasporic plants of *Rhodomela larix*. This life history factor was investigated further by Hornsey and Hide (1985), using representatives of the Rhodophyta, Phaeophyta, and Chlorophyta. Only three of the 10 red algae investigated and *Alaria esculenta* (brown) showed any variation. *Plocamium cartilagineum* (red) and *Gracilaria verrucosa* (red) produced cystocarpic plants with enhanced antibacterial properties, whilst the tetrasporophyte of *Bonnemaisonia hamifera* (red) possessed considerably more activity than the gametophyte phase.

Sieburth and Tootle (1981) attempted to investigate both quantitative and qualitative aspects of the bacterial colonization of four seaweeds, *A. nodosum, F. vesiculosus* (both brown), *P. lanosa*, and *C.crispus* (both red), using the SEM. Except for *P. lanosa* there was no statistically significant difference between the algal hosts in the mean yearly percentage of total microbial cover, or in the occurrence of filamentous bacteria. However, in the case of bacterial rods and cocci, the colonization was significantly lower only on *P. lanosa* compared with the other three seaweeds. Sieburth and Tootle (1981) suggested that this may have been due to the closer proximity to the surface of *Polysiphonia* of the smaller rods and cocci, compared with end-attached filamentous bacteria; thus they might have been more affected by antibacterial substances, such as the bromophenols, produced by *Polysiphonia*. The reduced total coverage of *A. nodosum* and *F. vesiculosus* by bacteria in the period May to October relative to that in April and November–December was inversely correlated with temperature. However, the lack of a similar correlation for the red algae suggested to Sieburth and Tootle (1981) that the inverse relationship was therefore caused by metabolic changes in the brown algae, because both species produce inhibitory polyphenols only in the summer months; these antibiotic substances may therefore account for the inverse correlation between total microbial cover and water temperature for both species. This conclusion was qualified, however, because in all cases there was no definite proof that antibiotic production caused this distribution. Sieburth and Tootle (1981) suggested therefore that other mechanisms may control the epiphytes on seaweeds, including physical factors because the slime-free rough stipes of *Laminaria claustoni* are very heavily coated with epiphytes. A scanning electron micrograph of *C. crispus* demonstrated that the sloughing-off of the cuticle containing the bacteria revealed an uncolonized surface; likewise, in *Ascophyllum*, the heavily fouled epidermis was possibly sloughed off or abraded to expose a non-colonized algal surface. Similarly, in *E. intestinalis*, McArthur and Moss (1977) demonstrated that the walls of the cells in contact with the sea were thick and layered; the outer wall layers were continually being worn away, thus removing the attached epiphytes. Moss (1982) examined *H. siliquosa* using both the SEM and transmission electron microscopy. As in other studies the microflora was always sparse on the young meristematic apices, whereas the remaining thallus was usually covered with varying densities of epiphytes. Pieces of thallus kept in culture were found to shed small pieces of 'skin' consisting only of cast-off outer layers of the cell wall. This was a repeated phenomenon and in static cultures layers of 'skin' accumulated. Similar shedding of 'skin' was also observed in *A. nodosum* and *Himanthalia elongata* in static culture, so Moss (1982) suggested that this continuous shedding of the outermost layer of the meristoderm cell walls and not of whole cells as claimed by Filion-Myklebust and Norton (1981) may be of general occurrence in the Fucales. Thus we have the possibility of both 'physical' and chemical control of epiphytes in these larger algae.

Obviously many factors besides antibiosis and 'skin' sloughing may well control the growth of epiphytic bacteria on algae. The failure of *Leucothrix mucor* to colonize some algae may indeed be related to the production of antimicrobial compounds, for example, *Desmarestia* has sulphuric acid in its cell sap at concentrations up to 0.44 M (Eppley and Bovell 1958), and Bland and Brock (1973) associated this feature with the rare occurrence of epiphytes on this alga. However, Bland and Brock (1973) also suggested that the absence of *Leucothrix* sp. on some algae was due to the host organism failing to produce utilizable carbon compounds. They postulated that the preferential attachment of *Leucothrix* sp. to *Bangia fuscopurpurea* rather than to a plastic support was related to the production of metabolizable mannose and mannans in the algal mucilage. Additionally, Bland and Brock (1973) suggested that the predominance of *Leucothrix* in the epiphyton of *Bangia* and *Porphyra* located at the top of the sea-shore may also be due to the water-retaining properties of the algae and the resistance to desiccation of the sheathed *Leucothrix* sp.

Conclusions

Amongst the various classes of algae there are many species which exhibit some sort of antimicrobial activity in laboratory test assays and a considerable number of compounds responsible for this activity have been isolated and chemically characterized. These include polysaccharides, derivatives of chlorophyll, fatty acids, nucleosides, cyclic polysulphide derivatives, small lactones, various phenolic substances and bromophenols, and a variety of other halogenated organic molecules, as well as diverse organic molecules, including unusual compounds like the diterpenoid monoacetates with the spatane skeleton. A number of classes of compounds to which these algal products belong have long been known in general pharmacology as effective external bacteriostatic substances. Some algal antimicrobials may have potential as antiseptics or may provide leads for the design of antiseptics and industrial inhibitory agents for use in industry (Fenical 1979; McConnell and Fenical 1977; Reichelt and Borowitzka 1984). It was concluded (Reichelt and Borowitzka 1984) that for other therapeutic purposes these algal antimicrobials are rather disappointing, very few showing *in vivo* activity in mice, being either inactivated in some way or conversely proving extremely toxic to the treated animal.

Alternatively, aquatic ecologists could find these molecules interesting because of their possible role in the chemical interactions that occur in communities of micro-organisms. It has also been suggested (Barbier 1981; Steinberg 1985) that such compounds could act as chemical defence mechanisms to deter phytophagous invertebrates. For example, Steinberg (1985) found that the preferred species of brown algae grazed by the intertidal gastropod *Tegula funebralis* had a markedly lower phenolic content than the non-preferred species. Barbier (1981) also suggests that in the vicinity of stands of algae there may be a sufficient accumulation of such active molecules for them to interfere with the growth of unicellular planktonic organisms. Glombitza (1979), however, could find no justification for definite statements on the ecological significance of these algal antimicrobials. In the case of the natural epiphytic flora of algae, it is difficult to relate in any precise way variations in microbial activity with variation in the microflora. This is because antimicrobial activity has often been tested against only a limited range of bacteria and these test organisms were often either the faecal indicator *E. coli* or animal pathogens such as *S. aureus*. There are some indications of antimicrobial activity towards indigenous aquatic bacteria and several algal extracts are inhibitory towards marine fungi and Gram-positive terrestrial and Gram-negative bacteria. Different (aqueous or organic) extracts from the same alga have sometimes suggested the presence of more than one antibacterial substance. Variation in the production of antibacterial activity seems also to occur in the different phases of growth of batch cultures of algae and of natural blooms of microscopic algae. Similarly, diverse patterns of seasonal variation in antibiotic production occur in various seaweeds. Different distributions of antibiotic activity in the various morphological regions of some seaweeds also occurs; and some show either no antibiotic activity or it is relatively uniformly distributed throughout the thallus. Although there is little direct evidence that antibiotics affect the abundance and distribution of bacteria on seaweeds *in situ*, the circumstantial evidence often seems to indicate that antibacterials may be important. Substances with antibacterial activity may also in part account for the wall-documented bactericidal nature of some seawaters, particularly if the antibiotic is one released on the decay of an algal bloom. Part of the problem resides in the absence of knowledge concerning the concentrations that such substances may reach in the environment, because the same substance may act towards a test bacterium as a chemotactic attractant at a particular concentration or as an antibacterial when presented at a higher concentration. It has been rather naively argued that evolutionary pressures would not have favoured the production of such a range of compounds with antimicrobial characteristics if they were not of some benefit to the producing organism. Indeed, some algal antibiotics may act as deterrents to potential invertebrate predators on the alga, as well as having antibacterial and, sometimes, antifungal, antiviral, and antineoplastic activity. However, these interactions may be purely fortuitous. The antibacterial effect of algal antibiotics to potential pathogens of large algae has still to be clearly demonstrated.

References

Aaronson, S., Berner, T., and Dubinsky, Z. (1980). In *Algal biomass* (ed. G. Shelef and C.J. Soeder) p. 575. Elsevier, Amsterdam.

Ackman, R.G., Tocher, C.S., and McLachlan, J. (1966). *J. Fish. Res. Bd. Can.* **23**, 357.

Allen, M.B. and Dawson, E.Y. (1960). *J. Bact.* **79**, 459.

Almodovar, L.R. (1964). *Bot. Mar.* **6**, 143.

Aubert, M. and Gauthier, M.J. (1967). *Rev. Int. Oceanogr. med.* **5**, 63.

Aubert, M., Aubert, J., and Gauthier, M. (1968). *Rev. Int. Oceanog. med.* **10**, 137.

Aubert, M., Aubert, J., and Gauthier, M. (1979). In *Marine algae in pharmaceutical science. Vol. 1* (ed. H.A. Hoppe, T. Levring, and Y. Tanaka) p. 267. De Gruyter, Berlin.

Aubert, M., Pesando, D., and Gauthier, M.J. (1970). *Rev. Int. Oceanogr. med.* **18–19**, 69.

Azam, F. and Ammerman, J.W. (1984*a*). In *Lecture notes on coastal and estuarine studies 8. Marine phytoplankton and productivity* (ed. O. Holm-Hansen, L. Bolis, and R. Gilles) p. 45. Springer-Verlag, Berlin.

Azam, F. and Ammerman, J.W. (1984*b*). Cycling of organic matter by bacterioplankton in pelagic marine ecosystems: microenvironmental considerations. In *Flows of energy and materials in marine ecosystems: theory and practice* (ed. M.J. Fasham) p. 345. Plenum, New York.

Ballantine, D.L. (1979). *Bot. Mar.* **22**, 107.

Barbier, M. (1981). In *Marine natural products, Vol. IV. Chemical and biological perspectives* (ed. P.J. Scheuer) p. 147. Academic Press, London.

Bell, W. and Mitchell, R. (1972). *Biol. Bull.—Mar. Biol. Lab. Woods Hole* **143**, 265.

Berdy, J., Aszalos, A., Bostian, M., and McNitt, K.L. (1982). CRC handbook of antibiotic compounds, Vol. IX. Antibiotics from higher forms of life: lichens, algae and animal organisms. CRC Press, Boca Wraton, Florida.

Berland, B.R., Bonin, D.J., Cornu, A.L., Maestrini, S.Y., and Marino, J.P. (1972), *J. Phycol.* **8**, 383.

Berland, B., Bonin, D., and Maestrini, S. (1974). Thèse Sc. Naturelles Université Aix-Marseille. Etude expérimentale de l'influence de facteurs nutritionnels sur la production du phytoplancton de Méditerranée, 239. p.

Blaauw-Jansen, G. (1954). *Proc. Koninke. Ned. Akad. Wetenschap. Ser. C* **57**, 498.

Bland, J.A. and Brock, T.D. (1973). *Mar. Biol.* **23**, 283.

Burkholder, P.R. and Burkholder, L.M. (1958) *Science* **127**, 1174.

Burkholder, P.R., Burkholder, L.M. and Almodovar, L.R. (1960). *Bot. Mar.* **2**, 149.

Bywood, R. and Challenger, R. (1953). *Biochem. J.* **53**, xxvi.

Caccamese, S., Azzolina, R., Furnari, G., Cormaci, M., and Grasso, S. (1980). *Bot. Mar.* **23**, 285.

Cantoni, G.L. and Anderson, D.G. (1956). *J. biol. Chem.* **222**, 171.

Challenger, F., Bywood, R., Thomas, P., and Hayward, B.J. (1957). *Arch. Biochem. Biophys.* **69**, 514.

Chan, E.C.S. and McManus, E.A. (1967). *Can. J. Microbiol.* **13**, 295.

Chesters, C.G.C. and Stott, J.A. (1956). In *The production of antibiotic substances by seaweeds. Proc. 2nd Int. Seaweed Symposium* (ed. T. Braarudad and N.A. Søorenson) p. 49. Pergamon Press, New York.

Chrost, R.J. (1972). *Acta Microbiol. Pol. Ser. B* **4**, 171.

Chrost, R.J. (1975*a*). *Acta Microbiol. Pol. Ser. B* **7**, 125.

Chrost, R.J. (1975*b*). *Acta Microbiol. Pol. Ser. B* **7**, 167.

Chrost, R.J. and Siuda, W. (1978). *Acta Microbiol. Pol. Ser. B* **27**, 129.

Ciereszko, L.S. (1962). *Trans. N.Y. Acad. Sci. Ser. II* **24**, 502.

Cole, J.J. (1982). *A. Rev. Ecol. Syst.* **13**, 291.

Conover, J.T. and Sieburth, J. McN. (1964). *Bot. Mar.* **6**, 147.

Cooper, S., Battat, A., Marsot, P., and Sylvestre, M. (1983). *Can. J. Microbiol.* **29**, 338.

Craigie, J.S. and McLaghlan, J. (1964). *Can. J. Bot.* **42**, 23.

Crato, E. (1892). *Ber. deut. bot. Ges.* **10**, 295.

Cundell, A.M., Sleeter, T.D., and Mitchell, R. (1977). *Microb. Ecol.* **4**, 81.

Debro, L.H. and Ward, H.B. (1979). *Planta Med.* **36**, 375.

Da Silva, S.S.M. (1982). *Phytochem.* **21**, 944.

Droop, M.R. and Elson, K.G.R. (1966). *Nature, Lond.* **211**, 1096.

Englemann, T.W. (1884). *Bot. Ztg.* **42**, 81.

Eppley, R.W. and Bovell, C.R. (1958). *Biol. Bull.—Mar. Biol. Lab. Woods Hole* **115**, 101.

Faulkner, D.J. and Fenical, W.H. (ed.) (1977). *Marine natural products chemistry*. Plenum, New York.

Fenical, W. (1975). *J. Phycol.* **11**, 245.

Fenical, W. and Paul, V.J. (1984). *Hydrobiologia* **116/117**, 135.

Fenical, W., McConnell, O.J., and Stone, A. (1979). In *Proc. 9th Int. Seaweed Symposium*, (ed. A. Jensen and J.R. Stein), p. 387. Science Press, Princeton, NJ.

Filion-Myklebust, C. and Norton, T.A. (1981). *Mar. Biol. Lett.* **2**, 45.

Fogg, G.E. (1962). In *Physiology and biochemistry of algae* (ed. R.A. Lewin) p. 475. Academic Press, New York.

Fogg, G.E. (1966). *Oceanogr. mar. Biol. A. Rev.* **4**, 195.

Fogg, G.E. (1971). *Arch. Hydrobiol.* **5**, 1.

Fogg, G.E. (1983). *Bot. Mar.* **26**, 3.
Franko, D.A. and Wetzel, R.G. (1980). *Physiol. Plant.* **49**, 65.
Franko, D.A. and Wetzel, R.G. (1981). *J. Phycol.* **17**, 129.
Gauthier, M.J. (1980). *Rev. Int. Oceanogr. med.* **58**, 41.
Gauthier, M.J., Bernard, P., and Aubert, M. (1978). *J. exp. mar. Biol. Ecol.* **33**, 37.
Glombitza, K-W. (1979). In *Marine algae in pharmaceutical science. Vol. 1* (ed. H.A. Hoppe, T. Levring, and Y. Tanaka) p. 303. De Gruyter, Berlin.
Glombitza, K-W. and Klapperich, K. (1985). *Bot. Mar.* **28**, 139.
Glombitza, K-W., Rosener, H.U., Vilter, H., and Raunwald, W. (1973). *Planta Med.* **24**, 301.
Harder, R. and Opperman, A. (1953). *Arch. Microbiol.* **19**, 298.
Haug, A. and Larsen, B. (1958). *Acta Chem. Scand.* **12**, 650.
Hellebust, J.A. (1974). In *Algal physiology and biochemistry* (ed. W.D.P. Stewart) p. 838. Blackwell Scientific Publications, Oxford.
Higa, T. (1981). In *Marine natural products* (ed. P.J. Scheuer) p. 93. Academic Press, London.
Hooper, R.G. and Tittley, I. (1985). *Br. phycol. J.* **20**, 186.
Hoppe, H.A. and Levring, T. (eds.) (1982). *Marine algae in pharmaceutical science. Vol. 2.* De Gruyter, Berlin.
Hoppe, H.A., Levring, T., and Tanaka, Y. (eds.) (1979). *Marine algae in pharmaceutical science. Vol. 1.* De Gruyter, Berlin.
Hornsey, I.S. and Hide, D. (1974). *Br. phycol. J.* **9**, 353.
Hornsey, I.S. and Hide, D. (1976*a*). *Br. phycol. J.* **11**, 63.
Hornsey, I.S. and Hide, D. (1976*b*). *Br. phycol. J.* **11**, 175.
Hornsey, I.S. and Hide, D. (1985). *Br. phycol. J.* **20**, 21.
Jensen, A. (1984). In *Lecture notes on coastal and estuarine studies 8. Marine phytoplankton and productivity* (ed. O. Holm-Hansen, L. Bolis, and R. Gilles) p. 61. Springer-Verlag, Berlin.
Jones, A.K. (1982). In *Microbial interactions and communities* (ed. A.T. Bull and J.H. Slater) p. 189. Academic Press, London.
Jones, A.K. (1986). In *Natural antimicrobial systems* (ed. G.W. Gould, M.E. Rhodes-Roberts, A.K. Charnley, R.M. Cooper, and R.G. Board) p. 232. Bath University Press.
Jones, A.K. and Cannon, R.C. (1986). *Br. phycol. J.* **21**, 341.
Jones, J.G. (1972). *J. Ecol.* **60**, 59.
Jorgensen, E.G. (1962). *Physiologia Pl.* **15**, 530.
Kamimoto, K. (1955). *Jap. J. Bact.* **10**, 897.
Katayama, T. (1962). In *Physiology and biochemistry of algae* (ed. R.A. Lewin) p. 467. Academic Press, New York.
Kazlauskas, R., Murphy, P.T., Quinn, R.J., and Wells, R.J. (1977). *Tetrahedron Lett.* **33**, 37.
Kogure, K., Simidu, U., and Taga, N. (1979). *J. exp. mar. Biol. Ecol.* **36**, 201.
Levina, R.T. (1961). In *The purification of waste-waters in biological ponds.*, Minsk. Akad, Nauk SSSR p 136 (Biol. Abst *42,* 1963).
McArthur, D.M. and Moss, B.L. (1977). *Br. Phycol. J.* **12**, 359.
McConnell, O. and Fenical, W. (1977). *Phytochemistry* **16**, 367.
McConnell, O.J. and Fenical, W. (1979) In *Marine algae in pharmaceutical science* vol. 1 (ed H.A. Hoppe, T. Leuring and Y. Tamaka) p. 403. De Gruyter, Berlin.

Matusiak, K., Chrost, R., and Krzywicko, A. (1971). *Acta Microbiol. Pol. Ser. B* **3**(4), 189.
Mautner, H.G., Gardner, G.M. and Pratt, R. (1953). *J. Am. Pharm. Assoc.* **42**, 294.
Metting, B. and Pyne, J.W. (1986). *Enzyme microb. Technol.* **8**, 386.
Mitchell, R.A. and Cundell, A.M. (1977). *The role of microorganisms in marine fouling and boring processes*. Technical Report, No. 3, US Office of Naval Research Contract N00014-76-c-0042 NR-104-967.
Moreth, C.M. and Yentsch, C.S. (1970). *J. exp. mar. Biol. Ecol.* **4**, 238.
Moss, B.L. (1982). *Phycologia* **21**, 185.
Nakatsu, T.B., Ravi, N., and Faulkner, D.J. (1981). *J. org. Chem.* **46**, 2435.
Ohta, K. (1979). In *Proc. 9th Int. Seaweed Symposium* (ed. A.T. Jensen and J.R. Stein) p. 401. Science Press, Princeton, NJ.
Olfers-Weber (1977) Antbakterielle Substanzen aus der Süβwesseralge *Hydrodictyon reticulatum*. Diss. Bonn.
Pande, B.N. and Gupta, A.B. (1977). *Phycologia* **16**, 439.
Patterson, G.M.L. and Harris, D.O. (1983). *Br. phycol. J.* **18**, 259.
Patterson, G.M.L., Harris, D.O., and Cohen, W.S. (1979). *Plant Sci. Lett.* **15**, 293.
Paul, V.J., Sun, H.H., and Fenical, W. (1982). *Phytochemistry* **21**, 468.
Pesando, D. (1972). *Rev. Int. Oceanogr. Med.* **25**, 49.
Pesando, D., Gnassia-Barelli, M., and Gueho, E. (1979). In *Marine algae in pharmaceutical science* (ed. H.A. Hoppe, T. Levring, and Y. Tanaka) p. 447. De Gruyter, Berlin.
Pettus, J.A., Wing, R.M., and Simms, J.J. (1977). *Tetrahedron Lett.* **31**, 41.
Phillips, D.W. and Towers, G.H.N. (1982*a*). *J. exp. mar. Biol. Ecol.* **58**, 285.
Phillips, D.W. and Towers, G.H.N. (1982*b*). *J. exp. mar. Biol. Ecol.* **58**, 295.
Pratt, R., Mautner, R.H., Gardner, G.M., Sha, Y., and Dufrenoy, J. (1951). *J. Am. pharm. Assoc. Sci. Ed.* **40**, 575.
Pratt, R., Daniels, T.C., Eiler, J.J., Gunnison, J.B., Kummler, W.D., Oneto, J.R., Spoehr, H.A., Hardin, G.J., Milner, H.W., Smith, J.H.C. and Strain, H.H. (1944). *Science New York* **49**, 351.
Ragan, M.A. (1976). *Bot. Mar.* **19**, 145.
Ragan, M.A. and Glombitza, K-W. (1986). In *Progress in phycological research*, Vol. 4 (ed. F.E. Round and D.J. Chapman) p. 130. Biopress, Bristol.
Ragan M.A. and Jensen, A. (1978). *J. exp. mar. Biol. Ecol.* **34**, 245.
Reichelt, J.L. and Borowitzka, M.A. (1984). *Hydrobiologia* **116/117**, 158.
Rinehart, K.L., *et al.* (1981). *Pure appl. Chem.* **53**, 795.
Roos, H. (1957). *Kiel Meeresforsch.* **13**, 41.
Saito, K. and Sameshima, J. (1955). *J. agr. Chem. Soc. Japan* **29**, 427.
Saz, A.K., Watson, S., Brown, S.R., and Lowery, D.L. (1963). *Limnol. Oceanogr.* **8**, 63.
Scheuer, P.J. (ed.) (1978–1981). *Marine natural products: chemical and biological perspectives*, Vol. 1–4. Academic Press, New York.
Scutt, J.E. (1964). *Am. J. Bot.* **51**, 581.
Shield, L.S. and Rinehart, K.L. (1978). *J. Chrom. Library* **15**, 309.
Sieburth, J.McN. (1959*a*). *J. Bact.* **77**, 521.
Sieburth, J.McN. (1959*b*). *Limnol. Oceanogr.* **4**, 419.
Sieburth, J.McN. (1960). *Science* **132**, 676.

Sieburth, J.McN. (1961). *J. Bact.* **82**, 72.
Sieburth, J.McN. (1964). *Dev. Ind. Microbiol.* **5**, 124.
Sieburth, J.McN. (1965). In *Pollutions marines par les microorganismes et les produits pétroliers (Symposium de Monaco, Avril 1964)* p. 217. Commission Internationale pour l'exploration Scientifique de la Mar Mediteranée, Paris.
Sieburth, J.McN. (1968). In *Advances in microbiology of the sea* (ed. M.R. Droop and J.F. Ferguson-Wood) p. 63. Academic Press, London.
Sieburth, J.McN. (1975). *Microbial seascapes*. University Park Press, Baltimore, Md.
Sieburth, J.McN. (1976). *Ann. Rev. ecol. Syst.* **7**, 259.
Sieburth, J.McN. (1979). *Sea microbes.* Oxford University Press.
Sieburth, J.McN. and Conover, J.T. (1965). *Nature, Lond.* **208**, 52.
Sieburth, J.McN. and Tootle, J.L. (1981). *J. Phycol.* **17**, 57.
Sieburth, J.McN., Brooks, R.D., Gessner, R.V., Thomas, C.D., and Tootle, J.L. (1974) In *Effects of the ocean environment on microbial activities* (ed. R.R. Colwell and R.Y. Morita) p. 418. University Park Press, Baltimore, Md.
Silva, M. and Bittner, M. (1979). In *Actas primer symposium sobre algae marinas Chilenas* (ed. B. Santelices) p. 235. Subsecretaria de Pesca Ministeria de Economia Fomento y Reconstruccion, Santiago, Chile.
Sjobald, R.D. and Mitchell, R. (1979). *Can. J. Microbiol.* **25**, 964.
Spoehr, H.A., Smith, J.H.C., Strain, H.H., Milner, H.W., and Hardin, G.J. (1949). *Carnegie Inst. Wash. Publ.* **586**, 1.
Starr, J.J., Deig, E.F., Church, K.K., and Allen, M.B. (1962). *Texas Rept. Biol. Med.* **20**, 271.
Steinberg, P.D. (1985). *Ecol. Monog.* **55**, 333.
Swain, T. (1977). *Plant Physiol.* **28**, 479.
Telitchenko, M.M., Davydova, N.V., and Fedorov. V.D. (1962). *Nauchn. Dokl. Vysshei. Shkoly. Biol. Nauki. Moscow* **4**, 157.
Vacca, D.D. and Walsh, R.A. (1954). *J. Am. pharm. Assoc. Sci. Ed.* **43**, 24.
Waksman, S.A., Stokes, J., and Butler, M.R. (1937). *J. mar. Biol. Assoc. UK* **22**, 359.
Wratten, S.J. and Faulkner, D.J. (1976). *J. org. Chem.* **41**, 2465.

15 Cyanobacterial toxins

G.A. CODD and G.K. POON

Department of Biological Sciences, University of Dundee, Dundee DD1 4HN, UK

Introduction

Rapid advances are occurring in the study of cyanobacterial toxins at the cellular and molecular levels. Although the properties of these compounds have largely emerged over the past decade and 'new' toxins are currently being characterized, the reported history of cyanobacterial poisoning episodes in aquatic environments extends back for over a century (Francis 1878). Some 12 cyanobacterial genera have been documented as including toxin-forming species. Individual genera, e.g. *Anabaena* and *Microcystis*, can include toxic and non-toxic species, and isolates of a single species may, or may not, produce toxin(s). However, it is of considerable interest that those cyanobacterial genera which typically dominate the heavy phytoplankton growths in many fresh and saline-waters belong to the group containing toxin-forming members. As a consequence, toxic cyanobacterial growths have been reported from at least 25 countries in Europe, Asia, the Americas, and Australasia.

Earlier reviews on toxin-producing cyanobacteria, poisoning incidents ascribed to these organisms, and of the toxins themselves include those of Gorham 1964; Schwimmer and Schwimmer 1964, 1968; Collins 1978; Gorham and Carmichael 1979; Moore 1981; Codd 1984; Skulberg *et al.* 1984; Stein and Borden 1984; and Carmichael *et al.* 1985.

The present review is not comprehensive but focuses on recent advances in understanding of the production of the toxins, their adverse effects on animals, new purification procedures for toxin isolation, and their molecular properties. Finally, we discuss the human health hazards posed by the toxins.

Occurrence of toxic cyanobacterial blooms in aquatic environments, and associated animal poisonings

Most investigations of cyanobacterial toxicity in aquatic environments have been stimulated by the occurrence of a poisoning incident affecting agricultural livestock, wild animals, birds, or fish. Deaths of large numbers of cattle, sheep, pigs, birds, and fish have been widely reported. Smaller numbers of horse, dog, rodent, amphibian, and invertebrate fatalities have also been ascribed to cyanobacterial poisoning (for primary references, see Carmichael 1981; Codd 1984; Skulberg *et al.* 1984). The planktonic cyanobacteria which often dominate fresh and brackish-water blooms

(*Microcystis, Anabaena, Aphanizomenon, Oscillatoria, Coelosphaerium, Gloeotrichia, Gomphosphaeria*, and *Nodularia*) have all been implicated in animal, bird, and fish deaths and the causative agents (the toxins) have been identified in many, but not all, cases.

A combination of biological and physical conditions has usually led to the animal poisonings. Death/illness occurs following oral ingestion of the cells or the released toxins. When toxin-forming cyanobacteria are suspended throughout the water column, then dilution of cells/toxins may occur so that an acute oral dose is not received during drinking. However, with the exception of *Synechocystis*, the planktonic cyanobacterial genera that include toxin-forming species (Table 15.1) can produce gas vesicles. These buoyancy devices (Walsby 1975) often result in the surface accumulation of the cells to form dense blooms (Reynolds and Walsby 1975) during calm weather. Surface bloom formation typically occurs during summer and autumn in Northern temperate latitudes and further concentration of the cells along the margin of a water body may occur by the action of a gentle on-shore breeze. This may result in concentration of toxic cyanobacteria so that an acute oral dose can be provided in considerably less than the daily water requirement of animals (Richard *et al.* 1983). Deaths of terrestrial animals have usually occurred in those circumstances where access to clearer water has not been possible and they have been obliged to drink in the vicinity of toxic blooms. Animal deaths may also occur following the bio-accumulation of cyanobacterial toxins via food chains.

Problems have arisen from inadequate recognition of the symptoms of cyanobacterial poisoning and from incomplete investigation and reporting of such incidents in the field (Schwimmer and Schwimmer 1964, 1968; Skulberg *et al.* 1984). These deficiencies, plus the finding of cyanobacterial toxins in many freshwaters for which no records of animal, bird, or fish poisonings are known (Richard *et al.* 1983; Codd and Bell 1985), indicate that these toxins are more widely present in waters containing cyanobacteria than may be inferred from poisoning reports alone. Indeed, cyanobacterial bloom samples collected from 15 British freshwaters over recent years have all been lethal according to intraperitoneal mouse bio-assay on numerous occasions (Table 15.2), although reports of deaths of animals after drinking at these sites are known for only two locations; Rostherne Mere, Cheshire (Reynolds 1980) and White Loch, Perthshire (T. Nicholson, personal communication). The high toxicities (low mouse bio-assay LD_{50}s) recorded on many occasions at these sites are as high as or higher than elsewhere in the world where livestock fatalities have occurred (Table 15.2). Toxic or potentially toxic blooms of cyanobacteria are also a common feature of those lakes in Norway, Sweden, and Finland which support cyanobacteria (Persson *et al.* 1984; Berg *et al.* 1986).

Cyanobacterial toxin levels per unit cyanobacterial biomass vary widely from week to week at individual locations (Codd and Bell 1985), and Carmichael and Gorham (1981) found a complex mosaic of high and low toxicity within an individual bloom sampled at intervals of a few metres on a

Table 15.1 Toxin-producing cyanobacteria and their toxins.

Cyanobacterium	Toxin name	Toxin structure	Reference
Marine and brackish water			
Lyngbya majuscula	lyngbyatoxin A	indole alkaloid	Moore 1981, 1984*a*
	aplysiatoxin	phenolic bislactone	
	debromoaplysiatoxin	phenolic bislactone	
Schizothrix calcicola	debromoaplysiatoxin	phenolic bislactone	
Oscillatoria nigroviridis	oscillatoxin A	phenolic bislactone	
Nodularia spumigena	nodularia toxin	peptide	J.E. Eriksson, personal communication
Freshwater		peptides	Runnegar and Falconer 1981;
Microcystis aeruginosa	microcystin or cyanoginosin		Santikarn *et al*. 1983;
			Botes *et al*. 1984.
Aphanizomenon flos-aquae	aphantoxins	alkaloids	Sasner *et al*. 1981;
			Ikawa *et al*. 1982
Anabaena flos-aquae	anatoxin *a*	alkaloid	Carmichael and Gorham, 1978
(different strains)	anatoxin *b*	unknown	
	anatoxin *c*	peptide	Krishnamurthy *et al*. 1986*a*, 1986*b*.
	anatoxin *d*	unknown	Carmichael and Gorham 1978
	anatoxin *a*(*s*)	unknown	
	anatoxin *b*(*s*)	unknown	
Oscillatoria agardhii	oscillatoria toxins	peptides	Skulberg *et al*. 1984;
			Krishnamurthy *et al*. 1986*b*;
			Eriksson *et al*. 1988
Gloeotrichia echinulata	gloeotrichia toxin	unknown	Codd and Bell 1985
Synechocystis sp.	synechocystis toxin	unknown	Lincoln and Carmichael, 1981
Cylindrospermopsis raciborskii	cylindrospermopsis toxin	unknown	Hawkins *et al*. 1985

Table 15.2 Toxicities of cyanobacterial blooms from British freshwaters.

Location	Date	Dominant cyanobacteria[a]	LD_{50}[b]
Scotland			
Loch Balgavies	Jul–Sep 1981–85	M	2–1512
Loch Balgavies	Jul–Oct 1982	Aph	116–290
Loch Rescobie	Jun–Oct 1981–85	M	1–1257
Monikie Island Pond	Jun–Dec 1981–85	G + Ana + Aph	22–909
Monikie Denfind Pond	Aug 1983	M + Ana	70–308
Monikie North Pond	May–Sep 1983–85	G	112–1593
White Loch	Sep 1982	Aph	1020
Fingask	Sep 1982	Aph	1150
Lindores	Sep 1984	M	85
Cameron Reservoir	Aug–Sep 1983–84	M + Ana	19–907
Loch Leven	Sep 1984	M	145
Loch Fad	June–Sep 1983–84	M	22–949
Loch Charn	Jul–Aug 1984	Ana	69–100
Clatto Reservoir	Aug 1983	Ana	116–173
England			
Rostherne Mere	Aug 1984	M + Osc	22–392

[a]M, *Microcystis aeruginosa*; Aph, *Aphanizomenon flos-aquae*; Ana, *Anabaena flos-aquae*; G, *Gloeotrichia echinulata*; Osc, *Oscillatoria agardhii*.
[b]Dose required to kill 50 per cent of population of Swiss balb *c* mice by intraperitoneal administration of freeze-dried cells. Values in mg dry wt cyanobacteria per kg body wt.

single occasion. Reliable morphological, physiological, or olfactory markers are not available to provide a ready indication in the field of the toxicity, or otherwise, of cyanobacterial blooms. The acknowledged need for new cyanobacterial toxin assay methods to complement the inadequate mouse bioassay for laboratory-based research clearly extends to the field, where a ready indication of cyanobacterial bloom toxicity is needed by veterinarians, park-rangers, aquaculturalists, and water amenity and treatment officials. In our laboratory we are developing chemical, immunological, and cytotoxity assay methods for cyanobacterial toxins for research and routine monitoring.

Purification and properties

Space permits discussion of only the toxins of selected freshwater cyanobacteria. For reviews of the toxins of marine species, see Moore 1981, 1984*a*; Carmichael *et al.* 1985).

Neurotoxins

Alkaloid neurotoxins are produced by strains of *Anabaena flos-aquae* and *Aphanizomenon flos-aquae* (Table 15.1). The anatoxin *a* of *Ana. flos-aquae*

NRC-44-1 has been the most intensively studied. Initial purification by solvent extraction, and then column and thin-layer chromatography, yielded a bicyclic secondary amine, 2-acetyl-9-azabicyclo (4.2.1) non-2-ene (Huber 1972; Devlin *et al.* 1977). This toxin, MW 165, is a structural analogue of cocaine and it can be synthesized from cocaine by ring expansion (Campbell *et al.* 1977).

Poisoning of animals by *Ana. flos-aquae* neurotoxins can occur after drinking water containing whole or lysed cells. Muscular tremors and staggering may occur 5–30 min after ingestion, and paralysis of peripheral skeletal muscles, then of the respiratory muscles, leads to convulsions and death from respiratory arrest. Anatoxin *a* acts as a post-synaptic neuromuscular blocking agent. The toxin binds to the nicotinic acetylcholine receptors of *Torpedo* electric tissue with high affinity and at muscarinic acetylcholine receptors of rat brain with low affinity and regional selectivity (Aronstam and Witkop 1981). No specific internal lesions are apparent in animals poisoned by anatoxins, although congestion of the brain, spinal cord, and meninges may occur and the lungs may be filled with frothy fluid. Opisthotonous (head bent over backwards) is a characteristic feature of anatoxin *a* poisoning in birds (Carmichael 1981). Several other physiologically distinguishable anatoxins have been recognized (Table 15.1), including neurotoxic factors, possibly anatoxins *b* and *d*, and other toxins which specifically cause salivation in addition to the neuromuscular effects of anatoxins, namely anatoxins *a*(*s*) and *b*(*s*). A procedure for anatoxin *a*(*s*) purification has been devised recently which involves extraction in acetic acid:ethanol, Sephadex G-15 and CM-Sephadex C-25 column chromatography, and high-performance liquid chromatography (HPLC) on a Cyano column (Mahmood and Carmichael 1986). The low molecular weight anatoxin *a*(*s*) molecule may be expected to differ structurally from anatoxin *a* since it shows a higher toxicity in intraperitoneal mouse bioassay (LD_{50}s, 50 μg/kg for anatoxin *a*(*s*) and *c*. 250 μg/kg for anatoxin *a*) and, in contrast to anatoxin *a*, anatoxin *a*(*s*) apparently acts as an anticholinesterase (Mahmood and Carmichael 1986). We have recently developed a rapid isolation and purification procedure for cyanobacterial toxins using high-performance thin-layer chromatography (HPTLC) (Poon *et al.* 1987) and have modified this to permit alkaloid extraction and purification from *Ana. flos-aquae*. By this means, we have purified anatoxin *a* plus two hepatotoxic peptides from a single cyanobacterial strain (K. Jamel Al-Layl, G.K. Poon, and G.A. Codd, unpublished observations).

Animal and fish deaths by *Aphanizomenon flos-aquae* blooms have been recorded in the eastern states of the USA. Signs of poisoning resemble those of anatoxins and marine paralytic shellfish poisons. The toxins responsible, aphantoxins, cause loss of co-ordination, twitching, irregular ventilation, gaping mouth, and death in mice by respiratory failure (Sasner *et al.* 1984). Two neurotoxins, which are similar in terms of molecular size and fluorescence, and electrophoretic and chromatographic properties, to saxitoxin and neosaxitoxin of the marine dinoflagellate *Gonyaulax tamarensis*, have been purified from a single *Aph. flos-aquae* isolate, strain NH-I (Ikawa *et al.*

1982). The aphanotoxins further resemble the paralytic shellfish toxins in their reversible blocking effect on sodium conductance of squid axon membrane.

Hepatotoxins

The toxic peptides of *Microcystis aeruginosa* have been the most intensively studied cyanobacterial hepatotoxins so far. Signs of poisoning in animals after drinking water containing toxic *M. aeruginosa* or in mice after intraperitoneal bio-assay usually appear after 30 min to 24 hours, depending on dose. Acute dosage of mice causes death between about 30 min and 2–3 hours. Symptoms include weakness, vomiting, pilorection, diarrhoea, cold extremities, pallor, and heavy breathing. The most obvious internal sign of microcystin poisoning is a dark mottled liver, swollen with blood to about twice its normal weight (Carmichael 1981; Codd and Carmichael 1982; Richard *et al.* 1983).

An intensive study of the clinical and pathological consequences of microcystin poisoning in animals has been made in Australia. Deaths of 27 kg sheep occurred 18–48 hours after intraruminal inoculation with a *M. aeruginosa* bloom, and were preceded by a sequence of depression, a transient rise in body temperature, increased heart rate, fluctuating respiration rate, cessation of ruminal sounds and cudding, persistant sternal then lateral recumbency, noisy laboured rapid breathing, muscle twitching, leg paddling, gasping respiration with periodic Cheyne–Stokes breathing cycles, mild nystagmus with eyelids widely opened, and finally loss of the eye preservation reflex (Jackson *et al.* 1983, 1984). The carcasses were jaundiced with yellow fluid in the body cavities and numerous haemorrhages. Lungs were slightly oedematous and the livers were swollen, with a blotchy pattern of congestion and petechial haemorrhages. Other organs appeared normal. Marked increases in serum activities of aspartate aminotransferase, lactate dehydrogenase, glutamate dehydrogenase, and alkaline phosphatase, an increased amount of bilirubin, and a decrease in serum glucose were observed. Massive hepatocyte necrosis was found in the livers of acutely poisoned sheep. The necrotic cells had eosinophilic, disintegrating cytoplasm and their nuclei showed pyknosis and karyorrhexis. The hepatocyte endoplasmic reticulum was aggregated and vacuolation was found in more severely affected cells. These observations are consistent with the *Microcystis* peptide toxins acting primarily on the liver (Heaney 1971; Skulberg 1978; Carmichael 1981; Codd and Carmichael 1982; Jackson *et al.* 1983, 1984).

An alternate hypothesis, that microcystin may primarily cause pulmonary congestion due to acclusion by thrombi-containing plates, has been proposed (Slatkin *et al.* 1983). Multiple pulmonary thrombi were found in mice after acute poisoning by microcystin from two South African *M. aeruginosa* isolates. This pulmonary thrombosis was not relieved by anticoagulants and was estimated to precede hepatomegaly (Slatkin *et al.* 1983). Although pulmonary thrombi have also been observed following the administration of microcystins from Australian and Scottish *M. aeruginosa* isolates, liver damage occurred in these studies before the appearance of the thrombi in the lungs, and measurements of venous and arterial blood pressures did not indicate congestion consistent with pulmonary thrombosis (Falconer *et al.* 1981; Theiss and Carmichael 1986).

Determinations of the distribution of radioisotope-labelled microcystins between the major organs of rodents have been performed recently which have confirmed that the liver is the main target organ for accumulation and excretion of the toxins. Thirty minutes after intraperitoneal administration of ^{125}I-labelled toxin to rats, 21.7 per cent of the radioactivity was in the liver and 9.4 per cent in the small intestine. The kidneys and urine contained only 5.6 and 2.9 per cent of the ^{125}I administered (Falconer *et al.* 1986). We have prepared ^{14}C-labelled microcystin and between 73 and 88 per cent of that administered to mice was recovered from the liver between 1 and 180 min after dosage (Brooks and Codd 1987). Levels of ^{14}C in the lungs, kidneys, and all other organs ranged from 1 to 10 per cent throughout. In addition to the microcystin-dependent changes in serum enzyme levels which are indicative of liver damage (Jackson *et al.* 1983, 1984), the toxins cause major changes in hepatic microsomal cytochrome levels: intraperitoneal administration of sub acute doses to mice resulted in 80 per cent decreases in cyt b_5 and P-450 and a fivefold increase in P-420 (Brooks and Codd 1987).

These changes were partly prevented by the administration of the microsomal enzyme inducers β-naphthoflavone, 3-methylcholanthrene, and phenobarbital 48 hours before the toxin. Furthermore, each of these inducers extended mouse survival time and partly prevented the microcystin-dependent increase in liver weight (due to blood pooling) in a dose-dependent manner if given before the toxin. Besides identifying the liver as the main target of microcystin(s), these findings indicate that the liver has a detoxifying function against cyanobacterial peptide poisoning (Brooks and Codd 1987).

The extension of rodent survival times by co-administration of the α-adrenoceptor and β-adrenoceptor blocking agents, phenoxybenzamine and acebutolol, respectively, has suggested that acute levels of microcystin may also cause stimulation of the sympathetic nervous system directly or indirectly with the contraction of vascular tissue (Oishi and Watanabe 1986). Whether this interpretation is correct or not, the efficacy of the compounds in extending survival and decreasing the potency of the toxin when co-administered offers prospects for further research into the mode of action of microcystins and for the treatment of poisoning cases.

Advances have been made over recent years in the purification of *Microcystis* peptide toxins. The usual procedure now involves solvent extraction from lyophilized cells, followed by adsorption by passage through C_{18} cartridges (e.g. Sep-Pak or Bond-Elut), then Sephadex gel filtration in some cases, and, invariably, reversed-phase HPLC (e.g. Siegelman *et al.* 1984; Brooks and Codd 1986; Krishnamurthy *et al.* 1986*a*). We have recently achieved microcystin purification without the need for gel filtration or HPLC by the use of HPTLC (Poon *et al.* 1987). Although only one or two toxic peptides are usually found in individual *M. aeruginosa* isolates, up to six different peptides have been found in some cases (Botes *et al.* 1982; Eloff *et al.* 1982; Poon *et al.* 1987). However, these peptides appear to conform to a basic pattern which is that of a cyclic heptapeptide of molecular weight about 1000. The general structure as elucidated by Botes, Williams, and colleagues is:

$$\boxed{\text{D-ala - }R_1\text{ -Masp - }R_2\text{ - Adda - D-glu - Mdha}}$$

where R_1 and R_2 are variable amino acids including leucine, alanine, arginine, tyrosine, and methionine; Masp is β-methylaspartate; Adda is a novel β-amino acid residue of 3-amino-9-methoxy-2,6,8-trimethyl-10-phenyldeca -4, 6-dienoic acid, and Mdha is methyldehydroalanine (Botes *et al.* 1982, 1984, 1985; Santikarn *et al.* 1983; Williams 1984; Krishnamurthy *et al.* 1986*a*, 1986*b*). Since this family of toxins shows similar toxicities in the standard intraperitoneal mouse bioassay (50–100 μg/kg), then it is inferred that the toxicity of the molecule resides in the invariable components. The effects of chemical modification of these components upon toxicity is under study in our laboratory.

Hepatotoxic peptides which cause similar signs of poisoning to those of *M. aeruginosa* have been isolated from strains of *Oscillatoria agardhii* and *Ana. flos-aquae* (Berg and Søli 1985*a*, 1985*b*; Krishnamurthy *et al.* 1986*a*, 1986*b*; K. Jamel Al-Layl, G.K. Poon, and G.A. Codd, unpublished observations; Eriksson *et al.* 1987). The peptide from *Ana. flos-aquae* strain S-23-g-1 contains equimolar amounts of alanine, arginine, glutamic acid, leucine, and β-methylaspartate in addition to the presumed Adda and methyldehydroalanine (Krishnamurthy *et al.* 1986*a*, 1986*b*). As with microcystins (Runnegar *et al.* 1981), treatment of isolated hepatocytes with the *O. agardhii* (Aune and Berg 1986) and *Ana. flos-aquae* peptide toxins (K. Jamel Al-Layl, G.K. Poon, and G.A. Codd, unpublished observations) causes gross morphological change (surface blebs and disruption), although toxicity of the *Oscillatoria* and *Anabaena* peptides is lower than that of the microcystins. The toxic principle(s) from the severely hepatotoxic *Cylindrospermopsis raciborskii* has not yet been reported. Although this may be a peptide, it is unlikely to be identical to the microcystins since, although mainly hepatotoxic, it differs in the pattern and onset of hepatocyte necrosis and haemorrhage *in vivo* (Hawkins *et al.* 1985).

Regulation of toxin production

Few studies on the regulation of cyanobacterial production have bene reported. Gorham (1964) found that the toxicity of *M. aeruginosa* to mice was greater after growth at 25 °C than at 20 °C, though it decreased at 30 °C to an intermediate value. Decreases in the toxicity of South African and Japanese *M. aeruginosa* isolates with increasing growth temperature have also occurred (van der Westhuizen and Eloff 1983, 1985; Watanabe and

Table 15.3 Toxicity of *Microcystis aeruginosa* 7813 grown in photoautotrophic batch culture.

Growth medium[a]	Temp (°C)	PFR[c]	LD_{50}[d]
Complete BG-11	25	5	16
	25	10	17
	25	15	22
	25	20	12
	25	50	29
	10	15	114
	34	15	87
BG-11 minus PO_4[b]	25	15	13
BG-11 minus NO_3[b]	25	15	118
BG-11 minus 'CO_2'[b]	25	15	139

[a]Cells grown to mid-log phase and harvested by centrifugation.
[b]Cells transferred at mid-log phase to medium minus added nutrient and harvested after approximately two weeks.
[c]Photon fluence rate, $\mu E\ m^{-2}\ sec^{-1}$.
[d]mg lyophilized cells per kg body of mice by intraperitoneal route.

Oishi 1985). Although the highest growth rate in batch culture occurred at pH 9.0, toxicity of the South African isolate was greater with growth at higher or lower pH values. Watanabe and Oishi (1985) obtained a four fold increase in *M. aeruginosa* toxicity by increasing the incident photon fluence rate (PFR) from 7.5 to 200 $\mu E\ m^{-2}\ sec^{-1}$, although a decline in toxicity has also been recorded at such very high light levels (van der Westhuizen and Eloff 1985). As shown in Table 15.3, the toxicity of *M. aeruginosa* 7813 was not greatly influenced by PFR between 5 and 50 $\mu E\ m^{-2}\ sec^{-1}$ when maintained at the optimum growth temperature of 25 °C. Growth at 10 °C and 34 °C, at near the lower and upper temperature limits for this strain, resulted in four- to fivefold decreases in toxicity of the cells. When such cultures were transferred from complete growth medium to medium minus added nutrients it was found that phosphorus removal did not influence toxicity, although nitrogen and inorganic carbon removal each caused an approximate tenfold decrease in toxicity (Table 15.3). Further studies on the regulation of microcystin production at the metabolic level are in progress.

The problem of the genetic regulation of cyanobacterial toxin production has received little attention. The plasmids of cyanobacteria, so far cryptic, have been considered as possible locations of genes involved in toxin production. Early studies on the effects of plasmid-curing agents on an *Ana. flos-aquae* (neurotoxin) and an Australian isolate of *M. aeruginosa* did not influence toxicity, although it is not clear whether these strains contained extrachromosomal DNA and, if so, whether it was eliminated by the plasmid-curing agents (Kumar and Gorham 1975; Runnegar *et al.* 1983). On the other hand, three plasmid-curing agents eliminated toxin formation by the South African isolate *M. aeruginosa* WR70, suggesting a role for plasmids in toxin production by this strain (Hauman 1981). The highly toxic *Microcystis* 7820 contains extrachromosomal DNA in 3–4 plasmids (Vakeria *et al.* 1985). Since prolonged growth in the presence of novobiocin eliminated the plasmids from this strain but did not influence the high level of toxin production, we have concluded that non-integrated plasmid DNA involvement in the genetic control of microcystin production by *Microcystis* 7820 is unlikely (Vakeria *et al.* 1985). The possibility has not been excluded, however, that plasmid DNA which may have become integrated into the chromosome may be involved. If the microcystin peptides were likely to be single gene products, then the use of a synthetic oligonucleotide, deduced from the amino acid sequence, may be expected to be a useful approach in the localization of the microcystin gene(s) by hybridization techniques. However, the presence of the unusual amino acid residues, Adda, Mdha, and Masp in microcystins, renders this straightforward approach questionable and it is likely that cyanobacterial peptide toxins require, and are controlled by, multiple genes.

Effects of cyanobacterial toxins on human activities and health

Much of the evidence implicating cyanobacterial toxins in human health problems is circumstantial, although some clear relationships have been established. The involvement of cyanobacterial toxins in Haff disease remains speculative: over a 1000 cases occurred in the 1920s and 1930s along the Baltic coast around Kaliningrad, USSR, among people eating fish, mainly burbot, from lagoons containing cyanobacterial blooms. Similar cases occurred along the Swedish coast (Berlin 1948) and the descriptions resemble that of the more recently described Yuksov–Sortlav disease (Birger *et al.* 1973). After eating fish, particularly the livers, from cyanobacterial bloom-containing locations, symptoms of severe muscular pain, respiratory distress, vomiting, and brownish-black urine persisted for several days, though, of the large numbers of people affected, few fatalities were reported. It is known that microcystin peptide toxin principally causes liver damage when administered intraperitoneally to fish (Phillips *et al.* 1985), and the possibility that cyanobacterial toxins accumulate at sub acute levels in fish requires further study.

Many cases of contact irritation ('swimmers itch') have occurred after bathing in marine and freshwaters containing toxic cyanobacteria. Symptoms of burning or itching of the skin, erythematous wheals, redness of the eyes and lips, often accompanied by sore throat, ear aché, and dizziness, have occurred among large numbers of bathers around the coasts of Hawaii, the Marshall Islands, Florida, and Okinawa (Moikeha and Chu 1971; Moikeha *et al.* 1971; Hashimoto *et al.* 1976). The main marine agent responsible is *Lyngbya majuscula*. The *Lyngbya* toxins, aplysiatoxin and debromoaplysiatoxin, cause inflammation of mucous membranes and skin irritation. These compounds, and the related phenolic bislactones of *Schizothrix calcicola* and *Oscillatoria nigroviridis*, also implicated in swimmers' itch, are potent tumour promoters in two-stage carcinogenesis in mouse skin (Moore 1981,

1984*a*, 1984*b*; Fujiki *et al.* 1984). Since the cyanobacteria which produce these toxins are eaten by marine animals, the presence of cyanobacterial tumour promoters in marine food chains requires investigation. Contact with blooms of toxic *Ana. flos-aquae, Aph. flos-aquae*, and *Gloeotrichia echinulata* may contribute to swimmers itch in freshwaters (G.A. Codd, W.W. Carmichael, and S.G. Bell, unpublished observations).

Gastroenteritis, coincident with skin irritation, and including diarrhoea, headache, nausea, vomiting, stomach cramps, and dizziness, has occurred among several thousand people at freshwaters in the USA containing *Ana. flos-aquae*, other *Anabaena* spp., and *Schizothrix calcicola* (Lippy and Erb 1976; Billings 1981; Keleti *et al.* 1981; Carmichael *et al.* 1985). The lack of evidence implicating enterobacteria and viruses in these cases has led these authors to suggest that cyanobacterial toxins may be (in part) responsible. In these cases, cyanobacterial lipopolysaccharide (LPS) endotoxins, a consistent feature of cyanobacteria since they are Gram-negative prokaryotes (Stanier and Cohen-Bazire 1977), may be responsible. (For details of cyanobacterial LPS, see Keleti *et al.* 1981; Keleti and Sykora 1982, and Raziuddin *et al.* 1983.)

A severe outbreak of hepato-enteritis which affected 148 inhabitants of Palm Island off Australia's Queensland coast occurred in 1979 (Bourke *et al.* 1983). The incident took place shortly after the treatment of their drinking water with copper sulphate to control a heavy cyanobacterial bloom, and cyanobacterial poisoning was suspected to have caused the 'Palm Island mystery disease'. Isolation of the highly hepatotoxic *Cylindrospermopsis raciborskii* has supported this possibility (Hawkins *et al.* 1985; Bourke *et al.* 1986; Runnegar and Jackson 1986), although it is also possible that illness may have been caused by elevated levels of copper sulphate used to control the bloom (Prociv 1986).

Epidemiological investigations of a population at Armidale, New South Wales, Australia, taking drinking water from a reservoir containing hepatotoxic toxic blooms of *M. aeruginosa*, have indicated significant links between toxic cyanobacteria and human liver illness (Falconer *et al.* 1983). In this study, a pattern of liver damage, evidenced by increased levels of hepatic γ-glutamyltranspeptidase and alanine aminotransferase in the plasma, was found by examination of the results of routine assays of patients admitted to a local hospital. A seasonal increase in toxic liver injury coincided with the peak of the toxic *M. aeruginosa* bloom and this was only detected in patients who had taken water from the affected reservoir. This possibility of chronic liver damage to humans requires increased awareness of cyanobacterial bloom toxicities, particularly in locations where the sole potable water source is known to contain blooms of gas-vacuolate cyanobacteria.

Acknowledgements

We are grateful to the Natural Environment Research Council, UK, and the Water Research Centre, UK, for financial support, and to Dr S.G. Bell for contributions.

References

Aronstam, R.S. and Witkop, B. (1981). *Proc. natn. Acad. Sci. USA* **78**, 4639.
Aune, T. and Berg, K. (1986). *J. Toxicol. environ. Hlth.* **19**, 325.
Berg, K. and Sφli, N.E. (1985*a*). *Acta Vet. Scand.* **26**, 363.
Berg, K. and Sφli, N.E. (1985*b*). *Acta Vet. Scand.* **26**, 374.
Berg, K., Skulberg, O.M., Skulberg, R., Underdal, B., and Willen, T. (1986). *Acta Vet. Scand.* **27**, 440.
Berlin, A. (1948). *Acta Med. Scand.* **129**, 560.
Billings, W. (1981). In *The water environment—algal toxins and health* (ed. W.W. Carmichael) p. 243. Plenum Press, New York.
Birger, T.L., Malyarevskaya, A.Y., and Arsan, O.M. (1973). *Hydrobiol. J.* **9**, 71.
Botes, D.P., Kruger, H., and Viljoen, C.C. (1982). *Toxicon* **20**, 945.
Botes, D.P., Tuinman, A.A., Wessels, P.L., Viljoen, C.C., Kruger, H., Williams, D.H., Santikarn, S., Smith, R.J. and Hammond, S.J. (1984). *J. chem. Soc. Perkin Trans.* **1**, 2311.
Botes, D.P. *et al.* (1985), *J. chem. Soc. Perkin Trans.* **1**, 2747.
Bourke, A.T.C., Hawes, R.B., Neilson, A., and Stallman, N.D. (1983). *Toxicon (Suppl.)* **3**, 45.
Bourke, A.T.C., Hawes, R.B., Neilson, A., and Stallman, N.D. (1986). *Med. J. Australia* **145**, 486.
Brooks, W.P. and Codd, G.A. (1986). *Lett. appl. Microbiol.* **2**, 1.
Brooks, W.P. and Codd, G.A. (1987). *Pharmacol. Toxicol.* **60**, 187.
Campbell, H.F., Edwards, O.E., and Holt, R. (1977). *Can. J. Chem.* **55**, 1372.
Carmichael, W.W. (1981). In *The water environment—algal toxins and health* (ed. W.W. Carmichael) p. 1. Plenum Press, New York.
Carmichael, W.W. and Gorham, P.R. (1978). *Mitt. Int. Ver. Limnol.* **21**, 285.
Carmichael, W.W. and Gorham, P.R. (1981). In *The water environment—algal toxins and health* (ed. W.W. Carmichael) p. 161. Plenum Press, New York.
Carmichael, W.W., Jones, C.L.A., Mahmood, N.A., and Theiss, W.C. (1985). *CRC Crit. Rev. environ. Control.* **15**, 275.
Codd, G.A. (1984). *Microbiol. Sci.* **1**, 48.
Codd, G.A. and Bell, S.G. (1985). *Water Poll. Control.* **84**, 225.
Codd, G.A. and Carmichael, W.W. (1982). *FEMS Microbiol. Lett.* **13**, 409.
Collins, M. (1978). *Microbiol. Rev.* **42**, 725.
Devlin, J.P., Edwards, O.E., Gorham, P.R., Hunter, N.R., Pike, R.K. and Stavric, B. (1977). *Can. J. Chem.* **55**, 1367.
Eloff, J.N., Siegelman, H.W., and Kycia, H. (1982). In *Abst. Int. Symp. on Toxins and Lectins*, p. 43. International Union of Biological Sciences, Pretoria, South Africa.
Eriksson, J.E., Meriluoto, J.A.O., Kujari, H.P., and Skulberg, O.M. (1988). *Comp. Biochem. Physiol.* (in press).
Falconer, I.R., Beresford, A.M., and Runnegar, M.T.C. (1983). *Med. J. Australia* **1**, 511.
Falconer, I.R., Buckley, T., and Runnegar, M.T.C. (1986). *Aust. J. biol. Sci.* **39**, 17.
Falconer, I.R., Jackson, A.R.B., Langley, J.V., and Runnegar, M.T.C. (1981). *Aust. J. biol. Sci.* **34**, 179.

Francis, G. (1878). *Nature, Lond.* **18**, 11.
Fujiki, H. *et al.* (1984). In *Cellular interactions by environmental tumor promoters* (ed. H. Fujiki) p. 37. Japan Science Society Press, Tokyo.
Gorham, P.R. (1964). In *Algae and man* (ed. D.F. Jackson) p. 307. Plenum Press, New York.
Gorham, P.R. and Carmichael, W.W. (1979). *Pure appl. Chem.* **52**, 165.
Hashimoto, Y., Kamiya, H., Yamazato, K., and Nosawa, K. (1976). *Anim. Plant Microb. Toxins.* **1**, 333.
Hauman, J.H. (1981). In *The water environment—algal toxins and health* (ed. W.W. Carmichael) p. 97. Plenum Press, New York.
Hawkins, P.R., Runnegar, M.T.C., Jackson, A.R.B., and Falconer, I.R. (1985). *Appl. environ. Microbiol.* **50**, 1292.
Heaney, S.I. (1971). *Wat. Treat. Exam.* **20**, 235.
Huber, C.S. (1972). *Acta Crystallog. Sect. B* **28**, 2577.
Ikawa, M., Wegener, K., Foxall, T.L., and Sasner, J.J. (1982). *Toxicon* **20**, 747.
Jackson, A.R.B., McInnes, A., Falconer, I.R., and Runnegar, M.T.C. (1983). *Toxicon* (Suppl.) **3**, 191.
Jackson, A.R.B., McInnes, A., Falconer, I.R., and Runnegar, M.T.C. (1984). *Vet. Path.* **21**, 102.
Keleti, G. and Sykora, J.L. (1982). *appl. environ. Microbiol.* **43**, 104.
Keleti, G., Sykora, J.L., Maiolie, L.A., Doerfler, D.L., and Campbell, I.M. (1981). In *The water environment—algal toxins and health* (ed. W.W. Carmichael) p. 447. Plenum Press, New York.
Krishnamurthy, T., Carmichael, W.W., and Sarver, E.W. (1986*a*). *Toxicon* **9**, 865.
Krishnamurthy, T., Szafraniec, L., Sarver, E.W., Hunt, D.F., Shabanowitz, J., Carmichael, W.W., Missler, S., Skulberg, O.M. and Codd, G.A. (1986*b*). *Proc. 34th Annual Conf. Mass Spec. Allied Topics, Cincinatti, Ohio*, p. 93.
Kumar, H.D. and Gorham, P.R. (1975). *Biochem. Physiol. Pflanzen* **167**, 473.
Lincoln, E.P. and Carmichael, W.W. (1981). In *The water environment—algal toxins and health* (ed. W.W. Carmichael) p. 223. Plenum Press, New York.
Lippy, E.C. and Erb, J. (1976). *J. Am. Wat. Works Assoc.* **88**, 606.
Mahmood, N.A. and Carmichael, W.W. (1986). *Toxicon* **24**, 425.
Moikeha, S.N. and Chu, G.W. (1971). *J. Phycol.* **7**, 8.
Moikeha, S.N., Chu, G.W., and Berger, L.R. (1971). *J. Phycol.* **7**, 4.
Moore, R.E. (1981). In *The water environment—algal toxins and health* (ed. W.W. Carmichael) p. 15. Plenum Press, New York.
Moore, R.E. (1984*a*). In *Seafood toxins* (ed. E.P. Ragelis) p. 369. American Chemical Society, Washington, DC.
Moore, R.E. (1984*b*). In *Cellular interactions by environmental tumor promoters* (ed. H. Fujiki) p. 49. Japan Science Society Press, Tokyo.
Oishi, S. and Watanabe, M.F. (1986). *Environ. Res.* **40**, 518.
Persson, P-E., Sivonen, K., Keto, J., Kononen, K., Niemi, M., and Viljamaa, H. (1984). *Aqua Fennica* **14**, 147.
Phillips, M.J., Roberts, R.J., Stewart, J.A., and Codd, G.A. (1985). *J. Fish Diseases* **8**, 339.
Poon, G.K., Priestley, I.M., Hunt, S.M., Fawell, J.K., and Codd, G.A. (1987). *J. Chromatog.* **387**, 551.

Prociv, P. (1986). *Med. J. Australia* **145**, 487.
Raziuddin, S., Siegelman, H.W., and Tornabene, T.G. (1983). *Eur. J. Biochem.* **137**, 333.
Reynolds, C.S. (1980). *J. Inst. wat. Eng. Sci.* **34**, 74.
Reynolds, C.S. and Walsby, A.E. (1975). *Biol. Rev.* **50**, 437.
Richard, D.S., Beattie, K.A., and Codd, G.A. (1983). *Environ. Technol. Lett.* **4**, 409.
Runnegar, M.T.C. and Falconer, I.R. (1981). In *The water environment—algal toxins and health* (ed. W.W. Carmichael) p. 325. Plenum Press, New York.
Runnegar, M.T.C. and Jackson, A.R.B. (1986). *Med. J. Australia* **145**, 486.
Runnegar, M.T.C., Falconer, I.R., and Silver, J. (1981). *Nauyn-Schmiedeberg's Arch. Pharmacol.* **317**, 268.
Runnegar, M.T.C., Falconer, I.R., Jackson, A.R.B., and McInnes, A. (1983). *Toxicon* (Suppl.) **3**, 377.
Santikarn, S., Williams, D.H., Smith, R.J., Hammond, S.J., Botes, D.P., Tuinman, A., Wessels, P.L., Viljoen, C.C. and Kruger, H. (1983). *J. chem. Soc. chem. Commun.* **275**, 652.
Sasner, J.J., Ikawa, M., and Foxall, T.L. (1984). In *Seafood toxins* (ed. E.P. Ragelis) p. 391. American Chemical Society, Washington, DC.
Sasner, J.J., Ikawa, M., Foxall, T.L., and Watson, W.H. (1981). In *The water environment—algal toxins and health* (ed. W.W. Carmichael) p. 389. Plenum Press, New York.
Schwimmer, D. and Schwimmer, M. (1964). *Algae and man* (ed. D.F. Jackson) p. 368. Plenum Press, New York.
Schwimmer, M. and Schwimmer, D. (1968). In *Algae, man and the environment* (ed. D.F. Jackson) p. 279. Syracuse University Press, New York.
Siegelman, H.W., Adams, W.H., Stoner, R.D., and Slatkin, D.N. (1984). In *Seafood toxins*, (ed. E.P. Ragelis), p. 407. American Chemical Society, Washington, DC.
Skulberg, O.M. (1978). In *Norsk Institutt for Vannforskning Arbok*, p. 73. Norwegian Institute for Water Research, Oslo.
Skulberg, O.M., Codd, G.A., and Carmichael, W.W. (1984). *Ambio* **13**, 244.
Slatkin, D.N., Stoner, R.D., Adams, W.H., Kycia, J.H., and Siegelman, H.W. (1983). *Science* **220**, 1383.
Stanier, R.Y. and Cohen-Bazire, G. (1977). *A. Rev. Microbiol.* **31**, 225.
Stein, J.R. and Borden, C.A. (1984). *Phycologia* **23**, 485.
Theiss, W.C. and Carmichael, W.W. (1986). In *Mycotoxins and phycotoxins*, (ed. P.S. Steyn and R. Vleggar), p. 353. Elsevier Amsterdam.
Vakeria, D., Codd, G.A., Bell, S.G., Beattie, K.A., and Priestley, I.M. (1985). *FEMS Microbiol. Lett.* **29**, 69.
Van der Westhuizen, A.J. and Eloff, J.N. (1983). *Z. Pflanzenphysiol.* **110**, 157.
Van der Westhuizen, A.J. and Eloff, J.N. (1985). *Planta* **163**, 55.
Walsby, A.E. (1975). *A. Rev. plant Physiol.* **26**, 427.
Watanabe, M.F. and Oishi, S. (1985). *Appl. environ. Microbiol.* **49**, 1342.
Williams, D.H. (1984). *Chem. Soc. Rev.* **13**, 131.

16 Alga–invertebrate symbiosis

A.E. DOUGLAS

John Innes Institute, Colney Lane, Norwich NR4 7UH, UK

Introduction

Algae have been reported in stable associations with a range of non-photosynthetic eukaryotes in both freshwater and marine environments. These associations are symbioses in the original meaning of the term as the living together of differently named organisms (de Bary 1879) and, by convention, the larger non-photosynthetic partner is known as the host and the smaller algal partner as the symbiont. The algal symbionts are within the body of the host and, in most associations, they are intracellular and enclosed within individual vacuoles, bounded by a membrane of host origin.

In virtually all symbioses, only one species or genus of alga is found in each host species. The most common algal symbionts in the marine environment are dinoflagellates of the *Symbiodinium microadriaticum* species complex (Blank and Trench 1986), while in freshwaters *Chlorella* species of the *vulgaris* group are the dominant symbionts (Reisser 1984; Douglas and Huss 1986). As reviewed by Droop (1963), Trench (1979), and Smith and Douglas (1987), these algae form associations with up to 200 genera of non-photosynthetic hosts, principally protists (e.g. ciliate protozoa, larger foraminifera) and 'lower' invertebrate animals (including sponges, coelenterates, and turbellaria).

For all associations examined to date, nutritional interactions between the partners play a central role. In this paper, recent developments in the study of two interactions will be reviewed: (a) the release of photosynthetic carbon from symbionts to the host; and (b) algal utilization of host nitrogenous waste products and the recycling of nitrogen within the association. The discussion will be restricted to alga–invertebrate symbioses, principally *Chlorella* associations with the freshwater coelenterate 'hydra' and neorhabdocoel turbellarians and associations between *S. microadriaticum* and marine coelenterates. The symbiosis between the marine acoel turbellarian *Convoluta roscoffensis* and prasinophycean alga *Tetraselmis convolutae* will also be considered in the section on nitrogen recycling.

Transfer of photosynthetic carbon from the algal symbionts to the host

Host tolerance of low food availability

Most animals which contain algal symbionts also feed holozoically. However, the algal symbionts contribute to the survival and growth of their hosts when the association is subjected to low food supply or starvation under photosynthesizing conditions. Most information is available for *Chlorella* symbioses, many of which can be maintained in the laboratory and for which techniques to obtain aposymbionts (i.e. symbiont-free hosts) are available. For example, illuminated cultures of green hydra survive for at least 10 weeks of starvation (although the animals decline in size), whereas symbiotic animals in the dark and aposymbiotic individuals in both light and darkness die within two weeks of starvation (Douglas and Smith 1984). Similarly, when the neorhabdocoel turbellarian *Phaenocora typhlops* is starved in the light, animals containing *Chlorella* cells decline in length more slowly than aposymbionts (Young and Eaton 1975), and symbiotic individuals of two further neorhabdocoels, *Typhloplana viridata* and *Dalyellia viridis*, exhibit a slower decrease in size when starved under illumination than in darkness (Douglas 1987). Comparable data are available for various marine associations (Muscatine 1973). The light-dependence of the symbionts' contribution to host survival and maintenance of size suggests that photosynthetic products of the algal cells are made available to the host tissues.

Demonstration of photosynthate release in the intact symbiosis

The photosynthetic rates of algal cells in intact symbiosis are comparable to or greater than values obtained for cells in culture or for related non-symbiotic species, and there is now overwhelming evidence that a substantial proportion of photosynthetic carbon is released from intact algal cells to the surrounding host tissues in a variety of associations (Smith *et al.* 1969; Hinde 1983; Smith and Douglas 1987).

Many studies of photosynthate release in alga–invertebrate symbioses have adopted ^{14}C-radiotracer techniques, originally applied to symbiotic systems by Smith and co-workers (Smith *et al.* 1969). In these experiments, the intact association is incubated in medium containing ^{14}C-bicarbonate for 0.5–3 hours and is then separated into 'host' and 'algal' fractions by homogenization and washing with centrifugation. In the marine coelenterate hosts of *S. microadriaticum*, over 50 per cent of the translocated ^{14}C is recovered from the lipid fraction, with up to 25 per cent in each of the protein fraction and the soluble fraction (Hinde 1983). The distribution of ^{14}C in the host fraction of the green hydra symbiosis, the only *Chlorella*-invertebrate symbiosis examined to date, is very different, with glycogen and sugars as the principal ^{14}C-labelled products; ^{14}C-labelled lipid has not been detected (Mews 1980).

Characteristics of photosynthate release

The characteristics of photosynthate release and identity of mobile products have been established largely from studies of the algal symbionts *in vitro*. Although both symbiotic *Chlorella* and *S. microadriaticum* have been isolated from a range of invertebrates into axenic culture, most cultured cells release little or no detectable photosynthetically fixed ^{14}C. As a result, virtually all studies are conducted on algal symbionts freshly isolated from the association. It is important to realize that most such preparations of algal symbionts are not pure, but contain host material. For example, the percentage of total host protein associated with symbionts isolated by standard procedures is up to 20 per cent for *Chlorella* from green hydra (Douglas and Smith 1983) and more than 50 per cent for *S. microadriaticum* from the sea-anemone *Aiptasia pulchella* (Steen 1986).

The characteristics of photosynthate release by freshly isolated *Chlorella* and *S. microadriaticum* are considered below.

Chlorella *Chlorella* cells freshly isolated from a variety of hosts and incubated in the commonly used culture media release little or no detectable photosynthate, but such release is induced by incubation in media of low pH. Under these conditions, 30–40 per cent of photosynthetically fixed ^{14}C is recovered from the medium, almost entirely in the form of the disaccharide maltose, with traces of alanine and glycollate (Cernichiari *et al.* 1969; Mews 1980). A range of ^{14}C-labelled products, but no detectable maltose, accumulates within the algal cells, indicating that symbiotic *Chlorella* possess a pH-dependent mechanism for the selective synthesis and efflux of maltose. Using a sensitive fluorimetric assay for maltose, Mews and Smith (1982) have demonstrated that freshly isolated symbionts release maltose at linear rates between 1–3 fmol $cell^{-1}$ $hour^{-1}$, dependent on pH. For all isolates of *Chlorella* so far investigated, including those derived from four strains of green hydra and two species of neorhabodocoel turbellaria, maximal release occurs at pH 4 (Fig. 16.1). It is widely assumed that maltose release from *Chlorella* cells in the intact association is also induced by low external pH and that the contents of the vacuole surrounding each algal cell are acidic.

The biochemical pathways of maltose synthesis and mechanism of release of the disaccharide from symbiotic *Chlorella* remain obscure. In the *Chlorella* symbionts of hydra, maltose is synthesized directly from intermediates of the Calvin cycle, probably triosephosphates, via gluconeogenesis under photosynthesizing conditions (Mews 1980). However, *Chlorella* cells continue to release maltose under non-photosynthesizing conditions, both over the short term (Mews 1980) and after incubation of the intact association for many days in darkness (Douglas and Smith 1983). Algal polysaccharide reserves are the principal source of carbon for maltose synthesis during short-term incubations in DCMU (3-(3,4-dichlorophenyl)-1,1-dimethyl urea)

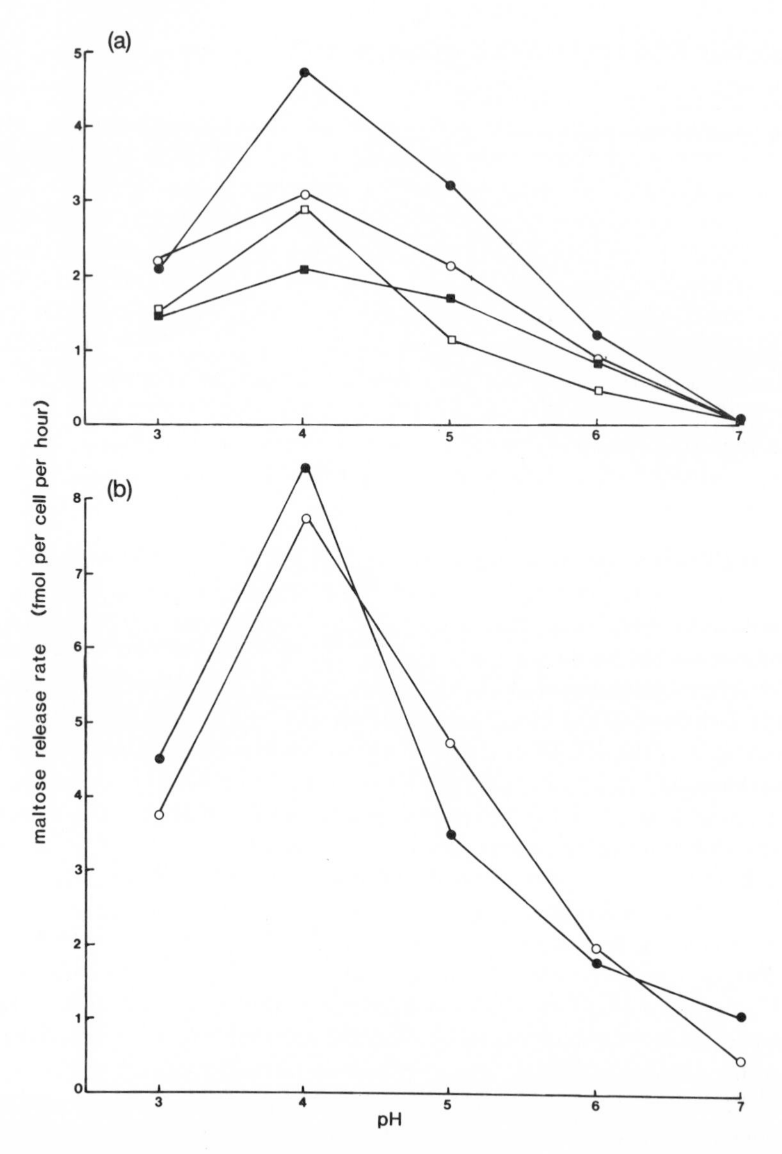

Fig. 16.1. pH-dependence of maltose release by freshly isolated *Chlorella* symbionts incubated in 20 mM phosphate–citrate buffer, pH 3–7. (a) Strains of green hydra: ○, Frome; ●, Florida; □, European; and ■, Jubilee. (b) Species of neorhabdocoel turbellaria: ○, *Dalyellia viridis*; ●, *Typhloplana viridata*. Details of assay procedure are described in Mews and Smith (1982) and Douglas and Smith (1984). [Data are from Douglas (1987) and unpublished work].

Table 16.1 Effect of light and feeding regime of green hydra on the maltose release rate by *Chlorella* symbionts.

Time since feeding (days)	Maltose release rate (fmol cell^{-1} hour^{-1})	
	L	D
1	1.80 ± 0.04	0.41 ± 0.03
3	1.33 ± 0.01	0.19 ± 0.02
6	1.15 ± 0.02	0.09 ± 0.05
14	0.62	<0.06

Green hydra of the European strain were maintained in continuous darkness (D) or under a 12 hour light: 12 hour dark regime (L) for 42 days and were fed thrice weekly until 14, 6, 3, or 1 days before assay, as indicated. The *Chlorella* symbionts were isolated from the symbiosis and maltose release rate scored at pH 4.5 in the presence of the photosynthetic inhibitor, DCMU, at 10^{-5} M. [Details of assay procedure are given in Mews and Smith (1982) and Douglas and Smith (1984)]. Data are the means of three experiments ± standard error of the mean.

(Mews 1980) and may also supply carbon for efflux from cells derived from associations maintained in darkness for long periods. However, in the latter case, the ultimate source of carbon is undoubtedly food ingested by the host (Douglas and Smith 1983); there is direct evidence for the translocation of food-derived host metabolites to the symbionts (Thorington and Margulis 1981). Feeding of the host has a stimulatory effect on maltose production, as is indicated by the decline in maltose release with duration of starvation both in darkness and under illumination (Table 16.1). The basis of this effect is not understood in detail. Possibly, food intake increases the capacity for maltose production by its direct contribution to the algal polysaccharide reserves and by a general enhancement of the nutritional status of the algal cells.

S. microadriaticum Early experiments conducted on *S. microadriaticum* freshly isolated from corals (e.g. *Pocillopora damicornis*, see Muscatine 1967) and anemones (e.g. *Anthopleura elegantissima*, see Trench 1971) and incubated with ^{14}C-bicarbonate during illumination demonstrated the release of water-soluble products of low molecular weight. Up to 90 per cent of the total ^{14}C released was glycerol; other released products included alanine, organic acids, and often glucose. The selectivity of release and concordance between the released products *in vitro* (glycerol, alanine, organic acids) and principal ^{14}C-labelled products in experiments conducted *in vivo* (lipid, protein; see earlier) have led to the conclusion that the characteristics of release by freshly isolated symbionts closely resemble those in the intact association. However, there is now strong evidence that intact lipid, as well as glycerol, is translocated from *S. microadriaticum* in various marine coelenterates (Kellogg and Patton 1983; Battey and Patton 1984). The fatty acid complement of lipids in the host tissues closely resembles that in the algal symbionts and, although over 90 per cent of total lipid is in host tissues, the symbionts and not the host are the principal site of lipid synthesis (Patton *et al.* 1977). The release of lipid from freshly isolated symbionts labelled with ^{14}C may not be apparent for two reasons. First, direct contact between the partners may be necessary for the release of lipid, but not of compounds of low molecular weight such as glycerol; and second, the pool of lipid precursors in the symbionts may not be labelled with ^{14}C during the short-term experiments used in studies on isolated symbionts (Crossland *et al.* 1980; Kellogg and Patton 1983). These investigations have drawn attention to the possibility that both short-term ^{14}C studies and the use of freshly isolated symbionts give an incomplete picture of the range of metabolites translocated in the intact association.

The release of photosynthate from *S. microadriaticum* declines rapidly with time after isolation from the association. However, when a portion of the host fraction is added to the symbiont preparation, the symbionts release substantial amounts of photosynthate (up to 60 per cent of ^{14}C fixed) over long periods. The stimulatory effect of the host homogenate is thermolabile, absent from aposymbionts, but is not species-specific (Hinde 1983). There is a widespread view that a single, chemically definable 'factor' is responsible for the increased photosynthate release and that this factor is active in the intact symbiosis. However, despite the considerable body of literature on the phenomenon *in vitro* (Hinde 1983), neither the identity of the factor nor its relevance to the intact symbiosis has been demonstrated.

The extent of photosynthate transfer in the intact symbiosis

Most quantitative estimates of transport of photosynthetically fixed carbon from algal symbionts to the host tissues have been obtained from short-term ^{14}C-fixation experiments on the intact symbiosis, followed by assay of ^{14}C content of the separated host and algal fractions. Published values of the percentage ^{14}C translocated vary between 20 and 50 per cent for various associations with *S. microadriaticum* and between 30 and 40 per cent for *Chlorella* symbioses (Hinde 1983; Smith and Douglas 1987). However, these values are probably underestimates. As described above, the algal fraction prepared by standard methods is often contaminated with host material. In a study of transport in the green hydra symbiosis (Douglas, unpublished work), the ^{14}C associated with the algal fraction was decreased from 63 to 42 per cent when the algal cells were freed of host material by treatment with sodium dodecyl sulphate (McAuley 1986*a*), suggesting that about 60 per cent of ^{14}C, and not 30 to 40 per cent, as previously reported, is translocated. However, ^{14}C methods may lead to an underestimation of the extent of translocation for two further reasons: first, certain mobile compounds may not be labelled in short-term experiments (e.g. lipids in *S. microadriaticum*, see earlier); and second, the disequilibrium between ^{14}C and ^{12}C in tracer experiments is

Table 16.2 Translocation of symbiont photosynthate to host tissues and its contribution to host respiration in the coral *Stylophora pistillata* [(data from Muscatine *et al*. 1984)].

Coral colony	Specific growth rate[a] (μ) (day^{-1})	Net carbon assimilated ($\mu g\ C\ cm^{-2}\ day^{-1}$)	Carbon specific growth rate[b] (μ_C) (day^{-1})	T[c] (%)	CZAR[d] (%)
Light-adapted	0.013	122.55	1.36	99	128–161
Shade-adapted	0.009	25.50	0.33	97	47–59

[a]Calculated from the proportion of dividing cells (f) by the formula $\mu = \ln(1+f)/t_d$, where t_d is the duration of cell division.
[b]μ_C—net carbon assimilated per day, per unit standing stock of symbiont carbon.
[c]Percentage of photosynthetic carbon translocated to the host: $T = (\mu_C - \mu)/\mu_C$.
[d]Contribution of symbionts to host respiration: CZAR = $P_z \cdot T/R_a$, where P_z is total carbon fixed by symbionts and R_a is total carbon respired by the host. The value of CZAR varies with the photosynthetic and respiratory quotients (PQ and RQ), and the range of CZAR shown is obtained for PQ = 1.0–1.1 and RQ = 0.7–0.8.

particularly pronounced in short-term experiments (Smith 1982).

Muscatine *et al.* (1981, 1984) have developed an alternative approach to quantify the translocation of photosynthate, dependent on the difference between the absolute growth rate of the algal cells in the intact association and the carbon-specific growth rate (i.e. total net carbon increment added per day) (Table 16.2). In a study of the coral *Stylophora pistillata*, it was estimated that more than 95 per cent of photosynthetically fixed carbon was translocated to the host, values 2–3 times greater than obtained by the ^{14}C method. From this value and data on the respiratory rates of the coral, Muscatine *et al.* (1984) derived estimates of the daily contribution of translocated photosynthate to the basal respiration of the host (CZAR) (Table 16.2). For coral colonies in the shade, the symbionts provide 50–60 per cent of the host's respiratory needs for carbon, and for colonies adapted to high light intensities, CZAR is greater than 100 per cent, i.e. symbiont-derived carbon is sufficient to fuel host respiration and potentially to contribute to host growth.

This approach has been adapted for several other associations containing *S. microadriaticum* (e.g. Hoegh-Guldberg *et al.* 1986), but, to date, for no *Chlorella* symbiosis. Despite the elegance of the method, it does incorporate various untested assumptions; e.g. the daily respiration rate of the coral is calculated from night-time respiration on the assumption that respiratory rate is unaffected by the photosynthetic activity (and enhanced translocation of photosynthate) during the day. Another area of uncertainty arises from the conversion of raw data on photosynthesis and respiration, obtained in terms of oxygen flux, into carbon equivalents using photosynthetic and respiratory quotients for which no direct measurements are available.

Nitrogen recycling

Nitrogen recycling in alga–invertebrate symbioses refers to the assimilation of host nitrogenous waste products (e.g. ammonia, uric acid) by the symbionts into compounds (e.g. amino acids) which are subsequently released to and utilized by the host tissues. For many years, nitrogen recycling has been considered to play a central role in various marine associations. In particular, it has been invoked as a primary explanation for the very high productivity of coral reefs in waters that are notoriously poor in mineral nutrients, especially nitrogen (Odum 1959; Muscatine 1973). Nitrogen recycling has also been proposed to occur in the symbiosis between the intertidal acoel *Convoluta roscoffensis* and the prasinophycean alga *Tetraselmis convolutae* (Boyle and Smith 1975), enabling the host to tolerate the nitrogen-poor conditions in which it has been claimed to thrive (Keeble 1912). However, as discussed below, recent data on the symbioses in both *C. roscoffensis* and coelenterates call into question the oft-repeated certitude that the algal symbionts utilize host nitrogeneous waste products.

Convoluta roscoffensis

Crystals of uric acid are conspicuous in the central tissues of *C. roscoffensis* and other acoel species, and they are generally regarded as an important nitrogenous waste product of the animal. Boyle and Smith (1975) have proposed that uric acid is translocated to the algal symbionts, which degrade it to carbon dioxide and ammonia; the ammonia is assimilated into amino acids, which are subsequently released to the host tissues. There is considerable evidence in support of this hypothesis. *Tetraselmis convolutae*, the usual algal symbiont, and all other *Tetraselmis* species examined to date, can utilize uric acid as sole nitrogen source for growth in axenic culture (Gooday 1970; Douglas 1983). Both cultured *Tetraselmis* cells and symbiotic *C. roscoffensis* possess urate oxidase, the enzyme which catalyses the degradation of uric acid (Gooday 1970; Boyle and Smith 1975). Further, the extent of solid uric acid in juvenile *C. roscoffensis* cells (which are aposymbiotic) increases with age, but declines to low levels when the animals are infected with *Tetraselmis* (Douglas 1983). Finally, uric acid utilization in *C. roscoffensis* is photosynthesis-dependent. In one experiment, animals were maintained in culture containing the photosynthetic inhibitor, DCMU, for 28 days, and the uric acid content increased from 0.97 nmol per worm to 2.43 nmol per worm, whereas, under photosynthesizing conditions, it decresed by 25 per cent over this period (Douglas 1983).

Although these data are consistent with the view that the algal cells are responsible for uric acid degradation in symbiotic *C. roscoffensis*, the case remains unproven. Two points should be borne in mind:

1. It is not justified to conclude from the presence of solid uric acid in juvenile aposymbiotic *C. roscoffensis* and other acoel species that these animals are incapable of uric acid utilization. In various animals, uric acid is known to have important functions other than (or in addition to) its role as a waste product, e.g. control of Na^+/K^+ balance (McNabb and McNabb 1980).

2. The photosynthesis-dependence of uric acid utilization in symbiotic *C. roscoffensis* does not demonstrate *algal* utilization of uric acid. Many *Tetraselmis* species can metabolize uric acid in the dark (Douglas 1983) and, when provided with a suitable carbon source, e.g. glucose, they can utilize it as a sole nitrogen source for growth under these conditions (Douglas, unpublished work). The net accumulation of uric acid in *C. roscoffensis* maintained in DCMU (see above) probably arises from the stringent requirement of urate oxidase for oxygen (Vogels and van der Drift 1976). The activity of this enzyme would be greatly decreased under the oxygen-deficient conditions

which occur in *C. roscoffensis* tissues under non-photosynthesizing conditions (Boyle and Smith 1975). Thus, although uric acid is utilized in the *C. roscoffensis* symbiosis, the evidence that it is metabolized by the algal cells and not the host remains circumstantial.

A further reservation concerning nitrogen recycling in the *C. roscoffensis* symbiosis relates to its contribution to the host tolerance of nitrogen-poor conditions. Contrary to the view of Keeble (1912), detailed studies of Doonan and Gooday (1982) indicate that the run-off water at one site of *C. roscoffensis* (Shell Beach on Herm in the Channel Islands, UK) is relatively rich in nitrogen at 8–35 μg atom N l^{-1}, comprising inorganic nitrate and organic nitrogen in approximately equal proportions. In laboratory culture, symbiotic *C. roscoffensis* is very sensitive to conditions of low external nitrogen. In one experiment, animals incubated in nitrogen-supplemented medium for 28 days increased in length by 80 per cent, whereas those in unsupplemented seawater or seawater with nitrogen-free enrichment declined in length by 5–25 per cent (Douglas 1983).

Coelenterate symbioses

Ammonia, and not uric acid, is generally agreed to be the principal nitrogenous waste product of coelenterates and it is excreted across the entire body surface. In symbiotic species, the host waste ammonia is also believed to represent a major source of nitrogen for the cells of *S. microadriaticum*. This view arises largely from the many studies which demonstrate that experimentally produced aposymbionts release substantial amounts of ammonia, whereas individuals containing *S. microadriaticum* release little or no detectable ammonia and may even assimilate exogenously supplied ammonia (Table 16.3a). Activities of the principal enzymes of ammonia assimilation, glutamine synthetase (GS) and glutamate dehydrogenase, have also been demonstrated in these associations (Wilkerson and Muscatine 1984). These enzymes are widely assumed to be located in the algal cells.

The ammonia exchange in the freshwater symbiotic coelenterate, hydra, has recently been demonstrated to follow a similar pattern to that in marine coelenterates; aposymbiotic hydra release more than five times the amount of ammonia than symbiotic individuals under illumination (Table 16.3) (Rees 1986). In parallel with these experiments, Rees (1986) demonstrated that the activity of GS was higher in the host tissues of symbiotic animals than of aposymbionts. Further, under normal culture conditions, the uptake of ammonia by algal symbionts was negligible (Rees 1986, and personal communication). These data suggest that the difference in the ammonia assimilatory capability of the host between aposymbiotic and symbiont-containing hydra is responsible for the greater release of ammonia from the aposymbionts.

Table 16.3 Flux of ammonia in symbiotic species of coelenterates.

Symbiosis	Mean ammonia flux (nmol g^{-1} wet wt)	
	Condylactis sp.	*Cassiopeia* sp.
Marine coelenterates and *S. microadriaticum*[a]		
Animals with symbionts	− 3.6	+ 3.6
Aposymbionts	+22.9	+25.7
Freshwater hydra and *Chlorella*[b]	Ammonia release (nmol animal^{-1})	
Animals with symbionts	$+1.31 \pm 0.59$	
Aposymbionts	$+7.15 \pm 0.36$	

[a]Net release of ammonia is shown as plus (+) and uptake as minus (−). Groups of 10 animals were maintained for 5 hours under illumination. (Data from Coates and McLaughlin 1976.)
[b]Groups of 50 animals were maintained for 48 hours under a 12-hour light: 12-hour dark regime. [Data from Rees (1986) is the mean of three experiments ± standard error of the mean.]

The enhanced capacity of symbiont-containing hydra to assimilate ammonia suggests that the interpretation of low ammonia release by symbiotic marine coelenterates exclusively in terms of algal utilization of ammonia may not be valid. There is a clear-cut need for study of the capacity of marine coelenterates to assimilate ammonia.

Concluding remarks

There is a wide divergence between the current status of the two classes of nutritional interactions traditionally believed to be of general significance in alga–invertebrate symbioses. On the one hand, the translocation of photosynthetically fixed carbon is now a firmly established phenomenon, and the main questions that remain are the identity of mobile products and the extent of transport in the intact symbioses, and the degree to which nutrient release by freshly isolated symbionts corresponds to that *in vivo*. On the other hand, the evidence for nitrogen recycling remains circumstantial and there is little information on its possible significance in host tolerance of low nitrogen conditions.

It is widely believed that the hosts benefit from the nutritional interactions with their symbionts. Many hosts exhibit enhanced survival and growth over aposymbionts, especially under conditions of low food supply, presumably due to the receipt of photosynthate from their symbionts. It is more difficult to assess whether the algal symbionts benefit from the association. They grow considerably more slowly in the symbiosis than in culture (e.g. Jolley and

Smith 1978; Muscatine *et al.* 1984) and a substantial proportion of photosynthetic carbon is lost to the host; these are undoubtedly 'costs' of the symbiotic habit. However, it has been argued that symbionts have access to a better nutrient supply in symbiosis than is available to non-symbiotic algae (e.g. Droop 1963; Hinde 1986). This has been propounded, particularly with respect to their nitrogen status, on the possibly erroneous assumption that host waste nitrogenous products are freely available to the algal cells. There is little direct information on the nutritional status of algal cells in symbiosis but a recent study on the amino-acid uptake kinetics of symbiotic *Chlorella* suggest that these algae may be nitrogen-deficient (McAuley 1986*b*; see also Rees 1986). The significance of symbiosis to the algal partner and the nutritional status of algal cells in the intact association are fascinating areas for future research.

Acknowledgements

I thank Dr P.J. McAuley, Dr T.A.V. Rees and Dr J.B. Searle for valuable comments on drafts of this article. Previously unpublished results were conducted while in receipt of research grants from SERC.

References

Battey, J.F. and Patton, J.S. (1984). *Mar. Biol.* **79**, 27.
Blank, R.J. and Trench, R.K. (1986). *Taxon* **35**, 286.
Boyle, J.E. and Smith, D.C. (1975). *Proc. R. Soc. Lond.* B **189**, 121.
Cates, J. and McLaughlin, J. (1976). *J. exp. mar. Biol. Ecol.* **25**, 1.
Cernichiari, E., Muscatine, L., and Smith, D.C. (1969). *Proc. R. Soc. Lond.* B **173**, 557.
Crossland, C.J., Barnes, D.J., Cox, T., and Devereaux, M. (1980). *Mar. Biol.* **59**, 181.
de Bary, A. (1879). *Naturf. Versamm. Cassel* **L1**, 121.
Doonan, S.A. and Gooday, G.W. (1982). *Mar. Ecol.—Prog. Ser.* **8**, 69.
Douglas, A.E. (1983). *J. mar. biol. Ass. UK* **63**, 435.
Douglas, A.E. (1987). *Br. phycol. J.* **22**, 157.
Douglas, A.E. and Huss, V.A.R. (1986). *Arch. Microbiol.* **145**, 80.
Douglas, A.E. and Smith, D.C. (1983). *In Endocytobiology, endosymbiosis and cell biology*, Vol. 2. (ed. H.E.A. Schenk and W. Schwemmler) p. 631. De Gruyter, Berlin.
Douglas, A.E. and Smith, D.C. (1984). *Proc. R. Soc. Lond.* B **221**, 291.
Droop, M.R. (1963). *Symp. Soc. gen. Microbiol.* **13**, 171.
Gooday, G.W. (1970). *J. mar. biol. Ass. UK* **50**, 199.
Hinde, R. (1983). *In Endocytobiology, endosymbiosis and cell biology*, Vol. 2 (ed. H.E.A. Schenk and W. Schwemmler) p. 709. De Gruyter, Berlin.
Hinde, R. (1986). *In Proc. 6th Int'. Cong. Parasitol* (ed. M.J. Howell) p. 383. Australian Academy of Sciences, Canberra.

Hoegh-Guldberg, O., Hinde, R., and Muscatine, L. (1986). *Proc. R. Soc. Lond.* B **228**, 511.
Jolley, E. and Smith, D.C. (1978). *New Phytol.* **81**, 637.
Keeble, F. (1912). *Plant-animals: a study in symbiosis*. Cambridge University Press.
Kellogg, R.B. and Patton, J.S. (1983). *Mar. Biol.* **75**, 137.
McAuley, P.J. (1986*a*). *Limnol. Oceanogr.* **31**, 222.
McAuley, P.J. (1986*b*). *New Phytol.* **104**, 415.
McNabb, R.A. and McNabb, F.M.A. (1980). *Comp. Biochem. Physiol.* **67A**, 27.
Mews, L.K. (1980). *Proc. R. Soc. Lond.* B **209**, 377.
Mews, L.K. and Smith, D.C. (1982). *Proc. R. Soc. Lond.* B **216**, 396.
Muscatine, L. (1967). *Science, New York* **156**, 516.
Muscatine, L. (1973). In *The biology and geology of coral reefs*, Vol. 2, p. 77. Academic Press, New York.
Muscatine, L., McCloskey, L.R., and Marian, R. (1981). *Limnol. Oceanog.* **26**, 601.
Muscatine, L., Falkowski, P.G., Porter, J.W., and Dubinsky, Z. (1984). *Proc. R. Soc. Lond.* B **222**, 181.
Odum, E.P. (1959). *Fundamentals of ecology*. W.B. Saunders, Philadelphia, Penn.
Patton, J.S., Abraham, S., and Benson, A.A. (1977). *Mar. Biol.* **44**, 235.
Rees, T.A.V. (1986). *Proc. R. Soc. Lond.* B **229**, 299.
Reisser, W. (1984). *Br. phycol. J.* **19**, 309.
Smith, D.C. and Douglas, A.E. (1987). *The biology of symbiosis*. Edward Arnold, London.
Smith, D.C., Muscatine, L., and Lewis, D.H. (1969). *Biol. Rev.* **44**, 17.
Smith, R.E.H. (1982). *Mar. Biol. Lett.* **3**, 325.
Steen, R.G. (1986). *Biol. Bull.* **170**, 267.
Thorington, G. and Margulis, L. (1981). *Biol. Bull.* **160**, 175.
Trench, R.K. (1971). *Proc. R. Soc. Lond.* B **177**, 225.
Trench, R.K. (1979). *A. Rev. plant Physiol.* **30**, 485.
Vogels, G.D. and van der Drift, C. (1976). *Bact. Rev.* **40**, 403.
Wilkerson, F.P. and Muscatine, L. (1984). *Proc. R. Soc. Lond.* B **221**, 71.
Young, J.O. and Eaton, J.W. (1975). *Arch. Hydrobiol.* **75**, 225.

Poster abstracts

Extracellular organic excretion by marine phytoplankton

I.T. Marlowe, R. Kaur, A.J. Smith, and L.J. Rogers
Department of Biochemistry, University College of Wales, Aberystwyth, Dyfed SY23 3DD, UK

Axenic isolates of three marine phytoplankton were batch cultured in an artificial seawater medium (900 ml) gassed with air. In cultures of the diatom *Phaeodactylum tricornutum* Bohlin (Gottingen 1090-1b) and the coccolithophorid *Hymenomonas carterae* (Braarud and Fagerl.) Braarud (Oban 254), dissolved inorganic carbon (DIC) uptake, measured radiochemically, per unit culture volume was roughly proportional to cell density during active growth. However, the rate per cell exhibited a pronounced maximum three days after inoculation, as the cultures entered early stationary phase. The patterns of dissolved organic carbon (DOC) production were generally similar. In contrast, a 'picoplankton' eukaryote (10/01 from P. Syrett, University College Swansea) showed a declining rate of DIC uptake per cell as the culture aged, whereas DOC production reached a maximum during the stationary phase, before declining. The proportion of DOC production relative to total DIC uptake remained consistently low in *P. tricornutum* (< 1 per cent), whereas for *H. carterae* and the picoplankton, values were routinely 5–10 per cent.

Samples of cell-free medium containing acid-stable ^{14}C-labelled DOC taken from batch and continuous cultures of *H. carterae* and the picoplankton were electrodialysed using ion-exchange membranes to separate anionic, cationic, and neutral species. For comparison, labelled cell material from the same cultures was extracted to simulate the effects of cell lysis. The distribution of ^{14}C in the electrodialysis fractions of the cell material was broadly similar for the two isolates (anion, $17 \pm$ SD 4 per cent; neutral, 77 ± 6 per cent; Cation, 6 ± 3 per cent). In contrast, when the culture media were fractionated that from the picoplankton possessed considerably more radioactivity in the anionic fraction (ca. 90 per cent) than did the medium from *H. carterae* (29 ± 10 per cent). This fraction would be expected to include acidic species, such as carboxylic acids. For both isolates, the distribution of ^{14}C in the medium electrodialysis fractions is quite different to that observed for the cell material. In such cases, the comparisons provide a method for distinguishing active algal extracellular production from a release of intracellular material by cell lysis, without the need for detailed chemical characterization.

Biologically active compounds from marine microalgae

S. J. Kellam and J. M. Walker
Algae Biotechnology Unit, Division of Biological and Environmental Sciences, The Hatfield Polytechnic, PO Box 109, College Lane, Hatfield, Herts, AL10 9AB, UK

Marine organisms are known to produce a wide range of novel chemical structures (Faulkner 1986), many of these having bio-activity. Marine algae have been widely examined for antimicrobial agents (e.g. Accorinti 1983; Richelt and Borowitzka 1984) but there are relatively few reports on screening for other possible activities. Many studies have used macro-algae due to local availability but these types lack the industrial potential of microalgae.

We have screened culture filtrates and organic solvent extracts from 132 marine microalgae for the production of antibacterial, antifungal, and enzyme inhibitory activities, and, in collaboration, for insecticidal and anti-*Herpes simplex* virus activity. Of the species surveyed, 11.4 per cent exhibited an effect against at least one test fungus, 21 per cent against at least one test bacterium, and 20 per cent inhibited one or more enzymes. Thiol enzymes in the screen were the most susceptible to inhibition, and, in accord with previous studies, the growth of Gram-positive bacteria was affected far more than Gram-negative. Five methanolic extracts were found to posses cytotoxicity and one of these had anti-herpes virus activity. Eight methanolic extracts of 21 screened had insecticidal activity. Two microalgal groups would seem to warrant particular attention, these being the Bacillariophyceae (diatoms) and the Prasinophyte genus *Tetraselmis*, which produced wide spectra of activities.

These results would indicate that the microalgae may have considerable potential as a source of new pharmacologically active compounds.

Accorinti, J. (1983). *Rev. Int. oceanog. Med.* **72**, 45.
Faulkner, D. J. (1986). *Mar. Nat. Prod.* **3**, 1.
Richelt, J. L. and Borowitzka, M. A. (1984). *Hydrobiologia*, **116/117**, 158.

Biologically active products from algae of the order Conjugales

R. J. P. Cannell and J. M. Walker
Department of Biological Sciences, The Hatfield Polytechnic, PO Box 109, Hatfield, Herts AL10 9AB, UK

Microbial culture filtrates, particularly those of actinomycetes, have been extensively screened for antibiotics and enzyme inhibitors. However, no such large-scale screening programmes for metabolites with pharmacological activity have yet been applied to microalgae. We therefore screened several hundred microalgal culture filtrates and organic solvent extracts for the presence of antibiotic and enzyme inhibitory activity.

The results included the identification of biological activity from four freshwater members of the order Conjugales. Inhibition of α-glucosidase by methanol extracts of *Spirogyra varians, Zygnema cylindricum, Mesotaenium caldariorum*, and *Mougeotia* sp. was detected. The *Zygnema* extract also possessed papain inhibitory activity and the *Zygnema* and *Spirogyra* extracts possessed antibacterial activity. The α-glucosidase inhibitors from the *Spirogyra* and *Zygnema* extracts have been purified by organic solvent extraction, Sephadex LH 20 column chromatography, and high-performance liquid chromatography gel filtration.

Biosynthesis of anatoxin *a*

J. R. Gallon, K. N. Chit, and E. G. Brown
Department of Biochemistry, University College of Swansea, Swansea SA2 8PP, UK

Anatoxin *a* [1-acetyl-9-azabicyclo (4.2.1) non-2-ene] is a highly toxic alkaloid with potent neuromuscular blocking properties in mammals. Structurally it is related to the tropane alkaloids and is produced by the cyanobacterium *Anabaena flos-aquae* NRC-44-1 (Devlin *et al.* 1977). Of various radioactive compounds tested, the most effective precursor of anatoxin *a* was L-[U-^{14}C]ornithine (0.021 per cent), followed by [1,4-^{14}C]putrescine (0.016 per cent). Isotope dilution studies further suggested that putrescine was an intermediate between ornithine and anatoxin *a*.

Extracts of *A. flos-aquae* NRC-44-1 were able to decarboxylate ornithine 100-fold more effectively than could extracts of *A. flos-aquae* CCAP 1403/13f, a non-toxic strain. Furthermore, ornithine decarboxylase activity was greatest in cultures 14 days after inoculation, coincident with maximum anatoxin *a* production. It is therefore likely that ornithine decarboxylase is a specific enzyme of anatoxin *a* biosynthesis.

Ornithine decarboxylase had a pH optimum of 8 and required pyridoxal phosphate for activity. It could be partially purified (up to 67-fold) by ion-exchange chromatography on DEAE-cellulose, after which two peaks of activity were found. Both of these enzymic activities gave equimolar amounts of [^{14}C]putrescine and $^{14}CO_2$ when incubated with L-[U-^{14}C]ornithine and had similar K_m values for ornithine of 1.25 mM and 2.5 mM, respectively.

Devlin, J. O. E., Gorham, P. R., Hunter, N., Pike, R., and Stavric, B. (1977). *Can. J. Chem.* **55**, 1367.

Common properties of cyanobacterial peptide toxins

J. E. Eriksson
Department of Biology, Åbo Akademi, SF-20500 Åbo, Finland

J. A. O. Meriluto
Department of Biochemistry, Åbo Akademi, SF-20500 Åbo, Finland

K. Fagerlund and L. Hällbom
Institute of Physiological Botany, University of Uppsala, Box 540, S-75121

The cyanobacterium *Microcystis aeruginosa* has previously been reported to contain a peptide toxin with the structure cyclo(-Ala–Leu–β-methyl–Asp–Arg–Adda–Glu–*N*-methyl–dehydro–Ala) where Adda refers to a novel β-amino acid with an unsaturated side-chain (Krishnamurthy *et al.* 1986). The structural and toxicological properties of toxins recently isolated from the cyanobacteria *Nodularia spumigena* and *Oscillatoria agardhii* were compared to those of the microcystis toxin. The toxins from all three species caused similar toxic effects in mice. The main target organ was the liver in which the toxins caused severe haemorrhages and cell necrosis. Both nodularia toxin (NT) and oscillatoria toxin (OT) were shown to be peptides with low molecular weights. The amino acids common to all three species were glutamic acid, arginine, β-methyl aspartic acid, an unknown amino acid, and *N*-methyl-dehydro-alanine. The latter escapes normal amino acid analysis but the presence of this amino acid in all three toxins could be deduced from UV spectra. The unknown component in the amino acid analysis could possibly be the amino acid derivative Adda. Neither OT nor NT contained leucine but OT contained alanine. IR spectra confirmed the differences between the toxins. The common structural feature or some part of this could represent the minimum requirements for the toxicological properties of these toxins.

Krishnamurthy, T., Carmichael, W. W., and Sarver, E. W. (1986). *Toxicon* **24**, 865.

Use of monoclonal antibodies to identify gamete receptors of *Fucus serratus*

J. L. Jones, J. A. Callow, and J. R. Green
Department of Plant Biology, University of Birmingham, Birmingham B15 2TT, UK

Fertilization in the marine brown alga *Fucus* involves an intriguing example of cell–cell signalling. Motile sperm, attracted by a pheromone, swarm around the naked egg and form a loose association with it (Evans *et al.* 1982). Following a species-specific recognition stage the egg plasma membrane fuses with that of one of the sperm cells (Callow 1985). Antisera raised to

Fucus serratus sperm flagellar antigens inhibited fertilization in a species-specific manner (Vithanage *et al.* 1983). Progress with the use of monoclonal antibodies (McAbs) to attempt to pinpoint the sperm antigens involved in egg–sperm recognition in *F. serratus* is reported.

McAbs have been raised against live *F. serratus* sperm and screened using ELISA and indirect immuno-fluorescence. Twelve McAbs have been selected on the basis of gamete-specific and species-specific binding and preferential binding to localized areas of cells (e.g. flagella). These McAbs are being used to screen for species-specific inhibition of fertilization in order to detect which are directed against antigens that form part of the sperm-bound egg recognition site.

Callow, J.A. (1985). *J. cell Sci. Suppl.* **2**, 219.
Evans, L.V., Callow, J.A., and Callow, M.E. (1982). *Prog. phycol. Res.* **1**, 67.
Vithanage, H.I.M.V., Catt, J.W., Callow, J.A., Callow, M.E., and Evans, L.V. (1983). *J. cell Sci.* **60**, 103.

Restriction mapping of five cyanophage

I. Bancroft and R. J. Smith
University of Lancaster, Department of Biological Sciences, Bailrigg, Lancaster LA1 4YQ, UK

Restriction maps of the genomic DNA of five cyanophage which infect the common host *Anabaena variabilis* have been constructed. Native DNA preparations followed improved techniques developed in our laboratory. Cloned DNA was prepared by standard techniques using Hind III and EcoRI fragments ligated into pBR322. The use of cloned DNA was particularly important in the preparation of a map of A-4L in which the native DNA preparation provides exceptionally poor yields due to the sensitivity of this cyanophage to Na^+ and the propensity of its DNA for binding to cell debris.

The similar restriction endonuclease cleavage patterns of two pairs of cyanophage (AN-13/AN-23 and A-1L/AN-10) suggest they are related. Neither pair show any relationship with A-4L or with each other. There is evidence in both related pairs of small insertions and deletions which have caused slight shifts in the positions of common cleavage sites as shown by the alignment of their maps. AN-10 and A-1L differ in one region of their maps which may reflect the event initiating their divergence. AN-10 and A-1L possess circular genomes of 67kb; AN-23 and AN-13 possess linear genomes of 46 kb, and A-4L possesses a linear genome of 40 kb.

The data provide a means of discriminating these cyanophage and provide data on their relatedness and evolution.

Sequence counter-selection in cyanophage

I. Bancroft and R. J. Smith
University of Lancaster, Department of Biological Sciences, Bailrigg, Lancaster LA1 4YQ, UK

An analysis of the cleavage of native and cloned DNA of five cyanophage which infect *Anabaena* 7120 by 34 restriction endonucleases provided evidence of sequence counter-selection similar to that present in T_7 (Kruger and Schroeder 1981). Seventeen restriction enzymes failed to cleave the DNA of all of the cyanophage. One group are isoschizomers of *Anabaena* 7120 endogenous restriction endonucleases. Another group contain the subsequence GATC; and a third group contain dicytosine residues. The fourth group have no common sequence structure and may represent isoschizomers of restriction endonucleases present in the host range of the five cyanophage.

Groups 1 and 4 represent counter-selection against endonuclease cleavage. Group 2 represent counter-selection against methylation by a dam-like methylase, and group 3 counter-selection against methylation by a cytosine methylase. The counter-selection in groups 2 and 3 confirm the presence of adenine and cytosine methylation by *Anabaena* species (Lambert and Carr 1984). Cyanophages AN-23, AN-13, and A-4L do not tolerate sequence methylation. A-1L and AN-10 tolerate adenine methylation, but differ in their tolerance of cytosine methylation. AN-10 appears able to prevent cytosine methylation by host enzymes. AN-10 and A-1L are closely related. Comparison of their restriction maps shows that counter-selection of some Hae III sites in AN-10 has occurred since divergence of the phage from their common ancestor.

Kruger, D.H. and Schroeder, C. (1981). *Microbiol. Rev.* **45**, 9.
Lambert, G.R. and Carr, N.G. (1984). *Biochim. Biophys. Acta* **781**, 45.

Part 6: Biotechnology

17 The biotechnology of microalgae and cyanobacteria

N.W. KERBY and W.D.P. STEWART

AFRC Research Group on Cyanobacteria and Department of Biological Sciences, University of Dundee, Dundee DD1 4HN, UK

Introduction

The algae are a diverse group of photosynthetic non-vascular plants which contain chlorophyll *a* and include unicellular and multicellular microscopic forms as well as large multicellular forms (such as 'seaweeds'). Cyanobacteria (blue–green algae) have been traditionally considered as algae but due to their prokaryotic nature they more closely resemble bacteria. Microalgae and cyanobacteria are found in a wide range of habitats including aquatic, terrestial, and extreme environments.

Photosynthesis is a unique means of utilizing CO_2, H_2O, and solar energy for the primary production of organic compounds. Microalgae are among the most prolific producers of plant biomass, having photosynthetic efficiencies broadly similar to those of higher plants (Aaronson and Dubinsky 1982). Theoretically the large-scale culture of microalgae is simpler and cheaper than that of bacteria and fungi since they do not require organic substrates. Consideration of microalgae as a source of food and chemicals began in the early 1940s, and in 1952 the first Algal Mass-culture Symposium was held (Burlew 1953). However, since that time very few commercial ventures have exploited microalgae although many potential uses for microalgal biomass have been discussed (see Shelef and Soeder 1980). The lack of exploitation may be due, in part at least, to the complexity of the systems required for cultivation. The commercial development of microalgae is largely dependent both on a sustained supply of adequate light and on the identification of high value products which will justify the necessary investment. Recently there has been renewed interest in the potential of microalgae and cyanobacteria as producers of high value metabolites (Borowitzka 1986; Curtin 1985; Gallon 1985; Klausner 1985, 1986; Lem and Glick 1985; Tucker 1985). This review discusses some recent developments in the applied potential of microalgae, emphasizing particularly our work at Dundee on the photoproduction of nitrogenous compounds by N_2-fixing cyanobacteria.

Microagal biomass

Microalgae and, in particular, cyanobacteria have served as a source of food in various parts of the world including Mexico, China, and Africa (Dangeard 1940; Farrar 1966). In the past the main focus has been on production of single cell protein (SCP) (Litchfield 1983) but other potential applications, which require the mass cultivation of microalgae, are becoming apparent. These include waste-water treatment, the production of commercial chemicals and fuels, the bioconversion of solar energy, and the cosmetic and health-food market.

Algae are currently grown either in batch culture in tanks or semi-continuously in ponds (Benemann *et al.* 1979; Goldman 1979; Litchfield 1983; Terry and Raymond 1985). The mass culture of algae in outdoor pond systems requires extensive land use, depends upon sunlight and climate, has large requirements for nutrients and water, and is threatened by contamination and invasion by weeds and herbivores. Species best suited to open cultivation are those which have rapid growth rates or which can withstand environmental extremes which their competitors cannot withstand e.g. high pH, high salinity, or high temperature. Species of *Chlorella, Scenedesmus, Spirulina*, and *Dunaliella* are, at present, so cultivated (see Borowitzka 1986; Goldman 1979; Terry and Raymond 1985). Because many algae require a temperature of above 20°C (eukaryotes) or 30°C (prokaryotes) for rapid growth, cultivation is particularly suited to tropical or sub tropical regions. However, the feasibility of microalgal treatment of agricultural wastes in temperate climates has been demonstrated (Fallowfield and Garrett 1985). Heating may be practical only in a few very specialized areas such as in the vicinity of power plants where waste heat may be utilized. During algal mass culture the CO_2 concentration in air is inadequate for growth (see Benemann *et al.* 1979) and additional CO_2 must be supplied from sources such as carbonates, bicarbonates, natural deposits, combustion gas, or from the decomposition of organic matter in sewage or industrial waste. Good mixing is required to prevent settling of the algae, thermal stratification, depletion of surface nutrients, and to ensure adequate light. Due to the dilute nature of algal cultures one of the major limiting economic steps is harvesting. Harvesting may be accomplished by centrifugation which is expensive, by chemical flocculation with lime or alum, or by sieving through fine mesh screens (25 μM), which is only suitable for large cells, colonial, or filamentous forms. Autoflocculation may be an economical method of harvesting. This phenomenon is associated with high pH, due to photosynthetic CO_2 depletion, and to the precipitation of minerals in the media which complex with the negatively charged surface of the cell (Sukenik and Shelef 1984). However, for practical use such a system requires further development and would require adjustment of pH and of calcium and orthophosphate concentrations.

Microalgal biomass productivities in large-scale cultures are estimated by Benemann *et al.* (1984) to be in the range of 60–100 t ha^{-1} yr^{-1}. In addition, they concluded that the production of fuels from microalgal and cyanobacterial biomass by fermentation could be considered to be feasible only when the microalgal biomass was derived from waste-water treatment. To compete with other sources of SCP and biomass, further development is required, including strain improvement to facilitate harvesting (e.g. buoyancy and large colony formation), genetic engineering of strains for desired attributes (e.g. increased content of amino acids or other high value components), and increasing the volumetric productivity by overcoming light limitation which is often due to self-shading.

The use of fermenters for culturing photosynthetic micro-organisms would minimize the problem of contamination and allow the control of environmental parameters. However, enclosed reactors are restricted in volume and, therefore, before their commercial exploitation with microalgae is economically feasible, suitable high value products will have to be identified. Enclosed systems at present include vertical glass columns, covered troughs, and rigid and collapsible tubular reactors (for a review, see Lee 1986).

Immobilization

The immobilization of microbial cells can be defined as a physical confinement or localisation of a microorganism that permits its economic reuse (Abbott 1977). Immobilization can confer a number of advantages, including: (1) physical stabilization of cells; (2) biomass retention, allowing the use of flow rates far in excess of those needed for maximum growth; (3) limitation of cell growth which may direct metabolism towards stationary phase activities, or prevent overgrowth of a desired genetically altered strain by revertants; (4) the use of higher biomass concentrations; (5) ease of product separation, obviating the need for separating cells from the product; (6) making possible multienzyme reactions, thus avoiding some of the disadvantages normally associated with enzyme immobilization (e.g. necessity for enzyme purification, loss of activity of isolated enzyme, and restriction to single step processes). A requirement for the immobilization of photosynthetic organisms is that the matrix must not limit the supply of light to the cells and, to be effective, the process of immobilization must not cause irreversible structural or physiological damage to the cells.

Immobilized algal cells have many potential applications including the following: biocatalysts for biotransformations and biosyntheses; waste-water treatment through bio-accumulation of, for example, nitrogen, phosphorus, and heavy metals; the production of energy; use in co-immobilized systems to supply O_2 and/or reductant. This subject has been reviewed recently by Robinson *et al.* (1986). Methods of immobilization have included entrapment in natural polymers such as agar, carageenans, and alginate or in synthetic polymers such as polyacrylamide, serum albumin cross-linked with gluteraldehyde, and polyurethane foam. In general, entrapment in natural polymers results in high cell viability. Alternatively, cells may be cultured in the presence of a matrix or support which can minimize contamination. Such techniques include invasion of preformed polyurethane and polyvinyl foams and adsorption onto glass beads.

Biotransformations by immobilized algae include the production of sulphated polysaccharides by *Porphyridium* (Gudin *et al.* 1984), the photoproduction of NH_4^+ and amino acids by N_2-fixing cyanobacteria (see Kerby *et al.* 1983, 1986*a*, 1987*a*), glycollate production from *Chlorella* (Day and Codd 1985), glycerol production from *Dunaliella* (Grizeau and Navarro 1986), hydrocarbon production from *Botryococcus* (Bailliez *et al.* 1985), and release of sugars and amino acids from immobilized *Synechococcus* and *Synechocystis* induced by osmotic shock (Reed *et al.* 1986).

Considerable interest has been shown in biological H_2 production and several reviews have dealt with its potential in cyanobacteria (Lambert and Smith 1981, Bothe 1982; Hallenbeck 1983). H_2 production by immobilized cyanobacteria has been demonstrated (Lambert *et al.* 1979; Smith and Lambert 1981; Rao *et al.* 1982; Muallem *et al.* 1983; Karube *et al.* 1986).

Immobilized algae have been employed in waste-water treatment; for example, immobilized *Scenedesmus quadricauda* efficiently removed nitrogen and phosphorus from urban secondary effluent (Chevalier and de la Noue 1985*a*, 1985*b*). Additionally, selective recovery of gold and other metal ions has been achieved using *Chlorella vulgaris* immobilized in polyacrylamide (Darnall *et al.* 1986). Using a packed column, an elution scheme was demonstrated for the binding and selective recovery of copper, zinc, gold, and mercury from an equimolar mixture.

Photoproduction of nitrogenous compounds

Photoproduction of ammonia

Free-living N_2-fixing cyanobacteria have been used as sources of biofertilizer in rice fields (De 1939; Watanabe 1956; Singh 1961; Stewart *et al.* 1979, Stewart 1980) and recently their potential in temperate agricultural soils has been identified (see Jenkinson 1977). Biological N_2 fixation has interesting prospects for biotechnological applications. Cyanotech (Woodinville, USA) is reported to be setting up a commercial facility for the development of strains of N_2-fixing cyanobacteria as agricultural fertilizer (Tucker 1985).

N_2-fixing cyanobacteria convert atmospheric N_2 to NH_4^+ which is assimilated by the primary NH_4^+-assimilating enzyme glutamine synthetase (GS).

Stewart and Rowell (1975) devised a mechanism whereby over 90 per cent of the fixed N_2 was released extracellularly as NH_4^+ when GS was inhibited by the glutamate analogue L-methionine-D,L-sulphoximine (MSX). Using this method Musgrave *et al.* (1982, 1983*a*) demonstrated the technical feasibility of continuous sustained NH_4^+ production by *Anabaena* ATCC 27893 immobilized in beads of Ca-alginate gel. A variety of continuous flow reactors of different configurations were investigated, including packed beds, fluidized beds, parallel plates, and air-lift reactors (Musgrave *et al.* 1982, 1983*b*; Kerby *et al.* 1983). Air-lift reactors were found to be the most suitable for the laboratory scale optimization of NH_4^+ photoproduction (Musgrave, unpublished thesis, 1985) due to their effective mixing, low shear forces, and high gaseous efficiencies. MSX was applied either continuously or intermittently to bioreactors and specific rates of NH_4^+ production were achieved of up to 40 μmol. mg chl $a^{-1}.h^{-1}$ for over 800 hours (Kerby *et al.* 1983, 1986*a*, Musgrave *et al.* 1983*b*; Musgrave, unpublished thesis, 1985). These rates compared favourably with those obtained for the photoproduction of NH_4^+ from nitrate by free-living *Anacystis nidulans* (Ramos *et al.* 1982*a*, 1982*b*) and from free-living N_2-fixing *Anabaena* ATCC 33047 (Ramos *et al.* 1984). Important factors which determine the duration of NH_4^+ production and the productivity of the system include: (1) biomass loading of the Ca-alginate beads; (2) MSX concentration and rate of supply (the optimum dose is that which allows the biomass concentration to remain constant, whilst pulses of MSX rather than a continuous dose prolong the duration of NH_4^+ release); (3) nitrogen limitation which induces high nitrogenase activity; (4) dilution rate which affects volumetric productivity and duration of production; and (5) environmental factors such as pH, light, and temperature (see Musgrave, unpublished thesis, 1985).

Other groups have recently used MSX to promote NH_4^+ production by immobilized cyanobacteria (Hall *et al.* 1985; Brouers and Hall 1986; Jeanfils and Loudeche 1986) with similar results to those obtained previously by this laboratory. However, the use of MSX has several disadvantages. It is difficult to sustain prolonged NH_4^+ production in bioreactors in its presence (see Musgrave, unpublished thesis, 1985), it is toxic and would have to be separated from the extracellular NH_4^+ prior to use of the latter as a biofertilizer, and it is very expensive relative to the cost of NH_4^+.

The alternative approach is to select mutant strains which are deficient or partly deficient in NH_4^+ assimilation. Such strains have been obtained by selecting for resistance to the NH_4^+ analogue ethylenediamine (EDA) (Polukhina *et al.* 1982, Kerby *et al.* 1986*a*). EDA is metabolized via GS to a glutamine analogue (aminoethylglutamine) which accumulates in the cell (Kerby *et al.* 1985). Unlike methylammonium, which is used as an NH_4^+ analogue in NH_4^+ transport studies (see Rai *et al.* 1984; Kerby *et al.* 1986*b*, 1987*b*), EDA is transported by passive diffusion in response to a pH gradient. Thus when EDA-resistant mutant strains are obtained at high pH values they generally prove to be GS-deficient and not (unwanted) transport mutants.

The EDA-resistant mutant strains ED81 and ED92 have decreased growth rates when grown either under N_2-fixing conditions or in the presence of NH_4^+, but show enhanced growth rates in the presence of glutamine (Sakhurieva *et al.* 1982). Nitrogenase activity was higher than in the parent strain and was derepressed in the presence of NH_4^+. The GS activity of ED92 was much decreased as measured by both the biosynthetic and transferase assay methods. However, the transferase activity of ED81 was comparable to that of the parent strain even though the biosynthetic activity was much decreased. ED92 had 25 per cent of the GS protein of the parent strain, as estimated by immuno-electrophoresis, whereas ED81 had a similar content to the parent strain (Kerby *et al.* 1986*a*). This implies that ED92 may be a regulatory mutant with less GS protein whilst ED81 may be a structural mutant. Photoproduction of NH_4^+ occurs at rates comparable to those obtained using MSX and is sustained for periods at least to 600 hours (Kerby *et al.* 1986*a*).

NH_4^+-liberating mutant strains have also been obtained by selection for resistance to MSX (Singh *et al.* 1983; Spiller *et al.* 1986). MSX-resistant strains which liberate NH_4^+, like those resistant to EDA, have derepressed nitrogenase, low levels of GS biosynthetic activity, and the ratio of biosynthetic activity to transferase activity may be decreased as compared to the parent strain (Spiller *et al.* 1986). Such strains enhanced the growth of rice plants grown in nitrogen-free medium, both the dry weight and nitrogen content of rice plants being increased by their presence, but not by the presence of the parent strain (Latorre *et al.* 1986; Kerby *et al.*, unpublished work). Because of their slow growth rates, GS-deficient strains, although suitable for NH_4^+-production studies in the laboratory, would be unlikely to compete in the field, except in certain specialized areas. However, one such area is the production of NH_4^+-liberating panels which are being evaluated for use in irrigation channels. It is also important to develop strains with elevated levels of N_2 fixation, including those with depressed nitrogenase which continue to fix N_2 in the presence of combined nitrogen. We have selected strains of *Anabaena variabilis* resistant to methylammonium, which show impaired NH_4^+ transport and which possess a nitrogenase which is less sensitive to exogenous NH_4^+ than is that of the parent strain (Kerby *et al.* unpublished work). The development of N_2-fixing strains for use in particular ecosystems is discussed by Stewart *et al.* (1987).

Photoproduction of amino acids

Studies on the photoproduction of NH_4^+ from air, light, and water by immobilized mutant strains of N_2-fixing cyanobacteria provided the fundamental basis for the production of other higher value metabolites. Amino acid liberation was obtained with mutant strains of *A. variabilis* selected for resistance to the tryptophan analogue 6-fluorotryptophan (FT) or the methionine analogue ethionine (ETH) (Kerby *et al.* 1987*a*). Some of these strains had increased heterocyst frequency and nitrogenase activity. Certain strains (e.g. FT2) liberated mainly the aromatic amino acids phenylalanine and tyrosine whereas others (e.g. FT7) liberated a broad range of amino acids. Free-pools of particular amino acids can be increased by selection for resistance to amino-acid analogues, for example mutant strains ETH9 and ETH10 had increased free-pools of methionine. Continuous photoproduction of amino acids was achieved by such strains immobilized in beads of Ca-alginate gel (Kerby *et al.* 1987*a*). FT-resistant mutant strains were found either to lack the phenylalanine-sensitive isoenzyme of 3-deoxy-D-arabino-heptulosonate-7-phosphate synthase (DAHP synthase) or to have a phenylalanine-insensitive DAHP synthase. The tyrosine-sensitive enzyme was unaffected (Niven *et al.* 1988). DAHP synthase is the first enzyme of the aromatic amino-acid pathway and is a key point of control (see Jensen and Hall 1982). However, such strains did not liberate tryptophan as the biosynthesis of this amino acid is tightly controlled by feedback inhibition of anthranilate synthase. Nevertheless, tryptophan liberation was observed in the presence of the detergent MYRJ45 (Niven *et al.* 1988) though this was possibly a result of uncoupling of specific amino acid transport systems in the plasmalemma, as has been shown for glutamate production by *Corynebacterium glutamicum* (Clement *et al.* 1984; Clement and Laneelle 1986). Hall *et al.* (1980) isolated amino-acid-producing mutants of *Anabaena* sp. 29151, *Synechocystis* sp. 29108, and *Synechococcus* 602 resistant to various amino acid analogues. Riccardi *et al.* (1981*a*, 1981*b*) have also demonstrated amino-acid overproduction by mutant strains of *Spirulina platensis*. In these studies amino acid overproduction was demonstrated by the ability of the mutant cyanobacterial strains to support the growth of heterotrophic bacteria which were auxotrophic for a particular amino acid. Future developments may include the possibility of improving the nutritional status of *S. platensis* by compensating for the low methionine content through the use of mutants which have high internal concentrations of this amino acid (Cifferi 1983).

Low molecular weight metabolites, including amino acids, may be released from free-living and immobilized cyanobacteria following osmotic shock (Reed *et al.* 1986). Transfer of *Synechocystis* 6714 and *Synechococcus* 6311 from a medium of high salt to one of low salt resulted in a transitory loss of plasmalemma integrity and release of organic compounds. This method could be repeated and after three cycles there was no evidence of long-term damage to the cells. Since this method does not involve harvesting or death of cells, it is a useful means of recovering metabolites from viable microorganisms and has been evaluated by the Ethyl Corporation at Baton Rouge, Louisiana, USA (Leavitt 1985) for the production of proline from *Chlorella* sp.; here proline accumulates as a compatiable solute in response to osmotic stress.

Other compounds and processes of potential value

Pigments

Cyanobacteria and microalgae are currently being evaluated for the production of other biological compounds, including pigments such as phycoerythrin, which has potential as a fluorescent probe (Glazer and Stryer 1984). Another biliprotein, phycocyanin, is being extracted from *S. platensis* for the health-food and cosmetic market in Japan. β-carotene, a vitamin A precursor which has uses as a food colourant, has attracted much attention. The commercial potential of *Dunaliella salina* for β-carotene and glycerol production has been well established (Williams *et al.* 1978; Ben-Amotz and Avron 1980; Ben-Amotz *et al.* 1982; Chen and Chi 1981; Borowitzka *et al.* 1984). At high salinities (250–320 g l^{-1}) *D. salina* contains high levels of β-carotene (approximately 10 per cent of the dry weight). Furthermore, *D. salina* has the advantage of remaining the dominant alga during outdoor mass culture and does not suffer readily from predation. Full-scale production of β-carotene from *Dunaliella* is expected to be under way during 1988 (Borowitzka 1986). A range of other carotenoids such as astaxanthin and canthaxathin are produced in significant quantities by certain species of *Euglena, Haematococcus*, and *Chlorella* and are also of commercial interest.

Toxins and cytostatic agents

Certain cyanobacteria, of the family Oscillatoriaceae, cause dermatitis in man and animals and promote tumours in laboratory animals. Debromoaplysiatoxin (a phenolic substance), along with its bromine-containing analogue bromoaplysiatoxin, and lyngbyatoxin A (an indole alkaloid with the same structure as teleocidin A) have been isolated from lipophilic extracts of *Lyngbya majuscula* and have all been reported as tumour promoters in mice (Fujiki *et al.* 1981, Fujiki and Sugimura 1983). However, both debromoaplysiatoxin and lyngbyatoxin A have also been shown to have activity against P-338 lymphocytic mouse leukaemia cells at sublethal doses (Mynderse *et al.* 1977). *Tolypothrix brysoidea* produces the cytotoxin tubercidin which has significant antitumour activity (Barchi *et al.* 1983). Tubercidin has previously been isolated from *Streptomyces tubercidus* (Suzuki and Marumo 1960) and has been used in human cancer therapy for the treatment of cutaneous neoplasms (Bisel *et al.* 1970; Grage *et al.* 1970). Moore (1982) has reviewed the properties and structures of anticancer agents and tumour promoters from marine cyanobacteria. An acidic polysaccharide extracted from *Chlorella pyrenoidosa* is claimed to show cytostatic activity against mouse leukaemia P-338 and S-180 tumours; additionally it induced interferon production in isolated spleen cells (Kitasato Institute 1985). Cystostatic activity has also been demonstrated for a polysaccharide, spirulinan, from *Spirulina subsala* (Seiko-Epson 1986) by a process which has been patented (Shinohara 1986). Similar activity has been obtained for an α-amylase hydrolysate of a *Chlorella* glycoprotein (Chlorella 1986).

Animal cell growth stimulants

Both microalgal and cyanobacterial extracts have been reported to stimulate the growth of cultured animal cells and to decrease or eliminate the requirement for animal sera, such as calf foetal serum, in the media. Hot water extracts of *Chlorella, Scenedesmus*, and *Spirulina* decrease the requirement for calf foetal serum from 10 per cent to 1 per cent (v/v), give accelerated growth rates and allow normal successive cultivation of animal cells (Chlorella 1984). A dialysate from *Synechococcus elongatus* added to basal medium enhanced the growth of three human cell lines to a greater extent than that obtained by addition of a mixture of insulin, transferrin, ethanolamine, and selenite (Shinohara 1985; Shinohara *et al*. 1986). We have tested these claims by examining the effects of hot water extracts of *S. platensis* on the growth of human hair cell fibroblasts (N.W. Kerby *et al*. unpublished work). Addition of freeze-dried extracts (10 $\mu g\ ml^{-1}$) to a basal medium substantially stimulated growth in the presence of 1 per cent (v/v) calf foetal serum. Higher concentrations were inhibitory and this was attributed to changes in solute concentration. Several advantages are derived from the use of low serum or serum-free media in the culture of animal cells, apart from the obvious economies. These include: (1) improved reproducibility of culture media; (2) reduced risk of contamination; (3) easier purification of culture products; (4) less protein interference in bio-assays; (5) avoidance of cytotoxicity; (6) prevention of fibroblast overgrowth in primary cultures. The algal compound(s) responsible for the stimulation of animal cell growth are currently being investigated.

Lipids and fatty acids

Microalgal lipids have been considered for commercial uses and as a possible source of renewable energy, especially when microalgal growth is coupled to waste-water treatment (Aaronson *et al*. 1980; Shifrin and Chrisholm 1980; Heller 1983). The colonial green alga *Botryococcus braunii* has received much attention because of its ability to accumulate up to 80 per cent of its dry weight as hydrocarbons, but its slow growth may make this organism unsuitable for mass culture (see Bachofen 1982; Wolf 1983). Eicosapentaenoic acid (EPA) is an unsaturated fatty acid and is important as a precursor in prostaglandin biosynthesis in mammals. EPA may decrease the concentration of cholesterol in the blood, lowering the risk of heart disease and atherosclerosis (Horrobin 1982), and is present in quantity in fish oil. However, the fish do not synthesise EPA but themselves acquire it from the algae they consume (Klausner 1986). By extracting EPA directly from algae, other undesirable attributes of fish oil could be avoided. EPA-rich algae are mainly marine species such as a marine *Chlorella* sp. in which 33 per cent of the fatty acids is EPA (Nisshin 1984). High levels of EPA have been found in a freshwater Xanthophyte *Monodus subterraneus* (Iwamoto and Sato 1986). γ-linolenic acid (GLA), which is also a precursor of prostaglandins, is a major component of fatty acids in certain microalgae. For example, *S. platensis* contains up to 11 per cent of its dry weight as lipid, of which 18 per cent is typically GLA (Materassi *et al.* 1980). Arachidonic acid, another essential fatty acid, is produced in significant quantities by certain microalgae (Nichols and Appleby 1969), including *Porphyridium cruentum* which may be a good potential source (Vonshak *et al.* 1985).

Antibiotics

Antibiotic compounds produced by microalgae are discussed by Metting and Pyne (1986) and Jones (1988, this volume, Chapter 14). As yet, algal antibiotics are mostly unidentified but are thought to include fatty acids, other organic acids, phenolic substances including bromophenols, tannins, terpenoids, polysaccharides, and other carbohydrates, and alcohols (Aubert *et al.* 1979). Such compounds are produced by a diverse range of microalgae and have been shown to be toxic to other microalgae, bacteria, fungi, viruses, and protozoans (see Metting and Pyne 1986). An algicide, cyanobacterin, produced by the cyanobacterium *Scytonema hofmanii* has been chemically characterized as a chlorine-containing γ-lactone (Pignatello *et al.* 1983; Gleason and Porwoll 1986). Cyanobacterin is highly toxic to cyanobacteria and algae (Mason *et al.* 1982) and has been shown to inhibit photosystem II electron transport both in cyanobacteria (Gleason and Paulson 1984) and in pea chloroplasts (Gleason and Case 1986). Screening studies have recently revealed that a marine cyanobacterium *Lyngbya aestuarii* possesses appreciable herbicide activity against *Lemna* (Entzeroth *et al.* 1985); the active compound was identified as 2, 5-dimethyldodecanoic acid. Majusculamide C, a depsipeptide from *L. majuscula*, controls the growth of fungal plant pathogens (Carter *et al.* 1984).

Plant growth stimulants

Filamentous cyanobacteria have also been shown to produce compounds that stimulate growth of plants. The following cyanobacteria or cyanobacterial extracts have been reported to stimulate plant growth: *Cylindrospermum muscicola* (Venkataraman and Neelakantan 1967); *Calothrix anomola* (Dadhich *et al.* 1969); *Anabaena* sp., *Nostoc* sp., *Oscillatoria* sp., *Plectomema* sp., and *Nodularia* sp. (Rodgers *et al.* 1979). Plant growth stimulants have also been reported in *Chlorella* sp. (Davis and Bigler 1973), and *Scenedesmus obliquus* stimulated the growth of *Rhizobium japonicum* (Fingerhut *et al.* 1984). To date, the characterization of any plant growth stimulant extracted from microalgae is lacking and it is hard to ascertain whether stimulation was caused by a particular plant growth regulator or solely by the nutritive value of such extracts. Additionally, cyanobacteria are important not only for their ability to fix N_2 but also for their soil-conditioning properties; for example, the presence of a thick mucilagenous sheath increases the water-holding capacity of soils.

Biochemicals

ATP can be generated photosynthetically using a living, thermophilic *Mastigocladus* sp. (Sawa *et al.* 1982). This system obviates the need for chromatophore preparation and immobilization for photosynthetic generation of ATP. The ATP-regenerating system has been employed to demonstrate photosynthetic glutathione production by living cells of *Phormidium lapideum* when glutamate, cysteine, and glycine were supplied (Sawa *et al.* 1986). Recently, biotransformations of pharmaceutical interest, such as the transformation of steriods, have been reported for microalgae (Abul-Hajj and Qian 1986). For example, certain microalgae, including cyanobacteria, transformed 4-androstenedione to testosterone with overall yields up to 80 per cent after seven days' incubation. Testosterone proved to be the major metabolite in most of the cultures, which were unable to perform the reverse reaction of oxidation of the 17β-hydroxyl group.

Restriction endonucleases

The cyanobacteria are a rich source of type II restriction endonucleases and some of these are currently marketed. Both filamentous forms (*Anabaena, Nostoc*) and unicellular forms (*Aphanothece, Dactylococcopsis, Synechocystis*, and *Anacystis* R2) display these activities (Murray *et al.* 1976; Reaston *et al.* 1982; Whitehead and Brown 1982, 1985; Calleja *et al.* 1985; Gallagher and Burke 1985).

Concluding remarks

The commercial potential of microalgae and cyanobacteria has yet to be fully realized and, indeed, is as yet scarcely tapped. However, there is currently considerable renewed interest in algae as potential sources of commercial products. Areas of particular importance include: further screening of strains to identify high value products; the purification and characterization of biologically active compounds which have been implicated in cell processes but not yet identified; strain improvement, first for rapid growth rates, second for the ability to withstand environmental extremes, and third for enhancement of synthesis of high value compounds; techniques for out door mass culture and harvesting; and fermentation technology for phototrophs. While potential markets for algal products exist these will require extensive development in the future.

Acknowledgements

This work was supported by the AFRC and we would like to thank Dr P. Rowell, Gordon Niven, and Hilary Powell for helpful discussions.

References

Aaronson, S. and Dubinsky, Z. (1982). Experientia **38**, 36.

Aaronson, S., Berner, T., and Dubinsky, Z. (1980). In *Algae biomass* (ed. G. Shelef and C. J. Soeder) p. 575. Elsevier, Amsterdam.

Abbott, B. J. (1977). *A. Rep. ferment. Process.* **1**, 205.

Abul-Hajj, Y. and Qian, X. (1986). *J. Natural Prod.* **49**, 244.

Aubert, M., Aubert, J., and Guathier, M. (1979). In *Marine algae in pharmaceutical science* (ed. H. A. Hoppe, T. Levring, and Y. Tanaka) p. 267. De Gruyter, Berlin.

Bachofen, R. (1982). Experientia **38**, 47.

Bailliez C., Largeau, C., and Casadevall, E. (1985). *Appl. Microbiol. Biotechnol.* **23**, 99.

Barchi, J. J., Norton, T. R., Furusawa, E., Patterson, G. M. L., and Moore, R. E. (1983). *Phytochemistry* **22**, 2851.

Ben-Amotz, A. and Avron, M. (1980). In *Algae biomass* (ed. G. Shelef and C. J. Soeder) p. 603. Elsevier, Amsterdam.

Ben-Amotz, A., Katz, A., and Avron, M. (1982). *J. Phycol.* **18**, 529.

Benemann, J. R., Weissman, J. C., and Oswald, W. J. (1979). In *Microbial biomass* (Economic Microbiology, Vol. 4) (ed. A. H. Rose) p. 177. Academic Press, London.

Benemann, J. R., Augenstein, D. C., Goebel, R., and Weissman, J. C. (1984). In *Energy from biomass and wastes viii* (ed. B. W. Feingold and L. Courtney) p. 133. Institute of Gas Technology, USA.

Bisel, H. F., Ansfield, F. J., Mason, J. H., and Wilson, W. L. (1970). *Cancer Res.* **30**, 76.

Borowitzka, L. J., Borowitzka, M. A., and Moulton, T. P. (1984). *Hydrobiologia* **116/117**, 115.

Borowitzka, M. A. (1986). *Microbiol. Sci.* **3**, 372.

Bothe, H. (1982). *Experientia* **38**, 59.

Brouers, M. and Hall, D. O. (1986). *J. Biotech.* **3**, 307.

Burlew, J. S. (1953). In *Algal culture from laboratory to pilot plant* (ed. J. S. Burlew) p. 3. Carnegie Institute of Washington, Washington, DC.

Calleja, F., Tandeau de Marsac, N., Coursin, T., van Ormondt, H., and de Waard, A. (1985). *Nucleic Acids Res.* **13**, 6745.

Carter, D.C., Moore, R.E., Mynderse, J.S., Niemczura, W.P., and Todd, J.S. (1984). *J. org. Chem.* **49**, 236.
Cehn, B.J. and Chi, C.H. (1981). *Biotech. Bioeng.* **23**, 1267.
Chevalier, P. and de la Noue, J. (1985*a*). *Enzyme Microb. Technol.* **7**, 621.
Chevalier, P. and de la Noue, J. (1985*b*). *Biotech. Lett.* **6**, 395.
Chlorella, (1984). US 4468-460: 15.06.81-JP-091733 (28.08.84) 21.05.82 as 380847. 83-00398K/01.
Chlorella, (1986). J6 1069-278: 13.09.84-JP-192043 (10.04.86) 13.09.84 as 192043. 86-133811/21.
Cifferi, O. (1983). *Microbiol. Rev.* **47**, 551.
Clement, Y. and Laneelle, G. (1986). *J. gen. Microbiol.* **132**, 925.
Clement, Y., Escoffier, B., Trombe, M.C., and Laneelle G. (1984). *J. gen Microbiol.* **130**, 2589.
Curtin, M.E. (1985). *Biotechnology* **3**, 34.
Dadhich, K.S., Varma, A.K., and Venkataraman, G.S. (1969). *Plant and Soil* **31**, 377.
Dangeard, P. (1940). *Actes Soc. Linn. Boréaux Extr. Procès-verbaux* **91**, 39.
Darnall, D.W., Greene, B., Henzl, M.T., Hosea, J.M., McPherson, R.A., and Sneddon J. (1986). *Environ. Sci. Technol.* **20**, 206.
Davis, C.H. and Bigler, E.R. (1973). *Agron. J.* **65**, 462.
Day, J.G. and Codd, G.A. (1985). *Biotech. Lett.* **7**, 573.
De, P.K. (1939). *Proc. R. Soc. Lond.* B **127**, 121.
Entzeroth, M., Mead, D.J., Patterson, G.M.L., and Moore, R.E. (1985). *Phytochemistry* **24**, 2875.
Fallowfield, H.J. and Garrett, M.K. (1985). *Agric. Wastes* **12**, 111.
Farrar, W.V. (1966). *Nature, Lond.* **211**, 341.
Fingerhut, U., Webb, L.E., and Soeder, C.J. (1984). *J. appl. Microbiol. Biotech.* **19**, 358.
Fujiki, H. and Sugimura, T. (1983). *Cancer Surveys* **2**, 552.
Fujiki, H., Mori, M., Nakayasu, M., Terada, M., Sugimura, T., and Moore, R.E. (1981). *Proc. natn. Acad. Sci. USA* **78**, 3872.
Gallagher, M.L. and Burke, W.F. (1985). *FEMS Microbiol. Lett.* **26**, 317.
Gallon, J.R. (1985). *Industrial Biotechnology* **5**, 44.
Glazer, A.N. and Stryer, L. (1984). *Trends biochem. Sci.* **9**, 423.
Gleason, F.K. and Case, D.E. (1986). *Plant Physiol.* **80**, 834.
Gleason, F.K. and Paulson, J.L. (1984). *Arch. Microbiol.* **138**, 273.
Gleason, F.K. And Porwoll, J. (1986). *J. org. Chem.* **51**, 1615.
Goldman, J.C. (1979). *Water Research* **13**, 1.
Grage, T.B., Rochlin, F.B., Weiss, A.J., and Wilson, W.L. (1970). *Cancer Res.* **30**, 79.
Grizeau, D. and Navarro, J.M. (1986). *Biotech. Lett.* **8**, 261.
Gudin, C., Bernard, A., Chaumont, D., Thepenier, C., and Hardy, T. (1984). In *Biotech '84*, p. 541. Online Conference Publications, Northwood.
Hall, D.O., Affolter, D.A., Brouers, M., Shi, D.J., Wang, L.W., and Rao, K.K. (1985). *A. Proc. Phytochem. Soc. Eur.* **26**, 161.
Hall, G.C., Flick, M.B., and Jensen, R.A. (1980). *J. Bact.* **143**, 981.
Hallenbeck, P.C. (1983). *Enzyme microb. Technol.* **5**, 171.
Heller, M.J. (1983). In *Energy from biomass and wastes vii* (ed. B.W. Feingold and L. Courtney) p. 1067. Institute of Gas Technology, USA.

Horrobin, D.F. (1982). In *Clinical uses of essential fatty acids* (ed. D.F. Horrobin) p. 1. Eden Press, London.
Iwamoto, H. and Sato, S. (1986). *J. Am. Oil Chem. Soc.* **63**, 434.
Jeanfils, J. and Loudeche R. (1986). *Biotech. Lett.* **4**, 265.
Jenkinson, D.S. (1977). In *Rothamsted Experimental Station Annual Report for 1976*, p. 103.
Jensen, R.A. and Hall, G. (1982). *Trends biochem. Sci.* **7**, 177.
Jones, A.K. (1988). In *Biochemistry of the algae and cyanobacteria*, (ed. L.J. Rogers and J.R. Gallon), p. 257. Clarendon Press, Oxford.
Karube, I., Ikemoto, H., Kajiwara, K., Tamiya, E., and Matsuoka, H. (1986). *J. Biotech.* **4**, 73.
Kerby, N.W., Rowell, P., and Stewart, W.D.P. (1985). *Arch. Microbiol.* **141**, 244.
Kerby, N.W., Rowell, P., and Stewart, W.D.P. (1986*b*). *Arch. Microbiol.* **143**, 353.
Kerby, N.W., Rowell, P., and Stewart, W.D.P. (1987*b*). *N.Z. J. mar. freshwater Res.* **21**, 447.
Kerby, N.W., Niven, G.W., Rowell, P., and Stewart, W.D.P. (1987*a*). *Appl. Microbiol. Biotech.* **25**, 547.
Kerby, N.W., Musgrave, S.C., Codd, G.A., Rowell, P., and Stewart, W.D.P. (1983). In *Biotech. '83*, p. 541. Online Conference Publications, Northwood, Middlesex.
Kerby, N.W., Musgrave, S.C., Rowell, P., Shestakov, S.V., and Stewart, W.D.P. (1986*a*). *Appl. Microbiol. Biotech.* **24**, 42.
Kitasato Institute (1985). US 4533-548: 02.12.81-JP-192899 (06.08.85) 15.11.82 as 441630. 83-25080K/14.
Klausner, A. (1985). *Biotechnology* **3**, 27.
Klausner, A. (1986). *Biotechnology* **4**, 947.
Lambert, G.R. and Smith, G.D. (1981). *Biol. Rev.* **56**, 589.
Lambert, G.R., Daday, A., and Smith, G.D. (1979). *FEBS Lett.* **101**, 125.
Latorre, C., Lee, J.H., Spiller, H., and Shanmugam, K.T. (1986). *Biotech. Lett.* **8**, 507.
Leavitt, R. (1985). *Abstr. Pap. Am. Chem. Soc.* **190**, 167.
Lee, Y.K. (1986). *Trends Biotech.* **4**, 186.
Lem, N.W. and Glick. B.R. (1985). *Biotech. Adv.* **3**, 195.
Litchfield, J.H. (1983). *Science* **219**, 740.
Mason, C.P., Edwards, K.R., Carlson, R.E., Pignatello, J., Gleason, F.K., and Wood, J.M. (1982). *Science* **215**, 400.
Materassi, R. Paoletti, C., Balloni, W., and Florenzano, G. (1980). In Algae biomass (ed. G. Shelef and C.J. Soeder) p. 619. Elsevier, Amsterdam.
Metting, B. and Pyne, J.W. (1986). *Enzyme microb. Technol.* **8**, 386.
Moore, R.E. (1982). *Pure appl. Chem.* **54**, 1919.
Muallem, A., Bruce, D., and Hall, D.D. (1983). *Biotech. Lett.* **5**, 365.
Murray, K., Hughes, S.G., Brown, J.S., and Bruce, S.A. (1976). *Biochem. J.* **159**, 317.
Musgrave, S.C. (1985). The technological use of cyanobacteria. Unpublished Ph.D. thesis, University of Dundee, UK.
Musgrave, S.C., Kerby, N.W., Codd, G.A., and Stewart, W.D.P. (1982). *Biotech. Lett.* **4**, 647.

Musgrave, S.C., Kerby, N.W., Codd, G.A., and Stewart, W.D.P. (1983*a*). *Eur. J. appl. Microbiol. Biotechnol.* **17**, 133.
Musgrave, S.C., Kerby, N.W., Codd, G.A., Rowell, P., and Stewart, W.D.P. (1983*b*). *Process Biochem.* (Suppl.), p. 184.
Mynderse, J.S., Moore, R.E., Kashiwagi, M., and Norton, T.R. (1977). *Science* **196**, 538.
Nichols, B.W. and Appleby, R.S. (1969). *Phytochemistry* **8**, 1907.
Nisshin (1984). J5 9196-086 21.04.83-JP-069205 (07.11.84) 21.04.83 as 069205. 84-314745/51.
Niven, G.W., Kerby, N.W., Rowell, P., and Stewart, W.D.P. (1988). *J. gen. Microbiol.*, **134**, 689.
Niven, G.W., Kerby, N.W., Rowell, P., and Stewart, W.D.P. (1988). *Arch. Microbiol.* (in press).
Pignatello, J.J., Porwoll, J., Carlson, R.E., Xavier, R.E., Gleason, F.K., and Wood, J.M. (1983). *J. org. Chem.* **48**, 4035.
Polukhina, L.E., Sakhurieva, G.N., and Shestakov, S.V. (1982). *Microbiology* **51**, 90.
Rai, A.N., Rowell, P., and Stewart, W.D.P. (1984). *Arch. Microbiol.* **137**, 241.
Ramos, J.L., Guerrero, M.G., and Losada, M. (1982*a*). *Biochim. Biophys. Acta* **679**, 323.
Ramos, J.L., Guerrero, M.G., and Losada, M. (1982*b*). *Appl. environ. Microbiol.* **44**, 1013.
Ramos, J.L., Guerrero, M.G., and Losada, M. (1984). *Appl. environ. Microbiol.* **48**, 114.
Rao, K.K., Muallem, A., Bruce, D.L., Smith, G.D., and Hall, D.O. (1982). *Biochem. Soc. Trans.* **10**, 527.
Reaston, J., Duyvesteyn, M.G.C., and de Waard, A. (1982). *Gene* **20**, 103.
Reed, R.H., Warr, S.R.C., Kerby, N.W., and Stewart, W.D.P. (1986). *Enzyme microb. Technol.* **8**, 101.
Riccardi, G., Sora, S., and Cifferi, O. (1981*a*). *J. Bact.* **147**, 1002.
Riccardi, G., Sanangelantoni, A.M., Carbonera, A., Savi, A., and Cifferi, O. (1981*b*). *FEMS Microbiol. Lett.* **12**, 333.
Robinson, P.K., Mak, A.L., and Trevan, M.D. (1986). *Process Biochem.* **11**, 122.
Rodgers, G.A., Bergman, B., and Henriksson, E. (1979). *Plant Soil* **52**, 99.
Sakhurieva, G.N., Polukhina, L.E., and Shestakov, S.E. (1982). *Microbiology* **51**, 258.
Sawa, Y., Kanayama, K., and Ochiai, H. (1982). *Biotech. Bioeng.* **24**, 305.
Sawa, Y., Shindo, H., Nishimura, S., and Ochiai, H. (1986). *Agric. Biol. Chem.* **50**, 1361.
Seiko-Epson (1986). J6 1158-926: 04.01.85-JP-000003 (18.07.86) 04.01.85 as 000003. 86-228982/35.
Shelef, G. and Soeder, C.J. (ed.) (1980). *Algae biomass.* Elsevier, Amsterdam.
Shifrin, N.S. and Chrisholm, S.W. (1980). In *Algae biomass.* (ed. G. Shelef and C.J. Soeder) p. 627. Elsevier, Amsterdam.
Shinohara, K. (1985). J6 0186-280: 07.03.84-JP-041996 (21.09.85) 07.04.84 as 041996. 85-273642/44.
Shinohara, K. (1986). J6 1031-095: 24.07.84-JP-152147 (13.02.86) 24.07.84 as 152147. 86-084802/13.

Shinohara, K., Okura, Y., Koyano, T., Murakami, H., Kim, E-H, and Omura, H. (1986). *Agric. Biol. Chem.* **50**, 2225.
Singh, H. N., Singh, R. K., and Sharma, R. (1983). *FEBS Lett.* **154**, 10.
Singh, R. N. (1961). *Role of blue-green algae in nitrogen economy of Indian agriculture*. Indian Council of Agricultural Research, New Delhi.
Smith, G. D. and Lambert, G. R. (1981). *Biotech. Bioeng.* **23**, 213.
Spiller, H., Latorre, C., Hassan, M. E., and Shanmugam, K. T. (1986). *J. Bact.* **165**, 412.
Stewart, W. D. P. (1980). In *Methods of evaluating biological N_2-fixation* (ed. F. J. Bergersen) p. 583. John Wiley, Chichester, Sussex.
Stewart, W. D. P. and Rowell, P. (1975). *Biochem. Biophys. Res. Commun.* **65**, 846.
Stewart, W. D. P., Rowell, P., Kerby, N. W., and Machray, G. C. (1987). *Phil. Trans. R. Soc. Lond.* B **317**, 245.
Stewart, W. D. P., Rowell, P., Ladha, J. K., and Sampaio, M. J. A. M. (1979). *Proc. Nitrogen and Rice Symposium*, IRRI, Manilla, p. 263.
Sukenik, A. and Shelef, G. (1984). *Biotech. Bioeng.* **26**, 142.
Suzuki, S. and Marumo, S. (1960). *J. Antibiot.* **13**, 360.
Terry, K. L. and Raymond, L. P. (1985). *Enzyme microb. Technol.* **7**, 474.
Tucker, J. B. (1985). *High Technol.*, February issue, p. 34.
Venkataraman, G. S. and Neelakantan, S. J. (1967). *J. gen. Microbiol.* **13**, 53.
Vonshak, A., Cohen, Z., and Richmond, A. (1985). *Biomass* **8**, 13.
Watanabe, A. (1956). *Bot. Mag.* **69**, 530.
Whitehead, P. R. and Brown, N. L. (1982). *FEBS Lett.* **143**, 296.
Whitehead, P. R. and Brown, N. L. (1985). *J. gen. Microbiol.* **131**, 951.
Williams, L. A., Foo, E. L. Foo, A. D., Kuhn, I., and Heden, C. J. (1978). *Biotech. Bioeng. Symp.* **8**, 115.
Wolf, F. R. (1983). *Appl. Biochem. Biotech.* **8**, 249.

18 Seaweed biotechnology—current status and future prospects

L.V. EVANS and D.M. BUTLER

Department of Pure and Applied Biology, University of Leeds, Leeds LS2 9JT, UK

Introduction

Advances in molecular biology and allied technologies over the past 10 years open up exciting new possibilities for the future with respect to exploitation of biological systems. Progress has been most rapid in micro-organisms and in mammalian systems, but work on plant systems is also progressing although the pace has been rather slower. Nevertheless, progress in the tissue culture and genetic engineering of plants now makes possible (1) large-scale and rapid propagation of genetically uniform plants from élite stock, (2) the

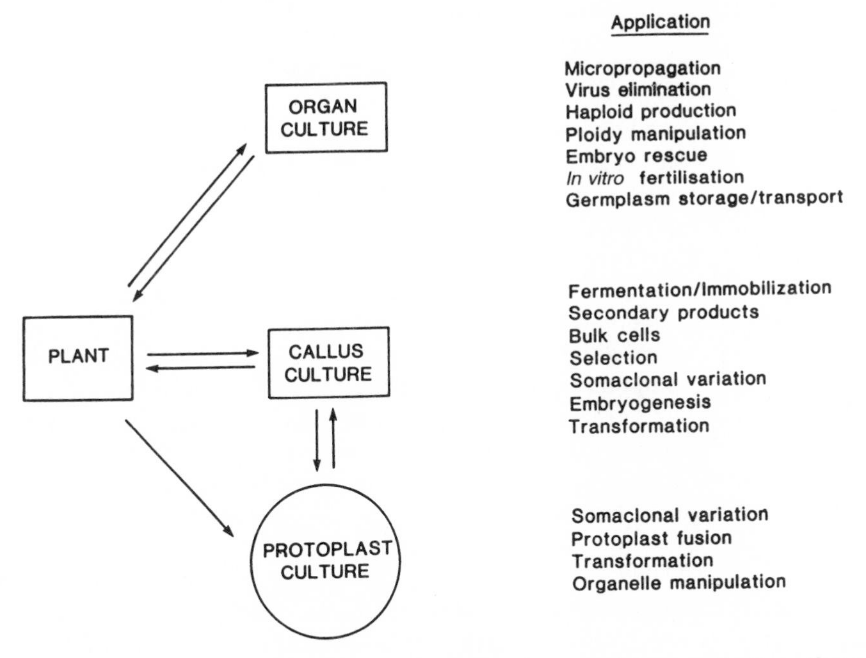

Fig. 18.1. Organizational levels in cultured plant tissues and their applications (from Jones 1985).

selection of novel and improved varieties using somaclonal variation techniques, (3) the development of new hybrids between different cultivars and species by protoplast fusion and cell culture techniques, and (4) the use of recombinant DNA technology to introduce new genetic material into plant cells (Ammirato *et al.* 1984). Excised plant tissues can be manipulated at different organizational levels (see Fig. 18.1) and the potential applications depend on which of these is utilized. In organ culture, tissue organization is maintained, whilst callus culture involves a disorganized callus phase. Except in the case of fermentation, secondary product formation, and bulk cell production, the aim is to regenerate whole plants from cultured tissues after manipulation (for further information, see e.g., Jones 1985; Butler and Evans 1988).

Not surprisingly, advances in macroalgal biotechnology lag far behind because of a low awareness of seaweeds and their commercial products, and because there are substantial gaps in our knowledge of the biochemistry and physiology of these plants. Before looking at the extent to which such techniques have been applied to macroalgae and the degree of success achieved, it is necessary to briefly consider the commercial applications and products involved.

Commercial applications and products

The most important use of seaweeds is as food. *Porphyra* (nori), *Laminaria* (kombu), and *Undaria* (wakame) are extensively cultivated for human consumption in the Far East, and seaweeds have a worldwide food value in excess of US$1 billion (Abbott and Cheney 1982). The second most important use of seaweeds is as a source of gel-forming polysaccharides (phycocolloids), notably agar and carrageenan from red seaweeds and alginic acid from brown algae. Seaweed phycocolloids are widely used in industry as emulsifying agents, gelling agents, stabilizers, thickeners, and suspension agents and have a combined annual market value in excess of US$250 million (Ryther 1984). Seaweeds are also a potential source of unique and pharmacologically useful compounds. Additional applications of seaweeds include use as animal food and fertilizers and extensive research is also being carried out aimed at commercially viable large-scale cultivation of seaweeds as a digestion feedstock (biomass) for energy (methane) production. For a recent review of seaweed bioproducts and their uses, see Evans (1986).

Agar and carrageenan

Agar and carrageenan are galactans consisting entirely of linear chains of galactose units which may be masked by modification or by substitution. Agar contains alternating 1,3-linked D-galactose residues and 1,4-linked L-galactose residues, but in carrageenan all the units are D-galactose. As well as differing in D- and L-galactose content the two types also differ in the extent to which the galactose residues are modified to the 3,6-anhydroderivative, and in the extent and position of sulphation and methylation (Percival 1978, 1979).

Agar Agar is made commercially from a number of red algal species of which the most important are *Gelidium* and *Gracilaria*. It can be fractionated into two components, *viz*. agarose (the gel-forming component) and agar-opectin (a mixture of variously sulphated molecules). Agarose [Fig. 18.2(a)] consists of chains of alternating 1,3-linked D-galactose residues (A units) and 1,4-linked 3,6-anhydro-L-galactose (B units). Agarose is either not sulphated or low in sulphate. Soluble in boiling water, agar forms a gel when cooled to 30 °C to 40 °C, and the more 3,6-anhydrogalactose residues present and the fewer sulphate and other charged groups, the better the gelling properties.

Because of the limited supply, its great importance as a bacterio logical medium, and its increasing use in biotechnology and in cell culture work of all kinds, agar and agar-derived agarose are the highest priced of the seaweed phycocolloids. Food-grade and industrial grade agar sell at US$8.75–9.25 per pound and top quality bacteriological grade agar at US$11–14 per pound

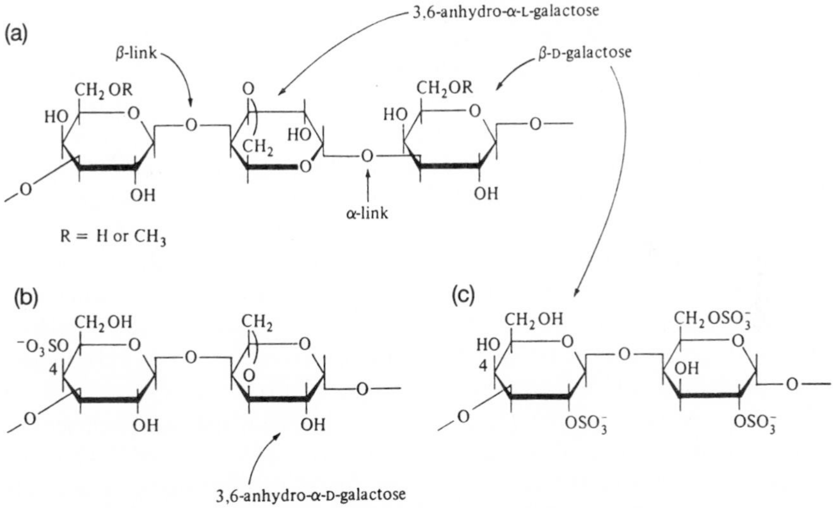

Fig. 18.2. (a) Chemical structure of agarose: 1.3 linked β-D-galactose (A unit) and 1,4 linked 3,6-anhydro-α-L-galactose (B unit). (b) Chemical structure of kappa-carrageenan: 1,3 linked β-D-galactose 4-sulphate (A unit) and 1,4 linked 3,6-anhydro-α-D-galactose (B unit). (c) Chemical structure of lambda-carrageenan: 1,3 linked β-D-galactose 2-sulphate (A unit) and 1,4 linked α-D-galactose 2,6-disulphate. (From Lobban *et al.* 1985.)

(Sandford and Baird 1983; Cheney 1984). Agarose is also used in immunodiffusion and electrophoretic separations and some specialized forms cost US$1600 per pound.

Carrageenan Carrageenans are mainly extracted from the red algae *Chondrus, Gigartina*, and *Eucheuma*. Carrageenan-type polysaccharides differ chemically from the agar type in that 3,6-anhydro-D-galactose [Fig. 18.2(b) and (c)] takes the place of the anhydro-L-galactose in agar, and they have a higher content (about 24 per cent) of ester sulphate. There are two main types of carrageenan, *viz*. kappa- and lambda-carrageenan, and the proportions of these (like agarose and agaropectin) vary in different seaweeds and, in the case of carrageenan, with season and habitat. In kappa-carrageenan [Fig. 18.2(b)] the repeating disaccharide units are D-galactose 4-sulphate (A units) and 3,6-anhydro-D-galactose (B units). In lambda-carrageenan [Fig. 18.2(c)] the A units are D-galactose 2-sulphate and the B units D-galactose 2,6-disulphate.

Kappa-carrageenan forms double helices during sol–gel transformation; lambda-carrageenan does not do this and does not form gels. Commercial carrageenan produces high viscosity solutions and gels and reacts with proteins such as casein in milk and so is widely used in the food industry as a thickening and stabilizing agent. Kappa-carrageenan is also used in biotechnology, e.g. in cell immobilization. Carrageenan sells for about US$3 per pound (Sandford and Baird 1983).

Alginic acid

Alginic acid (alginate) is extracted from brown algae, e.g. *Laminaria* and *Macrocystis*, where it accounts for 14–40 per cent of the dry weight. It is a linear polysaccharide composed of D-mannuronic and L-guluronic acids (Fig. 18.3) in varying proportions. These are arranged along the molecule in three types of blocks: mannuronic acid (M–M) blocks, guluronic acid (G–G) blocks, and mixed (M–G) blocks (Painter 1983). Alginate occurs within the plant as a mixed salt, the amount of bound calcium determining gel strength. Guluronic acid-rich alginates characteristic of older tissue have a much higher affinity for calcium ions than mannuronic acid-rich alginate typical of young tissues.

Alginate, unlike agar and carrageenan, forms gels not on cooling but by controlled addition of calcium ions. Alginate has extremely wide-ranging food and industrial uses, based upon its capacity to retain water, form gels, and stabilize emulsions. Calcium alginate is also used for cell encapsulation in bioconversion and immobilization systems (Painter 1983). Of the three commercially important seaweed phycocolloids, alginate is the lowest in price, selling at some US$1.86–3.49 per pound (Sandford and Baird 1983).

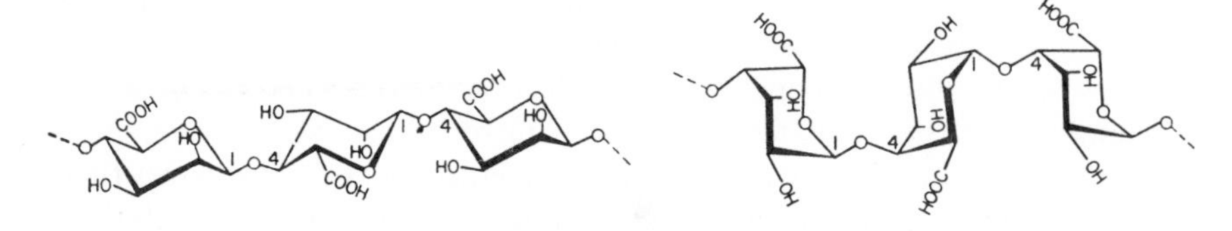

Fig. 18.3. Segment of polymannuronic acid chain (left) and polyguluronic acid chain (right) from alginic acid.

Pharmaceuticals

There is increasing evidence that seaweeds are a potential source of low-yield, high-cost pharmacologically useful compounds. Some of these are listed in Table 18.1.

Table 18.1 Pharmacologically useful compounds from seaweeds.

Compound	Source	Use
Fucosterol[1]	Phaeophyta	Base for sexual hormone synthesis/steroids
Sterols	*Fucus gardneri* *Sargassum muticum*	Reduction in blood cholesterol levels
Caulerpin Caulerpicin	*Caulerpa* spp.	Mild anaesthetics
Pachydictyol	*Pachydictyon coriaceum*	Fungicidal/bacteriocidal activity
Isozonarol/Zonarol	*Dictyopteris zonaroides*	Fungicidal/bacteriocidal activity
Squalene	*Fucus vesiculosus*	Fungicidal/bacteriocidal activity
Stypoldione[2]	*Stypopodium zonale*	Antitumour properties (inhibition of microtubule assembly)

[1]From Hoppe (1979); [2]from O'Brien *et al.* (1984). Rest of table based on data from Chapman (1979).

Research needs

Seaweeds are clearly an important resource with considerable potential. Since the demand for seaweed products is increasing annually it is necessary to carry out basic research on economically valuable species, aimed at creating genetically improved and novel strains, with increased yields and the capacity for producing new substances. This can be done to a certain extent by existing selective breeding methods but the possibility of exploiting the potential offered by new technologies such as somatic hybridization/somaclonal variation and recombinant DNA techniques must be kept in mind for the longer term. In order to do this it is essential to perfect basic techniques for protoplast isolation and fusion as well as axenic culture techniques based on haploid and diploid tissues. Hybrid selection studies can then be carried out, looking, for example, for strains showing resistance to disease or, in the longer term, for improved quality (or new) products. It is also necessary to be able to maintain actively dividing axenic callus tissue for possible production of secondary products and to regenerate plants from undifferentiated callus cells as well as from manipulated/fused protoplasts. Cell culturing techniques must also be developed to allow rapid and large-scale cloning of desirable new strains (which are likely to be sterile) and regeneration of new plants from these at will. To what extent have these objectives been achieved so far with respect to seaweeds?

Current status of seaweed biotechnology

Protoplast isolation

The starting point for application of the new technology to seaweeds is successful isolation of viable protoplasts from diploid and haploid tissues under aseptic conditions. Protoplast fusion gives a unique opportunity to produce new hybrids between related but sexually incompatible or sterile species. These somatic hybrids will have a combined genome and a novel cytoplasmic combination. The latter is known as cybridization (cytoplasmic transfer using protoplast fusion) and is a method of obtaining new cytoplasmic gene combinations (i.e. chloroplastic and mitochondrial) in one step. Protoplasts are also necessary for co-cultivation in the transformation of foreign or modified DNA into cells.

Vegetative tissues of at least some representatives of the green, brown, and red seaweeds have been dissociated into single cells which can be regenerated into complete plants (Polne-Fuller *et al.* 1986), and protoplasts have been isolated from at least 15 species of green, brown, and red seaweeds (Cheney 1986), although the yield is not high. Some success has also been achieved with fusing protoplasts, but regeneration from these has not been achieved (Saga *et al.* 1986).

In the case of the green seaweed *Enteromorpha intestinalis* enzymic digestion of the walls released viable protoplasts (see Fig. 18.4) which evolved oxygen in the light at a rate similar to that of vegetative cells. Some remained viable for several days and limited wall regeneration occurred (Millner *et al.* 1979). More recently, regeneration of protoplasts (and single cells) of *Enteromorpha* and *Ulva* into new thalli has been reported (Polne-Fuller *et al.* 1986; Polne-Fuller and Gibor 1987).

Viable protoplasts have been obtained from only two genera of red seaweeds so far, *viz. Porphyra* and *Gracilaria*. Four species of *Porhyra* have

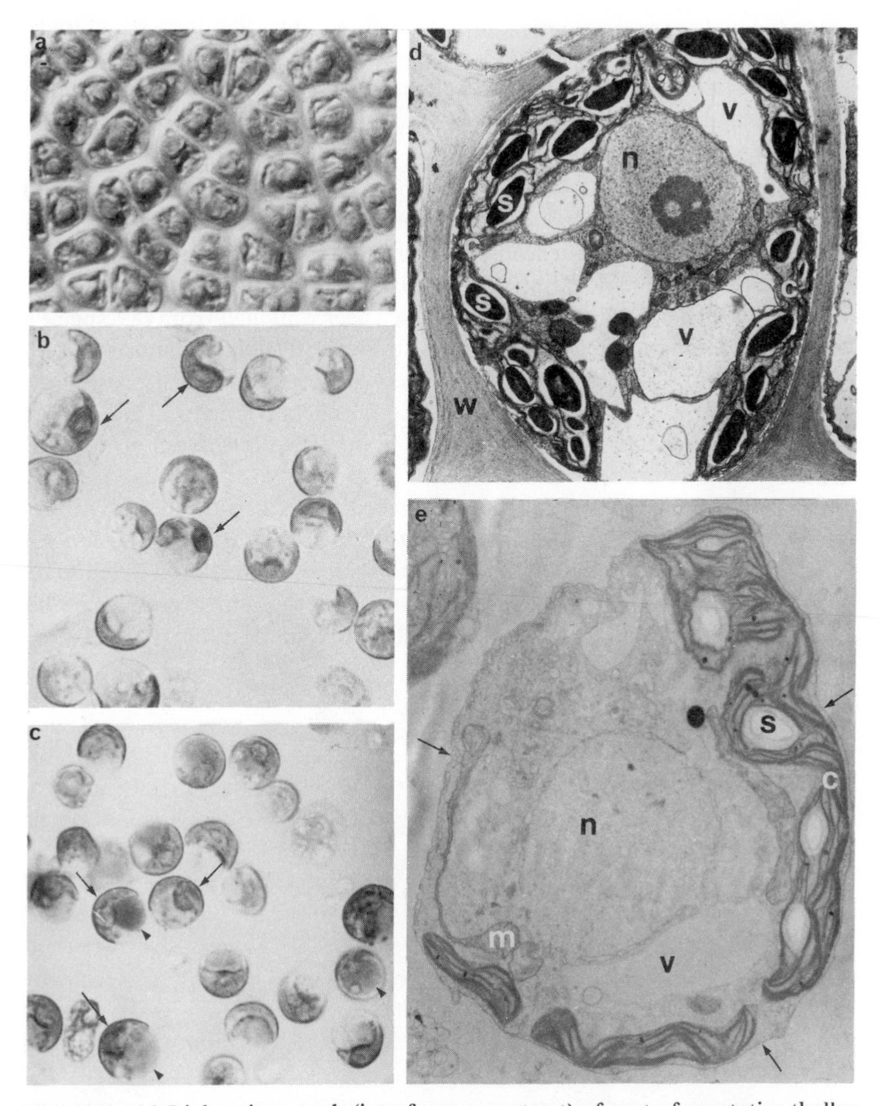

Fig. 18.4. (a) Light micrograph (interference contrast) of part of vegetative thallus of *Enteromorpha intestinalis* (× 625), (b) Light micrograph of isolated protoplasts showing apparent absence of cell walls. Arrows indicate the chloroplasts (× 625), (c) Same preparation as in (b) after neutral red staining. Vacuoles of viable protoplasts are stained red (arrowheads); also visible are the chloroplasts (arrows) (× 625), (d) Electron micrograph of vegetative cell surrounded be cell wall (w) and showing the nucleus (n), chloroplast (c) with starch grains (s), and vacuoles (v) (× 5 000), (e) Electron micrograph of isolated protoplast. Note absence of cell wall. The protoplast is bounded by a plasma membrane (arrows). Organelles as in (d) are also visible (× 5 500). (From Millner *et al.* 1979.)

yielded viable protoplasts which have proved capable of regeneration. They are *P. yezoensis* (Zhao and Zhan 1981; Saga and Sakai 1984; Fujita and Migita 1985), *P. suborbiculata* (Tang 1982), *P. perforata* (Polne-Fuller and Gibor 1984; Polne-Fuller *et al.* 1984; Saga *et al.* 1986; Polne-Fuller *et al.* 1986), and *P. miniata* (Chen 1986). Single cells and protoplasts from enzymically dissociated thalli of *Porphyra perforata* (Polne-Fuller *et al.* 1986) and *P. miniata* (Chen 1986) grew into normal plants. The growth pattern of regenerating cells was influenced by culture conditions, and, in the case of *P. perforata*, the original location of the cells on the blade. In the agar-producing genus *Gracilaria* viable protoplasts have been obtained from two species, *viz. G. tikvahiae* and *G. lemaneiformis*, following wall degradation (Cheney *et al.* 1986). Protoplast fusions were observed. Regeneration of protoplasts also occurred, followed by cell division, to give masses of up to 32 cells. More recently, whole plants of *Gracilaria* have been regenerated from protoplasts (Cheney, pers. comm.).

Isolation of protoplasts has been reported from at least six brown seaweeds, *viz. Laminaria japonica* (Saga 1984; Saga and Sakai 1984), *Macrocystis pyrifera* and *Sargassum muticum* (Saga *et al.* 1986; Polne-Fuller *et al.* 1986), *Sphacelaria* (Ducreux and Kloareg 1988), and *Fucus distichus* (zygotes) (Kloareg and Quatrano 1987). Recently, in the authors' laboratory

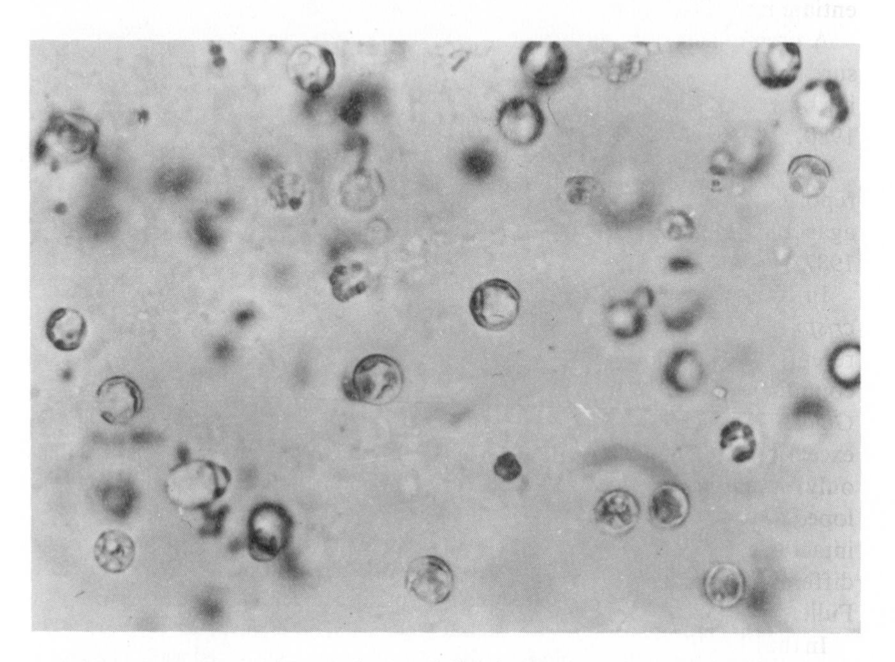

Fig. 18.5. Protoplasts of *L. saccharina*, following enzymic degradation of the thallus (× 550).

protoplasts have also been isolated from *Laminaria saccharina* (Fig. 18.5). Protoplasts (and single cells) from *Sargassum* spp. divided into small aggregates but did not regenerate into new plants, although erect shoots did develop from (pigmented) calluses (Polne-Fuller *et al.* 1986; Polne-Fuller and Gibor 1987). Apical cell protoplasts of *Sphacelaria*, however, regenerated into whole plants (Ducreux and Kloareg 1988).

Continued development of protoplast-derived cell masses into whole plants has therefore only been achieved in relatively few cases so far in the more complex red and brown seaweeds. Cheney *et al.* (1986) suggest this may be the result of a shortage of morphologically competent cells in the donor tissue or a lack of essential morphological substances in the (axenic) culture medium; the question of the nutritional requirements of seaweed tissue in culture will be considered later. Also, although there has been some success in achieving protoplast fusion (e.g. Zhang 1983), nuclear fusion followed by hybrid regeneration has not so far been convincingly demonstrated.

Callus formation

Formation of callus (undifferentiated cell masses on tissue explants) is an important phenomenon in higher plant culture. Calluses are used as a source of vegetative cells (or sometimes their products) or can be induced to differentiate into whole plants.

A prerequisite for callus formation is axenic tissue explants. Short exposure to betadine before culturing appears to be the best method for eliminating bacteria, preferably in conjunction with ultrasonic treatment (Polne *et al.* 1980; Gibor *et al.* 1981).

In green algae such as *Enteromorpha* and *Ulva*, calluses have been reported to develop regularly from isolated cells or protoplasts cultured on agar, but more rarely from sections of intact tissue (Polne-Fuller and Gibor 1987). Calluses had the capacity to regenerate into differentiated thalli.

In the red algae regeneration of medullary tissue explants of *Chondrus crispus* into new fronds was observed by Chen and Taylor in 1978. More recently, callus formation has been reported in *Agardhiella* (Cheney, pers. commun.) and at low frequencies from cortical cells of *Gracilaria, Gelidium, Gigartina, Prionitis*, and *Eucheuma* (Polne-Fuller and Gibor 1987). With the exception of *Eucheuma*, which could be cultured successfully, these survived only on their original tissue. In *Porphyra*, as in the green algae, calluses developed most readily from protoplasts or single cells on agar, rather than from intact tissue. These divided and regenerated new calluses on agar media, or differentiated into fronds in liquid media (Polne-Fuller *et al.* 1984; Polne-Fuller and Gibor 1987).

In the brown algae, Saga *et al.* (1978) first reported callus-like development from *Laminaria angustata* tissue, and single cells from this formed new sporophytes. Fries (1980) reported regeneration of small sporophytes (and meiospores) from axenic callus-like tissue formed from meristematic zone explants of *Laminaria digitata* and *L. hyperborea*. However, it is likely that these were derived at least to some extent from gametophytes (by normal fertilization, or apogamy) rather than from callus tissue. Saga *et al.* (1982) reported callus formation from the brown seaweed *Dictyosiphon* but redifferentiation of this did not occur. Callus masses formed from axenic medullary tissue from stipes of *L. angustata* (Saga and Sakai 1983). Tissue pieces of *L. japonica* and *Undaria pinnatifida* developed a callus which differentiated into new sporophytes in medium containing 'synthetic plant hormone' (Yan 1984). Lee (1985) reports regeneration of gametophytes from stipe explants of *L. saccharina*. Gametophyte colonies of *M. pyrifera*, in the form of round balls of fine branched filaments, have been found to develop callus-like cells 40 μm in diameter on agar media (Polne-Fuller *et al.* 1986). More recently (Polne-Fuller and Gibor 1987) colourless and pigmented callus formation has also been reported in sporophytic tissue of *Macrocystis, Egregia, Sargassum, Cystoseira*, and *Pelvetia*. Differentiation into plantlets occurred from pigmented callus in some cases, e.g. *S. muticum*. Protoplasts (which grew into calluses) were also produced from pigmented calluses of *Sargassum* and *Cystoseira*.

It can therefore be seen that success in obtaining actively dividing undifferentiated callus cell masses which can be subcultured (and grown in liquid culture or in fermenters) and regenerated is so far limited in seaweeds. Since some of the reports referred to do not involve an undifferentiated callus phase (and tissue organization is maintained), these presumably represent a form of organ culture. Work in the authors' laboratory has high lighted the difficulties in obtaining from sporophytic or gametophytic tissue of the more complex seaweeds undifferentiated callus cells which will grow at a rate comparable with those of higher plants. Further work is needed on nutritional requirements and, in particular, on carbon requirements, before such cells can be routinely and easily obtained under axenic conditions.

Nutritional requirements

Plant cells in culture require similar nutrients to those needed by whole plants. In addition, isolated tissues often require substances which would normally be synthesized in the intact plant, e.g. vitamins. Higher plant tissue culture media contain inorganic nutrients, vitamins, plant growth, substances, and a source of carbon. Complex organic mixtures such as casein hydrolysate may also be used to supplement the medium.

Existing seawater media are either enriched natural seawater or synthetic media. Natural seawater is complex and is the standard medium for the growth of marine plants; when enriched with additional nutrients it supports the growth of many seaweed species in culture. Synthetic media are fully defined and consist of the salts of the major ions found in seawater and mixtures of chelated metals and vitamins. The use of fully defined artificial seawaters is preferable as these are more suitable for nutritional experiments and less variable than natural seawater. However, defined media, being intrinsically simpler than seawater-based media, may not support good growth in some species. Formulae exist for several enriched and synthetic media (Provasoli 1963; McLachlan 1964, 1973) and these are the media most frequently used in the culture of seaweed tissues. However, these media were devised for the culture of marine microalgae and have been modified to support the growth of benthic species. The culture of isolated algal tissues will probably require further modification of the media, but so far there has been little research into this aspect.

Nutrient media composition (a) Inorganic nutrients. Seawater media, e.g. ASP-12-NTA, contain the macronutrients required for marine plant growth and these are similar to those required by higher plants although the ionic concentrations are in many cases quite different and the salinity much higher (Table 18.2). The inorganic requirements of seaweeds are diverse and the subject of reviews by O'Kelly (1974) and McLachlan (1982). Seaweed tissues have been successfully cultured on several media but the inorganic components have not yet been optimized for any species. Nitrate concentration has been shown to affect the morphology of cultured tissues, e.g. in *Cladosiphon zosterae* (Lockhart 1979), *Fucus* embryos (McLachlan 1977), and *Agardhiella subulata* (Cheney 1986). Detailed research on the trace metal nutrition of *Macrocystis* gametophytes (Kuwabara and North 1980) suggests that available metal formulations may be suboptimal for growth of seaweed cells and tissues. In summary, although the inorganic composition of existing media is adequate for the induction of tissue growth in some species, improvement will be necessary if growth rates are to be increased and the range of species cultured extended.

(b) Organic constituents.

(i) Carbon source. A carbon supply which can be metabolized is a necessary prerequisite for the rapid heterotrophic growth of cultured tissues, but the inclusion of a source of organic carbon is not reported in many of the published accounts of tissue culture in seaweeds (Chen and Taylor 1978; Fries 1980; Saga and Sakai 1983; Polne-Fuller and Gibor 1984) and discussed only in one report (Saga *et al.* 1982). Mannitol was found to be more effective than sucrose in the formation and growth of calli of the brown algae *Dictyosiphon foeniculaceus* (Saga *et al.* 1982). Mannitol is also the major form of soluble carbon in brown algae. However, Drew (1969) in an investigation of the uptake of carbon compounds by brown algae found that none of the species tested could metabolize exogenous glucose, fructose, galactose, mannose, or mannitol (see also Kingham and Evans 1986). Organic carbon has a central

Table 18.2 Inorganic salts: concentrations of ions in a seawater medium (ASP-12-NTA) and in two higher plant growth media (MS and B5).

Salt	Medium		
	MS[1]	B5[2]	ASP-12-NTA[3]
Macronutrients (mM)			
Na^+	0	1.0	483.0
Cl^-	6.0	2.0	548.0
SO_4^{2-}	1.5	2.0	28.4
K^+	20.0	25.0	9.5
NO_3^{3-}	39.4	25.0	1.2
PO_4^{3-}	1.25	1.0	0.1
NH_4^+	20.6	2.0	0
Mg^{2+}	1.5	1.0	48.0
Ca^{2+}	3.0	1.0	10.0
Total ionic conc.	93.3	59.5	1128.0
Micronutrients (μM)			
EDTA	100.0	50.0	26.9
Fe^{3+}	100.0	50.0	1.8
B^{3+}	100.0	50.0	185.0
Mn^{2+}	100.0	60.0	7.3
Zn^{2+}	30.0	7.0	0.8
Co^{2+}	0.1	0.11	0.17
Br^-	0	0	125.0
Sr^{2+}	0	0	23.0
Rb^+	0	0	2.3
Li^+	0	0	28.8
Mo^{6+}	1.0	1.0	5.0
I^-	5.0	4.5	0.1
Cu^{2+}	0.1	0.1	0

1. Murashige and Skoog (1962)
2. Gamborg *et al.* (1968)
3. Provasoli (1963)

role in supporting the growth of tissues in culture and the carbon nutrition of seaweeds requires urgent clarification.

(ii) Vitamins. Vitamin requirements in seaweeds have been reviewed by Provasoli and Carlucci (1974). Seawater media already include mixtures of vitamins thought to be required by algae. Axenic tissues may need additional vitamins previously available in the intact plant or supplied by the epiflora. Organic nutritional experiments can only be carried out in the absence of micro-organisms and axenic tissue cultures may therefore be useful in establishing the vitamin requirements of macroalgae.

(iii) Plant growth substances. Higher plant tissues usually require the addition of an auxin to induce cell division and callus formation. The addition of cytokinins is often necessary to maintain active cell division. Many seaweeds are complex multicellular plants and the presence of growth regulatory compounds is suggested by examples of apical dominance, localized meristems, tropisms, and cytodifferentiation. However, there is very little understanding of the role of plant growth substances in the growth and differentiation of marine plants (see Provasoli and Carlucci 1974; Buggeln 1981). Similarly, a clear action of plant growth substances in the induction and growth of callus, as occurs in dicotyledonous plants, is not yet evident in seaweeds. Explant development has been induced on media containing plant growth substances (Fries 1980, Chen and Taylor 1978; Fang *et al.* 1983) and on media without (Polne-Fuller and Gibor 1984; Lee 1985; Saga and Sakai 1983; Saga *et al.* 1982). Plant growth substances have been shown to enhance the rate of cell division in *Agardhiella* (Cheney 1985) but alternatively shown to have no effect on the formation and growth of callus in *Dictyosiphon* (Saga *et al.* 1982) and *Laminaria* (Saga and Sakai 1983). Knowledge of the action of growth substances in the intact marine plant is insufficient to form the basis for a tissue culture programme. However, tissue cultures provide good experimental systems for the study of the regulation of growth and differentiation, and algal tissue cultures may help provide important information on plant growth substances in seaweeds.

The development of specialized nutrient media is crucial to the advancement of tissue culture techniques in seaweeds. Of primary importance is an understanding of the requirements for plant growth substances and the metabolism of exogenous organic carbon compounds. Further research on other aspects of nutrition may ultimately lead to the development of a medium as widely applicable to marine algae as Murashige and Skoog medium (Murashige and Skoog 1962) is to higher plants.

Final comments

'We cannot yet point to a newly-produced crop variety as a product of recombinant DNA technology' (Day 1985). Indeed, production of protoplasts and regeneration of plants from these has only recently been achieved in monocotyledonous plants, and this so far in only a very limited number of species (e.g. see Abdullah *et al.* 1986; Srinivasan and Vasil 1986). In view of the extent of progress in higher plants it is hardly surprising that advance has been considerably more limited with respect to developing the new technologies for seaweeds. The cynic might say it is surprising there has been any advance at all, in view of the low-status image of seaweeds in most people's minds! However, greater progress is being made in developing genetic engineering techniques in microalgae such as *Chlamydomonas* (e.g. see Craig and Reichelt 1986) and this is encouraging.

Success can be achieved by exisiting technologies. For example, in *L. japonica* an intensive selective breeding programme resulted in increased yields, improved growth characteristics, and new warm water-tolerant varieties with an extended geographical range (Ryther 1984). What are the prospects for benefitting from the new technologies?

There are reports that agar can be successfully produced by callus culture of algae such as *Gelidium amansi* and *Gracilaria confervoides*; 100 g of dried callus cells were reported to produce 75 g of agar (Misawa 1977). This compares with 29 per cent and 26 per cent agar yields, respectively, from cysto carpic and tetrasporic plants of *Gracilaria verucosa* (Yao *et al.* 1984). It is difficult to judge whether it will be possible to produce phycocolloids such as agar and carrageenan cost-effectively by the new technologies because it is not yet known to what extent improved quality or yield can be achieved. The impetus for doing this may be lessened to some extent by the fact that phycocolloid substitutes are now being produced by bacterial fermentation processes (Sandford and Baird 1983). However, genetic engineering may eventually be useful in altering expression of existing genes, e.g. incorporation of high expression promotors with replicons involved in agar production could increase agar yields significantly. Production of speciality pharmaceuticals by seaweed tissue culture is likely to receive more interest in the future; only squalene (Table 18.1) can be obtained in useful quantities from whole plant extraction (Chapman 1979) and it is possible that greater yields of substances of this kind could be obtained by tissue culture. However, technical difficulties in isolation and characterization of such active compounds will probably prevent application of tissue culture to pharmaceutical production in the near future. On the other hand, the use of somatic hybridization for producing, for example, strains resistant to fungal diseases is a more immediate possibility, e.g. to generate *Porphyra* strains resistant to red rot caused by the fungus *Pythium* (Andrews 1979; Polne-Fuller and Gibor 1986).

As stated by Craig and Reichelt (1986), 'development of genetic engineering techniques will be as important for algae-based industries as for any other biotechnology'. Increased research effort must therefore be devoted to perfecting the underlying basic techniques, *viz.* protoplast isolation, somatic fusion, callus formation, and plant regeneration, because it is only through these that the potential offered by seaweeds can be realized to its fullest extent in the future.

References

Abbott, I.A. and Cheney, D.P. (1982). In *Selected papers in phycology* II (ed. J.R. Rosowski and B.C. Parker) p. 779. Phycological Society of America, Inc. Book Division, P.O. Box 368, Lawrence, Kansas 66044, USA.

Abdullah, R., Cocking, E.C., and Thompson, J.A. (1986). *Biotechnology* **4**, 1087.
Ammirato, V.A., Evans, D.A., Flick, C.E., Whitaker, R.J. and Sharp, W.R. (1984). *Trends Biotech.* **2**, 53.
Andrews, J.H. (1979). *Experientia* **35**, 429.
Buggeln, R.G. (1981). In *The biology of seaweeds* (ed. C.S. Lobban and M.J. Wynne) p. 627. Blackwell Scientific Publications, Oxford.
Butler, D.M. and Evans, L.V. (1988). In *An introduction to applied phycology* (ed. I. Akatsuka). SPB Academic Pub. Co. (in press).
Chapman, V.J. (1979). In *Marine algae in pharmaceutical science* (ed. H.A. Hoppe, T. Levring, and Y. Tanaka) p. 139. De Gruyter, Berlin.
Chen, L.C-M. (1986). *Bot. Mar.* **29**, 435.
Chen, L.C-M. and Taylor, A.R.A. (1978). *Can. J. Bot.* **56**, 883.
Cheney, D.P. (1984). In *Biotechnology in the marine sciences* (ed. R.R. Colwell, A.J. Sinskey and E.R. Pariser) p. 161. Wiley Interscience, New York.
Cheney, D.P. (1985). *Plant Phys.* (Suppl.) **77**, 69.
Cheney, D.P. (1986). *Beih. Nova Hedwigia* **83**, 22.
Cheney, D.P., Mar, E., Saga, N., and van der Meer, J. (1986). *J. Phycol.* **22**, 238.
Craig, R. and Reichelt, B.Y. (1986). *Tibtech*, November issue, p. 280.
Day, P.R. (1985). In *Biotechnology and its application to agriculture*, Monograph No. 32 (ed. L.C. Copping and P. Rodgers) p. 25. BCPC Publications, Croydon, Surrey.
Drew, E.A. (1969). *New Phytol.* **68**, 35.
Ducreux, G. and Kloareg, B. (1988). *Planta* **174**, 25.
Evans, L.V. (1986). *Sci. Prog., Oxf.* **70**, 287.
Fang, T., Yan, Z., and Wang, Z. (1983). *Science Bull.* **28**, 247.
Fries, L. (1980). *J. Phycol.* **16**, 475.
Fujita, Y. and Migita, S. (1985). *Bull. Fac. Fish. Nagasaki Univ.* **57**, 39.
Gamborg, O.L., Miller, R.A., and Ojima, K. (1968). *Exp. cell Res.* **50**, 151.
Gibor, A., Polne, M., Biniaminov, M., and Neushul, M. (1981). In *Proc. 10th Int. Seaweed Symp.*, (ed. T. Levring) p. 587. De Gruyter, Berlin.
Hoppe, H.A. (1979). In *Marine algae in pharmaceutical science* (ed. H.A. Hoppe, T. Levring, and Y. Tanaka) p. 25, De Gruyter, Berlin.
Jones, M.G.K. (1985). In *A. Proc. phytochem. Soc. Eur.*, Vol. 26, (ed. K.W. Fuller and J.R. Gallon), p. 215. Oxford University Press.
Kingham, D.L. and Evans, L.V. (1986). In *The biology of marine fungi* (ed. S.T. Moss) p. 177. Cambridge University Press.
Kloareg, B. and Quatrano, R.S. (1987). *Plant Sciences* **50**, 189.
Kuwabara, J.S. and North W.J. (1980). *J. Phycol.* **16**, 546.
Lee, T.F. (1985). *Botanica Marina* **28**, 179.
Lobban, C.S., Harrison, P.J., and Duncan, M.J., eds. (1985). *The physiological ecology of seaweeds* p. 128. Cambridge University Press.
Lockhart, J.C. (1979). *Am. J. Bot.* **66**, 836.
McLachlan, J. (1964). *Can. J. Microbiol.* **10**, 769.
McLachlan, J. (1973). In *Handbook of phycological methods* (ed. J.R. Stein) p. 25. Cambridge University Press.
McLachlan, J. (1977). *Phycologia* **16**, 329.
McLachlan, J. (1982). In *Synthetic and degradative processes in marine macrophytes* (ed. L.M. Srivastava) p. 71. De Gruyter, Berlin.
Millner, P.A., Callow, M.E., and Evans, L.V. (1979). *Planta* **147**, 174.

Misawa, M. (1977). In *Plant tissue culture and its biotechnological application* (ed. W. Barz, E. Reinhard, and M. Zenk) p. 17. Springer-Verlag, Berlin.
Murashige, T. and Skoog, F. (1962). *Physiologia Plantarum* **15**, 473.
O'Brien, E.T., White, S., Jacobs, R.S., Boder, G.B., and Wilson, L. (1984). *Hydrobiologia* **116/117**, 141.
O'Kelley, J.C. (1974). In *Algal Physiology and biochemistry* (ed. W.D.P. Stewart) p. 610. Blackwell Scientific Publications.
Painter, T.J. (1983). In *The polysaccharides*, Vol. 2 (ed. G.O. Aspinall) p. 195, Academic Press, London.
Percival, E. (1978). In *Carbohydrate sulfates* (ed. R.G. Schweiger) p. 213, ACS, Symposium Series No. 77, American Chemical Society, Washington, DC, USA.
Percival, E. (1979). *Br phycol. J.* **14**, 103.
Polne, M., Gibor, A., and Neushul, M. (1980). *Botanica Marine* **23**, 731.
Polne-Fuller, M. and Gibor, A. (1984). *J. Phycol.* **20**, 609.
Polne-Fuller, M. and Gibor, A. (1986). *Aquaculture* **57**, 117.
Polne-Fuller, M. and Gibor, A. (1987). In *Proc. 12th Int. Seaweed Symp.* São Paulo, Brasil (ed. M.A. Ragan and C.J. Bird) p. 131. Dr. W. Junk, Dordrecht.
Polne-Fuller, M., Biniaminov, B., and Gibor, A. (1984). *Hydrobiologia* **116/117**, 308.
Polne-Fuller, M., Saga, N., and Gibor, A. (1986). *Beih. Nova Hedwigia* **83**, 30.
Provasoli, L. (1963). In *Proc. 4th Int. Seaweed Symp.*, (ed. D. DeVirville and J. Feldmann) p. 9. Pergamon Press, Oxford.
Provasoli, L. and Carlucci, A.F. (1974). In *Algal physiology and biochemistry* (ed. W.D.P. Stewart) p. 741. Blackwell Scientific Publications, Oxford.
Ryther, J.H. (1984). In *Biotechnology in the marine sciences* (ed. R.R. Colwell, A.J. Sinskey, and E.R. Pariser) p. 123, Wiley Interscience, New York.
Saga, N. (1984). *Bot. Mag. Tokyo* **97**, 423.
Saga, N. and Sakai, Y. (1983). *Bull. Jap. Soc. Sci. Fish.* **49**, 1561.
Saga, N. and Sakai, Y. (1984). *Bull. Jap. Soc. Sci. Fish.* **50**, 1085.
Saga, N., Motomura, T., and Sakai, Y. (1982). *Plant and Cell Physiol.* **23**, 727.
Saga, N., Polne-Fuller, M., and Gibor, A. (1986). *Beih. Nova Hedwigia* **83**, 37.
Saga, N., Uchida, T., and Sakai, Y. (1978). *Bull. Jap. Soc. Sci. Fish.* **44**, 87.
Sandford, P.A. and Baird, J. (1983). In *The polysaccharides*, Vol. 2 (ed. G.O. Aspinall) p. 411. Academic Press, New York.
Srinivasan, C. and Vasil, I.K. (1986). *J. plant Physiol.* **126**, 41.
Tang, Y. (1982). *J. Shandong Coll. Oceanography* **12**, 37.
Yao, S.S., Xia, Z.Y., En, L.Z., and Qing, L.W. (1984). *Hydrobiologia* **116/117**, 551.
Yan, Z. (1984). *Hydrobiologia* **116/117**, 314.
Zhang, D. (1983). *J. Shandong Coll. Oceanography* **13**, 57.
Zhao, H. and Zhan, X. (1981). *J. Shandong Coll. Oceanography* **11**, 61.

19 Making mutants and influencing genes—the genetic exploitation of algae and cyanobacteria

R.A. LEWIN

Scripps Institution of Oceanography, A-002, University of California, La Jolla, California 92093, USA

Introduction

Phytochemists study the chemical nature of plant products, paying particular attention to the more useful ones like the constituents of wood, sugars, dyes, and drugs. These are manufactured by plant cells for their own purposes, and the plants make them by sequences of biological processes which may be quite complicated. Although many phytochemists have been concerned more with the finished products than with their biosyntheses, for many plant biochemists it is the step-by-step creation of complicated substances, starting from CO_2, that appears more fascinating and more challenging. Many plants produce biochemicals which unquestionably repel or poison animals that would otherwise eat them, and perhaps the greater part of phytochemical effort has been devoted to characterizing such products from higher plants. But vascular plants are not the only ones to synthesize secondary metabolites. To the experimental phycologist, there is interest particularly in the so-called lower plants, the algae. Many are normally single-celled, and reproduce rapidly: they can accordingly be grown in culture and subjected to analyses, both biochemical and genetic, more readily—and more quickly and cheaply—than sugar-cane or pine trees. Consequently the algal geneticist can envision potentialities for studying certain phytochemical problems by biochemical genetics, using convenient kinds of single-celled algae. For such studies we need to compare normal cells with mutants, so that we may be able to chart the normal processes that go on in wild-type cells by using information from studies of mutant strains which in one way or another are genetically impaired. For these reasons I have summarized here a few of the genetic techniques that we have used in experimental phycology and that you might find useful in future investigations of biosynthetic phytochemistry. I shall confine my attention to mutants of small algae: how to obtain them and keep them, and why this might be useful.

We will deal with the last topic first: why would anyone want mutant algae? According to Murphy's principal law, if anything can go wrong, it will, sooner or later. Rephrased in a genetic context, this means that every metabolic or biosynthetic pathway in every kind of organism can go wrong, and presumably often does so. If the mistake is genetic and potentially heritable, we call it a mutation. If by any means we can keep the changed, mutant strain alive and self-perpetuating, we have a handy way of studying the impaired pathway because, if we can find some chemical that will get into the cells and repair the genetic damage, we may postulate that the substance, or at least something very like it, is a normal component of the normally functioning machinery of the wild-type cells. The same kind of reasoning and experimentation—with appropriate modifications—might help us understand other natural processes, such as those involved in cell division, wall formation, maturation, movements, mating processes, and so on.

The growth of applied phycology has brought out clearly the need for expansion of algal genetics, so that we may more logically develop strains that can grow better in specified condition, or are resistant to diseases that occasionally reduce productivity, or produce more abundant or more useful products. So a study of algal mutants may be of practical as well as fundamental significance.

In this paper I propose to outline some general methods that have been successfully employed for obtaining mutants of microscopic algae (especially *Chlamydomonas*) and other micro-organisms, and to give illustrative examples of the strategies employed. Many, perhaps most, have been adopted or adapted from laboratories where heterotrophic microorganisms such as *Escherichia*, yeasts, or *Neurospora* species are under study. I hope this review will stimulate investigators to use their own ingenuity to develop other systems and other tricks for finding and using algal mutants. In all such cases, of course, the more we know about the details of the life cycle of the organism we want to work with, the more possibilities present themselves for specific mutagenesis. I shall discuss, first, methods for the induction of mutation by using chemical mutagens or ionizing radiation, touching only lightly on the theory involved; our object here is to deal with more practical problems. I shall go on to discuss briefly the processes by which mutations may be reversed, and ways to prevent mass reversals. Since mutation at any locus is a rare event, even under the most drastic mutagenesis, an outline of the ways to enrich for the rare kinds we are seeking and to discriminate against the much commoner wild types is presented. All kinds of selective media and devices have been developed for this purpose; their need becomes obvious when one considers potential savings in media, manpower, and laboratory resources. If we can increase the frequency of a rare cell type 100-fold, say from 10^{-5} to 10^{-3}, we might be able to find a mutant of the kind we need by use of less than ten Petri plates, instead of having to hunt through the colonies in a thousand. I shall then suggest ways of isolating desired mutants, freeing them from potentially contaminant cells of other types, and keeping mutant stocks, unchanged in so far as this can be controlled, for whatever study we had originally envisaged.

Induction of mutations

Anything that interferes specifically with the ability of a DNA chain to replicate itself accurately may act as a mutagen. Ionizing radiations, such as X-rays, do this by something akin to mechanical rupture of the nucleotide chain, which may incorporate one or several errors at the damaged site before it 'heals'. Non-ionizing, short-wave irradiation, notably ultraviolet light in the range of 260 nm, is absorbed fairly specifically by nucleotides and is thus also mutagenic, though with less drastic results than those of ionizing radiations. It tends to induce point mutations. Homologues of the nitrogenous bases may be even more specific in their effects, replacing natural nucleotides by, e.g., bromo-uracil and thereby changing the correct sequence to an incorrect one after one round of replication. Some poisons like fluorodeoxyuridine, by inhibiting the synthesis of certain nucleotides, may produce similar mutagenic effects. Changes, if they are to result in stable mutations, have to be self-perpetuating, otherwise the nucleotide chain may revert to its natural state. Too high a dose of any of these potential mutagens causes so much havoc among the nucleotides (i.e. so many changes in the nucleotide sequences of which genes are composed) that the cells all die; with too low a dose, on the other hand, the induced mutants may be too rare to find. One has therefore to compromise. A dose-response curve, plotting effect against dose—this is exposure time with a standard irradiation intensity (for radiation) or concentration (for a chemical mutagen)—or, alternatively, plotting different intensities or concentrations presented for a standard time, gives a survival curve. Conveniently, data are plotted on semi-log paper, since the effects are more or less exponential. It is probably best to select a treatment that kills about 99 per cent of the cells, which should put one into an effective mutagenic range. There are biological agents, too, notably certain viruses, that can sometimes induce mutations, but I know of none that have been used with algae.

Other things being equal, it is harder to obtain mutants in single-celled algae that have double or multiple copies of their genes (e.g. because of diploidy) because there is only a very small chance that both or all of the genes at the same locus will be mutated simultaneously. Normally one may expect that the wild-type gene will dominate and the phenotype of the heterozygous clone will therefore show no evidence of mutation. For the same reason, cells that occur in pairs or filaments, even if mutated, would not show their changed nature until separated from their normal, wild-type kin, which may anyway be expected to overgrow them eventually unless steps are taken to prevent this. A diploid, polyploid, or multicellular type generally reveals this condition by the form of its survival curve, which at low mutagen doses exhibits a plateau (because at any locus a single damaging 'hit' impairs only one of the two or more replicates).

X-ray damage, which may affect many genetic sites in a single region, tends to be irreversible, but UV mutations, being typically single-hit effects, are more likely to revert to the wild-type condition. Reversals may be spontaneous, although the rate may be affected by chemical and physical factors. In the case of UV mutagenesis, reversion is promoted by visible light due to the activity of a light-dependent repair enzyme, so one generally keeps irradiated cells for a few hours in darkness, to allow time for the errors to be stabilized, before exposing the cells to the light that they may need for photosynthesis and growth. Some lethal damage is also photo reversible, and visible light can be expected to 'rescue' some cells that might otherwise not have survived. One has to consider net effects in the balances of mutagenesis, survival, and death. Only by experimentation can one tell what combination of factors is likely to yield the most mutants.

Enrichment cultures

Mutagenesis, being essentially a random process, produces many kinds of gene changes. Most of them result in retarded or arrested growth, the latter being operationally lethal. Usually an experimenter with a specific objective seeks mutants of a particular class; but how, in the plethora of normal and variously crippled types, is he to find the special kinds he needs? The simplest way, though inefficient and tedious, is to separate surviving cells and grow clones among which a tiny proportion may show the desired characteristic. There are thousands of different genes in every organism, each capable of being damaged in several ways, and we might have to examine 10 000 or even 100 000 such clones before finding one of specific interest. Hence the need to use, whenever practicable, enrichment cultures after mutagenesis. Essentially, the idea is to make cells of the desired class of mutant proportionally less rare by creating conditions that favour them and thereby, after some cell division cycles, increase their relative frequency in the population. That is really all that one can hope for: it is usually impossible to effect anything like a 100 per cent separation of mutant cells from wild types. I shall give several examples of how this could be done.

Mutants of the flagellate *Chlamydomonas*, cells of which normally swim towards the light, can be selected in various ways. Those that cannot swim at all, or that swim weakly due to some impairment of the flagellar mechanism, tend to accumulate at the bottom of liquid cultures, while their wild-type counterparts accumulate at the top, nearest the light, thereby allowing us easily to effect a partial separation. Likewise, those that swim at normal speeds but have a changed response to the direction of light can be enriched by pipetting samples from the appropriate side of a culture vessel illuminated from one direction. Capillary tubes, in which turbulence is reduced, can be particularly useful for such separations. Some of these aberrant strains may be regarded as behavioural rather than biochemical mutants, but I have little doubt that the bases of their changed behaviour are, in the broad sense, phytochemical.

Biological selection of mutant types goes on all the time in Nature, being particularly evident in host–pathogen systems. Among fungi we could quote many examples from laboratory studies. Pathogenic algae, however, are rare, and few pathogens of algae have been critically studied. But we could easily seek a mutant alga resistant to a normally lethal agent, such as a lytic phage or predatory protozoan, by mixing mutagenized cells with the pathogens (or predators) and illuminating them in the hope of turning up resistant strains. Even a single cell, if it survives, could produce a clone with the desired resistant characteristic.

A variety of physical factors can be employed to enrich for other specific kinds of mutants. Obvious examples include incubation under conditions selective for cells tolerant of temperatures, salinities, pressures, or pH values outside the normal range.

A similar system is that used to enrich for mutants resistant to some inhibitory or toxic agent. Antibiotic-resistant mutants, obtained in this way, may be of special biological interest since many antibiotics are specifically inhibitory to a single process or step in normal biosynthesis. When we know what this is, investigation of different kinds of resistant mutants can be particularly illuminating. A rather useful scheme for selecting oil-rich strains involves flotation: cells that float may include mutants containing more lipid than normal. Various filtration devices, which have been employed for seeking mutants of fungi, are equally applicable to algal genetic research, permitting us to enrich for small-celled or large-celled strains, or strains producing longer or shorter filaments than the wild type. Here again, it should be emphasized that the discrimination cannot be expected to be more than partial, but even a 10 per cent 'edge' (advantage) in each of a sequence of filtrations could enormously increase the chances of ultimately finding what we are looking for.

Chemical selective factors are among the most widely used in such investigations: they can be aimed either at a requirement (affecting the lower part of the tolerance range) or an inhibitor (affecting the upper part). If a wild type requires for growth a specific organic nutrient, such as vitamin B_1 (thiamine), we could theoretically enrich for autotrophic mutants without this requirement, although our chances of success would be slight because such a mutation necessitates generating a 'new' biosynthesis, which is highly unlikely. Generally such mutants are recognized (though not selected for) by their failure to grow under standard conditions. The reverse—finding mutants with a new (unnatural) requirement, e.g. for a vitamin or an amino acid—is much more probable, since it presumably entails the impairment of a normal biosynthetic pathway. But at best this would mean that growth of wild-type cells is no worse than that of the mutants. Could we somehow create conditions that would reverse the expected effects of this bias? Possibly, if we could arrange for it to make a toxic product. A neat example where this has been achieved involves the enrichment for cells unable to reduce nitrate, presumably because they lack nitrate reductase. This enzyme, as it happens, does not clearly discriminate between its natural substrate, NO_3^-, and an unnatural one, ClO_3^-, and it can mediate the reduction of either. In the latter case, this reduction produces ClO_2^-, an anion so toxic that it may kill wild-type cells while deficient mutants tend to survive. Another trick is to incorporate into the medium an agent such as azaguanidine, which under some conditions is taken up by normal growing cells. Wild-type cells are thereby poisoned, whereas non-growing cells, which do not take up this substance to the same extent, consequently have a better chance of survival. Of course, as in all enrichment cultures, one has to work between certain concentration limits to achieve maximum differential survival rates.

Selection of mutants

Few enrichment procedures can be expected to be strictly selective; most confer only a relative advantage, in survival or rate of reproduction, over wild-type cells. To separate and recognize mutants from suspensions of cells of different genotypes, the most convenient system is to disperse them in or on a solid medium (generally 1 per cent agar) and in due course to seek colonies with distinguishing features. In suitable concentrations of an inhibitory agent or a lytic virus suspension, only resistant strains will survive and give rise to colonies. Other kinds of mutants are immediately recognizable by their abnormal colony colour, diameter (indicative of different, usually decreased, rates of growth), form (flat or heaped, according to the ease with which cells slip apart and settle down or aggregate in a heap with minimum surface area), and texture (smooth or rough, slimy or rubbery), and other characteristics indicative of changed factors at the cell surface, such as capsule production. Cell types unable to produce a specific metabolite, such as a vitamin, generally produce only pin-point colonies, the slight growth being attributable to intracellular or experimental carry-over or to traces of the required substance in the growth medium. The difference can sometimes be enhanced by incorporating very small amounts of a specific antimetabolite into the medium, just enough to suppress the growth of the mutants while allowing the wild types to grow almost normally. Care must of course be exercised here: too much antimetabolite would kill all of the cells. One may expect that cell types with impaired production of a photosynthetic pigment would fail to thrive and die out unless supplied with an assimilable organic substrate, and that mutants lacking photoprotective pigments (such as carotenoids) would be abnormally sensitive to light. However, many mutant strains of multicellular red algae, producing little or no phycoerythrin, have been obtained, and, at least in the laboratory, they seem to be as viable and robust as the wild type under ordinary conditions of illumination. Mutant strains producing less secondary carotenoids might become distinguishable from the wild type only as the colonies age and the wild-type cells turn yellow, orange, or red while the mutants remain green.

Other kinds of mutations become apparent because of changes in the viscosity of the agar around the colonies. Some diatoms, for instance, produce agar-digesting enzymes: mutants that lack this ability are revealed by the absence of peripheral liquefaction. A little phenylalanine or tyrosine, incorporated into the medium (either as the pure compound, or as a component of tryptone or some other protein hydrolysate) may cause brown haloes to appear around colonies that secrete tyrosine oxidase; colonies of mutants lacking this characteristic can be recognized by the absence of such a halo.

In such cases, the amino acid acts as a discriminatory indicator. The presence or absence of certain esterases or glycosidases can be revealed particularly neatly by incorporating into the medium some synthetic dye-conjugated substrate which, when broken down, liberates around the colonies a coloured compound, preferably as an insoluble precipitate (e.g. tetrazolium). Even a pH change can sometimes be detected by an indicator dye, though usually, because the rate of growth of an alga is slow relative to the rate of diffusion of hydrogen ions, such differential effects tend to be obscured.

Few unicellular algae are known to produce agents inhibitory or lethal to other kinds of micro-organisms, but the use of indicator organisms (e.g. bacteria or yeasts) can serve to reveal mutants that tend to 'leak' metabolites into the medium. For example, if we incorporate into the medium large numbers of innocuous bacterial cells lacking the ability to synthesize lysine, then (provided that all other requirements of both algae and bacteria are met) algal colonies leaking lysine will reveal themselves by associated bacterial haloes.

For some purposes we may wish to seek mutants that are deficient in certain biosynthetic pathways, or specifically sensitive to certain agents or treatments. On an agar medium prepared with a discriminating agent of this sort, colonies of such strains would die while the wild-type cells would survive to form colonies. A neat method for selecting the mutants involves the use of a technique known as replica plating. A set of 'master' plates on which both mutants and wild types grow and survive well is prepared, the colonies being either randomly dispersed or arranged in a set pattern (e.g. on a rectangular grid). We then replicate the colony patterns on two other sets of plates, one discriminatory and the other not, and allow the colonies to develop. (When the technique was originally developed, for bacteria, the patterns were replicated by a sort of printing from velveteen, but for unicellular algae like

Chlamydomonas it has been found that the use of filter paper generally gives better results.) By comparing the respective growth of different colonies on these two daughter replicates we can detect which represent deficient or sensitive mutants, and for further study can subculture these from the original 'master' plates. By multiple screening on plates of different compositions one can seek several different kinds of mutant in one experiment.

To keep biochemical or other physiologically impaired mutant strains alive, it is generally necessary to supply the cells with whatever essential nutrient factors they can no longer synthesize for themselves. But a manipulation whereby such supplementation of the medium is unnecessary may be feasible. It involves cases where the mutant is temperature-sensitive (TS). A TS mutant for the synthesis of thiamine, for example, might manifest its mutant condition only at temperatures above 30 °C, whereas in cooler (permissive) conditions it may behave autotrophically like wild-type strains. So although it can normally be maintained alive and well at 25 °C, whenever we wish to study its impairment for thiamine biosynthesis all we have to do is to grow cultures at an elevated (restrictive) temperature. It is not hard to devise methodologies for the specific selection of such TS mutants.

One should always bear in mind the possibility that, even among normally haploid algae, an apparently integral colony may have arisen from an effectively heterozygous cell, in which a mutation had occurred in only one of a pair of chromatids produced by DNA replication, or from two or more genetically dissimilar cells that chanced to be next to one another. So one should always re-streak colonies of selected strains, at least once, to ensure clonality. Without such purification, one may find oneself unwittingly working with mixed cultures, with consequently variable and irreproducible results. And one should also be aware of the possibility of pleiotropic effects, in which a single mutation can indirectly produce more than one biochemical, physiological, or morphogenetic effect.

Maintenance of mutants

Lastly, having found and isolated our mutant clones, how do we keep them alive? The possibility of partial or complete reverse mutation has always to be considered, because a switched nucleotide may switch back to the wild-type coding. Alternatively, a suppressor gene may arise. This is generally a result of a mutation in cells in which, perhaps on a completely different metabolic pathway, a substance accumulates in quantities sufficient to effectively nullify the manifestation of the original mutation. Moreover, under natural conditions a return to the wild phenotype—a recovery from genetic impairment—is almost certain to confer a selective advantage, so that, after a few cell generations, an ostensibly mutant clone that we had worked so hard to find and isolate loses its distinctive characteristics. As a rule, the slower the rate of cell division (in maintenance cultures), the slower the rate of back mutation, but the possibility always exists. If the cells can be stored frozen, under liquid nitrogen, no such back mutation can occur, but unfortunately only a few species of algae have so far been shown to remain viable when frozen to -195 °C. If they cannot be kept alive in liquid nitrogen, then at least we should try to keep the mutant cells viable in cool conditions, and reduce their selective disadvantage relative to potential revertants. A strain requiring a vitamin or streptomycin, for instance, should be maintained with an adequate supply of the required substance in the medium. It is a good idea to keep at least two or three replicated sets of mutant cultures, and to test these periodically, so that, if reversion occurs in one, the others can still constitute a source of mutant cells. And if valuable cultures are maintained in more than one laboratory, so much the better.

Phytochemistry takes over

If we have chosen the right species of alga, and if, through a combination of ingenuity, perseverance, and serendipity we succeed in finding one or more strains with the kind of mutation we wanted, the next thing is to try to discover exactly the nature of the defect. Most metabolic impairments can be attributed to a genetic inability of the cells to produce a certain end-product. Something else may accumulate, either on its direct biosynthetic pathway or as a result of a deviant biosynthesis. Such substances are to be sought in the growth medium, or in cell extracts or residues. Obviously, for comparative purposes, wild-type material has to be studied under similar conditions. The cells would have to be grown in pure culture, in quantities sufficient to provide phytochemists with enough material for analysis. Fortunately, as the field of analytical chemistry has progressed in recent years, technologies have been extensively refined and the quantities of material required have been very substantially decreased.

So, phytochemists, the next steps of the research programme are up to you. We biologists look forward to learning what you discover.

Acknowledgements

I am indebted to my colleagues Drs Karen van Winkle-Swift and Christopher Wills for many suggestions which helped to repair deficiencies in this article.

Poster abstracts

Ammonia production by foam-immobilized *Anabaena azollae*

H. de Jong, M. Brouers, D.J. Shi, D.O. Hall
King's College London, Department of Biology, University of London, Campden Hill Road, London W8 7AH, UK

Studies on the bioenergetic characteristics of photosynthetic cells and membranes entrapped in polymer matrices show that these immobilized photosynthetic systems have the capacity for sustained production of fuels and chemicals. Heterocystous cyanobacteria can fix atmospheric nitrogen, producing ammonia, and evolve hydrogen. Our aim is to develop a closed system for continuous production of ammonia by immobilized cyanobacteria.

Anabaena azollae cells have been immobilized in preformed polyvinyl and polyurethane foams and have also been entrapped in the polyurethane pre-polymer HYPOL 2002. Immobilization of *Anabaena azollae* results in the stabilization of photosynthetic electron transport and enzyme activities. Low temperature scanning electron microscopy shows that mucilage acts to adhere the cells to the matrix surface.

Ammonia production is induced by supplying L-methionine-D,L-sulphoximine (MSX) to the growth medium, which inhibits glutamine synthetase activity. Immobilization itself already results in increased rates of MSX-induced ammonia production in batch reactors, compared to the ammonia production by free-living cells. Continuous ammonia production, lasting for more than a month, using 12 hours light/dark cycling, giving intermittent MSX pulses (once every seven days), was observed in packed-bed continuous flow reactors. Recently a new type of bioreactor has been designed in which problems resulting from limited gaseous diffusion in the conventional packed-bed bioreactor have been overcome. In this type of photobioreactor an equilibrium concentration of 500 μM NH_4^+ was reached (in continuous light) at a dilution rate of 2.0 hour^{-1}; this compares to 200–300 μM at 0.4 hour^{-1} in the conventional packed-bed reactor.

Nitrogen fixation by immobilized cultures of cyanobacteria

M.S.H. Griffiths, J.R. Gallon and A.E. Chaplin
Department of Biochemistry, University College of Swansea, Swansea SA2 8PP, UK

The behaviour of cyanobacteria immobilized in polyvinyl foam, alginate, or agar or encapsulated in liquid membranes has been examined, though the latter system was intrinsically unstable. In many experimental situations, immobilized cultures are easier to handle and less sensitive in experimental manipulation than free-living cultures.

Cyanobacteria maintained in liquid culture under alternating light and darkness display one of three patterns of N_2 fixation; they either fix N_2 mainly in the light, e.g. *Anabaena cylindrica*, or mainly in the dark, e.g. *Gloeothece*, or both in the light and the dark, e.g. *Scytonema javanicum*. These patterns were not significantly altered by immobilization and, particularly in polyvinyl foam, could be maintained for many weeks.

Mixed cultures of *Gloeothece* and *Anabaena* immobilized in foam or alginate beads were very stable and displayed two peaks of acetylene reduction under an alternating 12 hour light/12 hour dark cycle; one in the light due to *Anabaena*, and one in the dark due to *Gloeothece*. This contrasts to mixed liquid cultures which soon became dominated by the faster growing *Anabaena* and consequently then fixed N_2 only in the light.

N_2 fixation by *Gloeothece* immobilized in agar blocks or alginate beads or encapsulated in liquid membranes was more sensitive than liquid cultures to inactivation by O_2. On the other hand, cultures immobilized in polyvinyl foam showed the same sensitivity to O_2 as free-living cultures.

Urea as a nitrogen source for immobilized *Chlorella*

A.L. Mak and M.D. Trevan
School of Biological Sciences, Hatfield Polytechnic, PO Box 109, Hatfield, Herts, UK

A problem encountered with many immobilized cells is their peripheral distribution in gel beads, suggesting diffusional limitations within the matrix. In the case of photosynthetic algae, CO_2 limitations appear to be particularly important. Calcium alginate-immobilized cells of *Chlorella emersonii* (CCAP 211/8a) were therefore provided with urea as a dual source of nitrogen and carbon, and co-immobilized with the enzyme urease to increase the rate of urea hydrolysis.

This co-immobilization allowed improved cell penetration of the gel matrix and increased biomass retention in media of high urea concentration (37.5 mM and 75 mM). Leakage of cells from the beads was markedly decreased. In contrast, the addition of urease to free cultures of *C. emersonii*, incubated at equivalent urea concentrations, resulted in severe cell debilitation and loss of chlorophyll. Possible causes of this apparent difference in tolerance to these conditions were investigated. An important influence appeared to be an upward shift in pK_a of ammonium ions in the beads due to the presence of alginate. Thus, at the high media pH values resulting from urea hydrolysis, significantly more molecules exist in the toxic ammonia form in free solution than within the alginate microenvironment. As with free cell cultures, cells which leak from alginate beads, or contaminating cells, are debilitated by ammonia. Their growth and multiplication are, therefore, selectively inhibited. Thus co-immobilization of urease with micro-algae is proposed as a suitable system not only to encourage biomass retention but also as a potential sterility control in immobilized cell reactors.

Cell culture of multicellular red algae

M.I. Tait, A.M. Milne, J.A. Somers, W.F. Long, F.B. Williamson and S.B. Wilson
Department of Biochemistry, University of Aberdeen, Marischal College, Aberdeen AB9 1AS, UK

A range of Rhodophyta cell culture lines has been established. A number of these lines grow as 'undifferentiated' aggregates, typically 0.5–2.0 mm in diameter. These have been successfully grown in batch and chemostat culture, in shake-flasks and an air-lift fermenter, respectively. Other cell lines grow as differentiated aggregates, up to 5 mm in diameter, formed from 3–4 monostromatic sheets of cells attached to a central region.

Growth and polysaccharide production by cell cultures of the red alga *Porphyridium cruentum*

M. Iqbal, D. Grey, G. Stepan-Sarkissian and M.W. Fowler
Wolfson Institute of Biotechnology, The University, Sheffield S10 2TN, UK

The extracellular polysaccharides of red algae have long been of interest because of their unique gelling properties, and consequently have been used extensively as colloidal stabilizers. These polysaccharides are best known commercially as agar and carrageenan. The present work is concerned with optimization of culture conditions and polysaccharide production.

Porphyridium cruentum was grown in artificial seawater (ASW) (Jones *et al.* 1963) and M11 medium (Asher and Spalding 1982). Maximum growth and polysaccharide yield were seen in M11. In ASW, biomass yield and polysaccharide production were lower by 35 per cent and 25 per cent, respectively. The lag phase of cultures was shortened when they were shaken at 100 r.p.m.; however, maximum growth rates were unaffected. At the termination of growth (53 days) cultures grown on M11 with shaking had the highest cell count (8.76×10^9 cells dm^{-3}) and absorbance (1.52 at 760 nm). The mean generation times ranged from 41.9 hours in M11 without agitation to 44.7 hours for ASW with agitation. In M11 the culture produces 3.15 g dm^{-3} of polysaccharide with a medium viscosity of 2.23 centistokes. Unlike growth, polysaccharide production appears to be adversely affected by agitation.

Asher, A. and Spalding, D.F., eds. (1982). *Culture centre of algae and protozoa: list of strains*. Institute of Terrestrial Ecology, 68, Hills Road, Cambridge CB2 1LA, UK.

Jones, R.F., Speer, H.L., and Kury, W. (1963). *Physiol. Plant.* **16**, 636.

Recovery by solvent extraction of extracellular hydrocarbons from *Botryococcus braunii*

J. Frenz and C. Largeau
Laboratoire de Chimie Bioorganique et Organique Physique, UA CNRS 456, ENSCP 11, rue P. et M. Curie, 75231 Paris Cedex 05, France

A.J. Daugulis and F. Kollerup
Department of Chemical Engineering, Queen's University, Kingston, Ontario K74 3N6, Canada

E. Casadevall
Laboratoire de Chimie Bioorganique et Organique Physique, UA CNRS 456, ENSCP 11, rue P. et M. Curie, 75231 Paris Cedex 05, France

Botryococcus braunii produces long-chain hydrocarbons that are insoluble in the culture medium and are stored extracellularly in outer walls. Solvent extraction is a promising unit operation for permeabilizing the outer walls and recovering the hydrocarbon product from continuous cultures. Solvent selection for this process is guided by two main requirements: high recovery of the product of interest and biocompatibility with the micro-organism.

Potential solvents were screened on the basis of published data for biocompatibility towards other organisms, low boiling point, immiscibility with water, low density and ready availability. Nineteen potential solvents were tested both for short- and long-term extraction of hydrocarbon and for impairment of photosynthetic activity.

Non-aromatic, non-polar, relatively high-boiling-point solvents proved most biocompatible with *B. braunii*. Recoveries by six non-toxic solvents were, however, very low. Evidently, direct contact of solvent with the algal culture is impractical since the large amount of water surrounding the cells acts as a barrier inhibiting transfer of hydrocarbon into the solvent phase.

By filtering the algae prior to solvent contact, thereby removing the bulk of the water in the culture, high recoveries of hydrocarbon were achieved without seriously impairing cellular activity. In this manner, extraction with hexane yielded 50 per cent or more of the hydrocarbon product, and obviated the need to search for a more complex solvent system.

Problems encountered in cloning *Chlorella* genomic DNA

N.A.R. Urwin and D.S.T. Nicholl
Biology Department, Paisley College of Technology, Paisley PA1 2BE, Scotland

High molecular weight genomic DNA was isolated from the eukaryotic alga *Chlorella fusca*, essentially by the method of Lichtenstein and Draper (1985). Digestion of the DNA with restriction endonucleases revealed that *Chlorella* DNA contains an abnormally large number of sites for the enzymes *Pst*I and *Pvu*II, possibly indicating the presence of repetitive sequences. Using the restriction enzymes *Hpa*II and *Msp*I we have found Chlorella DNA to be modified, probably by methylation of the base cytosine at some CpG sequences (see also Van Etten *et al.* 1985).

Methylated and repetitive sequences in eukaryotic DNA are thought to inhibit cloning (Blumenthal 1986; Cox 1986). *Escherichia coli* K802 lacks the modified cytosine restriction (*mcr*) system and can therefore enable cloning of some methylated sequences when used as a host. *E. coli* CES200 lacks recombination exonucleases I and V, the products of the *sbcB* and *recBC* genes, and can allow propagation of some repetitive sequences.

Genomic DNA from *Chlorella* has so far proved refractory to cloning in the lambda replacement vector EMBL3 using both normal hosts and *E. coli* strains K802 and CES200, although genomic DNA from *E. coli* C-la was successfully cloned in control experiments. It is hoped that further research into the structure of *Chlorella* DNA will resolve this problem.

Blumenthal, R.M. (1986). *Trends Biotech.* **4**, 302.

Cox, R.A. (1986). In *The molecular biology of* Physarum polycephalum (ed. W.F. Dove) p. 301. Plenum, New York.

Lichtenstein, C. and Draper, J. (1985). In *DNA cloning, a practical approach* (ed. D.M. Glover) p. 67. IRL Press, Oxford.

Nader, W.F. (1986). In *The molecular biology of* Physarum polycephalum, (ed. W.F. Dove) p. 291. Plenum Press, New York.

Van Etten, J.L., Schuster, A.M., Girton, L., Burbank, D.E., Swinton, D., and Hattman, S. (1985). *Nucleic Acids Res.* **13**, 3471.

Index of organisms

Subject index